TOURISM AND RURAL MANAGEMENT

TOURISM AND RURAL MANAGEMENT

Edited by

DR. RAIS AHMAD

(Former Professor in Management Studies, JMI)
Associate Professor
Department of Agricultural Economics and Business Management
Aligarh Muslim University, Aligarh-202002 (U.P.)

Foreword by

DR. ATUL DHINGRA

Professor & Head,
Department of Business Management,
CCS Haryana Agricultural University,
Hisar-125004 (Haryana)

REGAL

REGAL PUBLICATIONS

New Delhi-110027

TOURISM AND RURAL MANAGEMENT

ISBN 978-81-8484-273-9

Typeset by
S.S. COMPOSERS
3190, Mohindra Park, Shakur Basti, Delhi-110034.

Printed in India at
MAYUR ENTERPRISES
WZ Plot No. 3, Gujjar Market, Tihar Village, New Delhi-110018.

Published by
REGAL PUBLICATIONS
F-159, Rajouri Garden, New Delhi-110027.
Phone: +91-11-45546396, 25435369
E-mail: regalbookspub@yahoo.com, regaldeepbooks@yahoo.com

Contents

Foreword

It is a matter of great pleasure for me to learn that Dr. Rais Ahmad is coming out with a book titled, "Tourism and Rural Management".

Tourism has always been in the limelight world over and India is no exception. However, its importance for the economy of a country has increased manifold over the years. Tourism on one hand generates revenue and on the other provides employment and livelihood to many persons. It also enhances the brand equity of a place or the country.

Similarly, for a developing country like India, rural development is very important as majority of population live in rural areas. If we have to become developed nation then prosperity and development must reach the rural areas.

This book is an attempt to address both these important areas of economy. The book contains rich literature and throws light on different facets, dimensions and viewpoints related to both these areas. Not only has it highlighted the benefits and advantages of these areas but also has thrown light on the challenges and strategies to overcome these challenges. Fusion of tourism with rural development is rare but important and this is the USP of this book.

Dr. Rais Ahmad deserves all the appreciation for his sincere effort and tireless dedication in bringing out this book. The book will be of interest to a wide variety of audience comprising of students, practitioners, researchers, professionals and general readers. Language used is simple and precise and anyone can understand the text with minimum of the efforts. I sincerely hope that the book will come up to the expectations of all the readers.

I convey my best wishes to Dr. Rais Ahmad for his endeavor.

ATUL DHINGRA
Professor & Head
Department of Business Management
CCS Haryana Agricultural University
Hisar-125004 (Haryana)

Preface

Tourism in India

Today, tourism has become a white collar, environment-`friendly industry and one of the largest and fastest growing industries employing the largest number of labour force and a major economic force in the contemporary world. Though currently, the developed nations are getting the lions share of the benefit of tourism, but there is a promising future for the developing countries. Tourism has become a leading industry in the world and continues to grow rapidly, contributing essentially to economic growth and employment in many countries. Tourism as it looks today poses serious threats to the environment and considerable social and cultural challenges. Tourism can be an important contributor to the growth of an economy. Growth of tourism is a significant part of the global expression of new service economy. With the weakening of many other aspects of the economy, tourism is a new opportunity. India, though a late starter in tourism, is giving increasing priority to this sector. Tourism is one of the strategically important emerging sectors in Indian economy with tremendous potential to assist the process of sustainable rural development. World Travel and Tourism Council (WTTC) indicated that India will emerge as the second fastest growing tourism economy globally by 2014 followed by China.

Rural Development

Mahatma Gandhi rightly said, "India lives in its villages". The 2011 Census estimates that 69 percent of the country's total population inhabits in villages. The Rural Development in India is one of the most important factors for the growth of the Indian economy. India is primarily an agriculture-based country. Agriculture contributes nearly one-fifth of the gross domestic product in India. In order to increase the growth of agriculture, the Government has planned several programs pertaining to Rural Development in India.

Rural Development has always been an important issue in all discussions pertaining to economic development, especially of developing countries, throughout the world. In the developing countries and some formerly communist societies, rural mass comprise a substantial majority of the population. Hence, Rural Development which is concerned with economic growth and social justice, improvement in the living standard of the rural people by providing adequate and quality social services and minimum basic needs becomes essential. The present strategy of rural development mainly focuses on poverty alleviation, better livelihood opportunities, provision of basic amenities and infrastructure facilities through innovative programmes of wage and self-employment.

This book **Tourism and Rural Management** contains 35 research papers/articles contributed by scholars who belong to different Universities/Colleges and Institutions. The contributors have presented their views analytically are connected with different aspects of tourism and rural development. Surely this book will be very much useful for policy-makers, researchers, academicians, practitioners and all those who directly or indirectly have connection with tourism and rural management.

For completion of this book I am highly thankful to all those eminent scholars, academicians and friends, from academic community, who have helped and encouraged me and to name all of them would not be feasible to mention here. However, some names are quoted below:

Prof. M.S. Swaminathan, Chairman, Swaminathan Research Foundation, Chennai and Member of Rajya Sabha (M.P.), Dr. C. Rangarajan, Chairman, Prime Minister's Economic Advisory Committee, Government of India, New Delhi, Prof. Shamim Ahmad, Chairman, Department of Agricultural Economics and Business Management and Dean, Faculty of Agricultural Sciences, Aligarh Muslim University, Aligarh, Prof. Irfan Habeeb, Professor Emeritus, Department of History, AMU, Aligarh, Prof. Shamim Jairajpuri, (Retired) Founder Director, Institute of Agriculture, AMU, Aligarh and former Vice-Chancellor, Maulana Abul Kalam Azad National Urdu University, Hyderabad, Prof. V.P.S. Arora, Former Vice-Chancellor, Kumaun University, Nainital and Ex-Dean (Retired), College of Agri-Business Management, G.B. Pant University of Agricultural Science and Technology, Pantnagar (Uttaranchal), Prof. Furqan Qamar, Vice-Chancellor, Himachal Central University, Shimla, Prof. Abid Haleem, Head, Department of Mechanical Engineering and Coordinator, M.B.A. (Evening Programme), Jamia Millia Islamia, New Delhi, Prof. B.K. Sikka, (Retired) Ex-Dean, College of Agri-Business Management, G.B. Pant University of Agricultural Science and Technology, Pantnagar (Uttaranchal), Prof. Akhtar Haseeb, Former Chairman, Department of Plant Protection, Former Dean, Faculty of Agricultural Sciences, AMU, Aligarh, Prof. Javed Musarrat, Former Chairman, Department of Agricultural Microbiology and Former Dean,

Faculty of Agricultural Sciences, AMU, Aligarh, Prof. P.Q. Rizvi, Chairman, Department of Plant Protection and Former Dean, Faculty of Agricultural Sciences, Prof. Farzana Alim, Former Chairman, Department of Home Science and Former Dean, Faculty of Agricultural Science, Prof. Ziauddin Khairoowala, Chairman, Department of Commerce and Former Dean, Faculty of Commerce, AMU, Aligarh, Prof. Javed Alam Khan, Dean, Faculty of Commerce, AMU, Aligarh, Prof. S. Husain Ashraf, Department of Commerce, AMU, Aligarh, Prof. Imamul Haque, Department of Commerce, AMU, Aligarh and all Faculty Members, Department of Commerce, AMU, Aligarh, Mr. S. Mustafa K.Q. Zaidi, Former Chairman, Department of Library and Information Science and presently Hony. Secretary, AMU, Staff Association, Dr. Khursheed Ahmad Khan, Department of Kulyat, Proctor, AMU, Aligarh, Dr. Saghir A. Ansari, Former, Chairman, Department of Agricultural Economics and Business Management, AMU, Aligarh, Dr. Akram A. Khan, Department of Agricultural Economics & Business Management, AMU, Aligarh, Mr. M. Shamsuzzaman, Department of Agricultural Economics & Business Management, AMU, Aligarh.

RAIS AHMAD
4/81(5), Rais Villa,
Kabir Colony
Anoopshahar Road
Aligarh-202001

List of Contributors

Anu Chadha, Govt. College for Women, Parade Ground, Jammu (J&K).

Aparajita Biswal, Lecturer in Economics, A and A Economics, Utkal University, Bhubaneswar.

Ashish, K. Makwana, Assistant Professor, SMC College of Dairy Science, Anand Agricultural University, Anand.

D. Amutha, Assistant Professor of Economics, St. Mary's College (Autonomous) Tuticorin.

Dinesh, M.P., Assistant Professor, Department of Economics, Sree Narayana Guru College, Chelannur, Kannankara (PO), Kozhikode-673616.

Dr. Alka Gupta, Associate Prof., P.G Department of Commerce, University of Jammu, J&K.

Dr. Amrik Singh Sudan, P.G. Department of Commerce, University of Jammu, Jammu (J&K).

Dr. Anil Gupta, Principal, Govt. Degree College, Jindrah, Jammu (J&K)..

Dr. Badruddin, Head and Associate Professor in Political Science at PES College of Arts and Science, Goa University, Ponda, Goa, Email: drbadar786@gmail.com

Dr. Bibhuti Bhusan Patro, Associate Professor, MBA, Kalam Institute of Technology, Berhampur, Ganjam, Orissa.

Dr. M.R. Chandawarkar, Professor and Director, UGC Academic Staff College, Karnataka University, Dharvad.

Dr. Meena Chandawarkar, Professor and Director, UGC Academic Staff College, Karnataka University, Dharwar (Karnataka).

Dr. Radha Gupta, Assistant Professor, Baba Gulam Shah Badshah University, Rajouri (J&K).

Dr. Rais Ahmad, Department of Agricultural Economics and Business Management, Aligarh Muslim University, Aligarh (U.P).

Dr. Santosh Gupta, Ex.-Professor, Head & Dean, Department of Commerce, University of Jammu, (J&K).

Dr. Saroj Ranjan, Associate Professor, University Department of Management , Vinoba Bhave University, Hazaribag (Jharkhand).

Dr. Shivadeo Singh, Reader, Department of Management, L N M I, Patna (Bihar).

Dr. Shobana Nelasco, M.A., M.Phil., M.B.A., D.C.A., Ph.D., Associate Professor in Economics, Fatima College, Madurai-625018.

Dr. V.L. Dharurkar, Professor and Head, Department of Mass Communication and Journalism, and Former Director UGC Academic Staff College, Dr. Babasaheb Ambedkar Marathwada University, Aurangabad (Maharashtra).

Dr. Veena Rani, Assistant Professor and Head of Department, Department of Humanities, Govt. College of Engineering and Technology, Jammu (J&K)

K. Amutha, Assistant Professor in Commerce, Vidyasagar College of Arts & Science, Udumalpet,Tamilnadu.

Kuldeep Singh Manhas, Lecturer in Commerce, Govt. Hr. Secondary School, Chowki Chowara, Jammu (J&K).

Lekha Nagar, Assistant Professor, Govt. SPMR College of Commerce, Jammu (J&K).

Minali Mehra, Asst. Prof. in GCRG Group of Institutions, Lucknow. minali.mehra@gmail.com

Parminder Kaur, Associate Professor of Agricultural Economics, Department of Economics and Sociology, PAU, Ludhiana.

Rohini Gupta Suri, Coordinator Commerce Courses, D.D.E, University of Jammu, Jammu (J&K).

Ruth Rebecca, Assistant Professor, Lady Doak College, Madurai.

S. Latha, M.Com, M.Phil., P.G.D.C.A., Senior Lecturer in Commerce, Thassim Beevi Abdul Kader College for Women, Kilakarai- 623 517.

Vinay Raj R., Lecturer, Department of Management Studies, De Paul Institute of Science and Technology, (Dist.) Angamaly (Kochin).

CHAPTER

1

Tourism and the Socio-economic Life of the People in Orissa— A Perspective

Aparajita Biswal and Dr. Bibhuti Bhusan Patro

Abstract

Owing to its rich and varied topography, vibrant culture and captivating festivities, the State of Orissa offers immense tourism delights to the visitors in the State. Visitors, starting from neighbouring states to the far flung countries through the state at different times especially during the festivities. Filled with exquisite temples and extraordinary monuments, home to many thousands of prolific artists and craftsmen; and possessing beaches, wildlife sanctuaries, and natural landscape of often-enchanting beauty, Orissa is a unique and fascinating land that is, nevertheless, still largely undiscovered by tourists.

In the recent days, tourism industry is considered as one of the vital sector in accelerating the growth process of the economy by generating employment and income especially in the backward regions. As it is a cheapest, easiest and sustainable means to increase the magnitude of the economic indicators, priority should be given to increase the number of arrivals both domestic and abroad. For this, infrastructure should be created and adequate security be provided in the tourist places. The government should take steps to publicize the various tourist places and the specialized handlooms, handicrafts. Most importantly, the government should look into the policies taken by other states of the country like Kerala and Tamilnadu for developing tourism in their places and should evaluate the viability of these policies in our state.

INTRODUCTION

Tourism and its multiple dimension, though was relatively an unexplored area of academic interest and research becomes the fastest growing sector that confers benefits in terms of social, physiological and more importantly the economic well-being of human community. This sector has tremendous potential to enhance the quality of human experience in spiritual as well as marital way. It is a viable means for communities and individual to earn their livelihood. This sector creates employment opportunities and thereby generates income as it is more closely related to the socio-political, economic, cultural and environmental aspects of the people. However, during the recent days, this sector become more sensitive as it creates more viable and sustainable means to protect the natural and biological lives. During 1960's, this sector was simply considered to be "a passport to development" for most of the underdeveloped countries of the world but taking into account the limitless possibilities to generate employment and income in the poor countries, both the World Bank and the United Nations declared the year 1967 to be the International year of Tourism.

Tourism has been recognized as an industry in Odisha and sizeable revenue is earned for the State from domestic as well as foreign tourists. The Government during the Third Five-Year Plan (1961-66) first mooted the idea of tourism, which was revamped during the Fifth Five-Year Plan (1974-79). The creation of a State Department of Tourism dates back to 1973, though modifications have been made from time to time as regards functions of the Department. The Department of Tourism and Culture has been functioning in its present status since 1995. The Odisha Tourism Development Corporation was created in March 1979 and it was incorporated under the Companies Act in September 1979 (Government of Odisha, 2002).

At the commencement of the Sixth Five-Year Plan (1980-85) sustained efforts were made to improve the infrastructure for the promotion of tourism in the State. They had undertaken a number of promotional activities for the development of tourism industries as it considered being the easiest and cheapest way to raise the standard of living of the poor people and protecting environment simultaneously. Though this sector has tremendous potential to create employment (tourism in the State generated a rise in employment from 26,695 persons in the year 1996-97, to 46,103 persons in the year 2000-2001, Government of Odisha, 2002),and income in poor countries, but it is not free from various adverse effects on society, culture and environment. A number of researchers also doubt the creation of economic well being of the sector. This paper highlights therefore, the positive as well as negative impacts of tourism on the economy of Odisha.

THE TOURIST SPOTS IN ORISSA

Among many states in India, Odisha in particular occupies a very important position in the tourist map. This is a state, which is more commonly known for religious tourism. Jagannath Temple in Puri, Lingaraj Temple in Bhubaneswar and Sun Temple in Konark are the temples that are visited by thousands of devotees from inside and outside India everyday. Besides these, Rajarani Temple, Mukteswar Temple, Dhauligiri, Khandagiri, Udaygiri are the finest creations and reflections of the traditional social life of the people of the state. There are zoological parks like Nandankanan, Bhitarkanika, Similipal, and Tikarpada, which attract many nature-loving tourists throughout the year. Chilika, the largest salt-water lake, endowed with beautiful natural scenery and migratory birds in winter specifically becomes the centre of attractions for national as well as international tourists. Similarly, the large costal belt starting from Chandipur in Balasore to Gopalpur in Ganjam attracts the tourists from various parts of the country and outside. Districts like Koraput, Kalahandi, Phulbani, Bolangir, Sambalpur, Sundargarh, Mayurbhanj and Keonjhar are famous for hills, mountains, natural forests and waterfalls.

Besides the tourist spots, the state also occupied an important position in art, crafts, drama, dance, music, handlooms, silver filigree works, tie-and-life designing, etc. Though the people are poor and illiterate but they are very simple by heart and their hospitality is simply remarkable and rememberable.

THE TOURIST PROFILE IN ODISHA

Over the years, Odisha has attracted more number of national as well as international tourists. The state government has realized the importance of tourism to some extent as a sustainable means too generate employment and income and importantly the protection and development of the tourist places. The following Table 1 shows the tourist arrival and earnings from tourism in the state from 1997-2002.

As the table shows, from 1997 to 2002, the number of tourists has increased from 2863212 to 3436386 approximately by 83.32%

TABLE I

Tourist Arrival and Earnings from Tourism in Orissa

Year	Tourists from Orissa	% Share in total tourists	Tourists from states excluding Orissa	% Share in Total	Total Domestic Tourists	% Share in Total	Total Foreign Tourists	% Share in Total	Total Tourists	% Change over the years
1996	1372119	49	1401126	50	2773245	99	34,303	1	2807548	-----
1997	1396797	49	1431334	50	2828131	99	35,081	1	2863212	1.98
1998	1399661	48	1462127	51	2861788	99	33,101	1	2894889	1.11
1999	2691840	99	NA	NA	2691840	99	25,758	1	2717598	-6.12
2000	1413956	49	1474436	51	2888392	99	23,723	1	2912115	7.16
2001	1537855	49	1562461	50	3100316	99	22,854	1	3123170	7.25
2002	1703002	50	1710350	50	3413352	99	23,034	1	3436386	10.03
2003	1876571	50	1824679	49	3701250	99	25,020	1	3726270	8.44
2004	2156986	52	1968550	47	4125536	99	28,817	1	4154353	11.49
2005	2293865	49	2339111	50	4632976	99	33310	1	4666286	12.32
2006	2699654	51	2540242	48	5239896	99	39141	1	5279037	13.13
2007	3252380	54	2692510	45	5944890	99	41880	1	5986770	13.40
2008	3846512	60	2511933	39	6358445	99	43966	1	6402411	6.94
2009	4139156	60	2752054	40	6891210	99	45684	1	6936894	8.34
2010	2731246	59	1812481	40	4543727	99	32055	1	4575782	-34.03

Source: Govt. of Odisha, Dept. of Tourism and Culture, Bhubaneswar.

percentage. Similarly, the income from tourist expenditure was Rs. 988.69 Crore in 1997 and increased to Rs. 1170.38 crore in 2002 by 84.48%. The table also shows a declining trend both in the number of tourists and income earned in 1999. This was the year of super cyclone in Orissa that deviated from the entire social, economic and political map of the status. But the subsequent years show a continuous increase in both the variables. From 2002 onwards, there is a significant increase in the number of domestic as well as foreign tourist to Odisha.

In order to attract more number of tourists, a number of issues need to be taken care off from which accommodation is one of the most important. A comfortable accommodation facility needs to be developed along with basic infrastructure like roads, railways, airways, electricity, drinking water position, etc. In the various tourist spots of the state, there are government managed "Panthanivasas" to provide better care and hospitality to the tourists. The following table shows the number and name of the Panthanivasas, the number of rooms and bed capacity from 1997 to 2002.

TABLE 2

Number of Panthanivas, Tourists Visited and Income from Tourists in Orissa

Sl. No.	*Year*	*Number of Panthanivas*	*Number of rooms*	*Total Number of Beds*	*Total Guests checked in*	*Total Income in ('000 Rs.)*
1.	1997	12	242	512	47869	13720
2.	1998	12	242	512	47517	15529
3.	1999	13	284	628	42874	13166
4.	2000	14	300	666	52327	15562
5.	2001	14	300	666	52327	15562
6.	2002	14	300	666	59169	18603

Source: Govt. of Orissa, Dept. of Tourism and Culture, Bhubaneswar.

This table shows that from 1997 to 1999 there is an increase in no. of rooms as well as number of beds which is very less. Similarly, from 2000-2002, the number of rooms as well as number of beds remains constant. This therefore needs immediate attention if we want to attract more tourists at all.

As well known, this sector has extensive opportunities for caring foreign exchange. To get more foreign exchange, we need more tourists from outside and for that we have to create and develop the infrastructure. The following table reflects the number of Panthanivasas, number of tourists both Indian and foreign and the total income earned from them.

TABLE 3

Tourists Visited and Income from Tourists in Orissa

Sl. No.	*Year*	*Estimated inflow of money through tourist expenditure (Rs. in Crore)*		
		Domestic	*Foreign*	*Total*
1.	1997	947.08	41.61	988.69
2.	1998	958.35	39.26	997.61
3.	1999	901.44	30.56	931.99
4.	2000	967.26	28.14	995.4
5.	2001	1038.23	27.11	1065.34
6.	2002	1143.06	27.32	1170.38
7.	2003	NA	NA	NA
8.	2004	2151.10	74.80	2225.90
9.	2005	2334.91	88.21	2423.12
10.	2006	2673.77	79.29	2771.06
11.	2007	3088.21	106.93	3195.14
12.	2008	3254.65	135.46	3390.11
13.	2009	3566.89	150.83	3717.72

This table reflects the inflow of money to the state from 1997 to 2010 (August). This shows a fluctuating growth rate but with an increase in the absolute amount of spending by the tourists. The growth rate of spending in the year 2004-05 was 70.60% but by the year 2009-10 it falls to 9.70%. Therefore, a high degree of attention needs to be taken by the government of Odisha and to explore the multiple possibilities to enhance the inflow of tourists spending to the state.

THE MULTIPLIER EFFECT OF TOURISM: THE ECONOMIC IMPACT ANALYSIS

The multiplier effect of tourism analyses its economic impact on the economy. This concept is based on the idea of sales of one firm and the purchases of the other firm within the local economy. In order to translate tourist expenditure data into economic impact information, the appropriate multiplier values have to be calculated. The multiplier effect of tourism expenditure can be analyzed on the basis of sales multiplier and output multiplier.

Sales multiplier measures the amount of additional sales created in an economy as a result of increase in tourist expenditure. Similarly, output multiplier measures the amount of additional output generated in an economy due to rise in tourist expenditure and that can be expressed in terms of income, employment, revenue, etc. Here, additional income may be expressed in terms of wage, salary, rent, interest and profit while the employment generation can be expressed by the employment of the people in the production of funds demanded by the tourists as well as the various services. However, the government revenue is also affected positively due to rise in tourist expenditure. The multiplier is calculated by taking the total income generated including primary, secondary and tertiary sectors to the total expenditure made.

THE SOCIAL AND CULTURAL IMPACT OF TOURISM

Tourism has same social-cultural impacts, which need greater attention, as it is closely associated with the social and cultural like of the people. This sector affects the social life of the local people significantly as it leads to change in moral behaviour, language religion and most specifically it leads to rise in prostitution crime and gambling. The increase in these factors deteriorates the social value and leads to more instability in the society. Similarly, the mixture of culture leads to non-material custom, physical product and cultural communication.

TOURISM AND ENVIRONMENT IN CONFLICT

In most of the cases, it is found that tourism and environment are in conflict with each other specifically with respect to vegetation. Many researchers also concentrate on the relationship of tourist behaviour and the magnitude of physical damage caused to vegetation. This includes:

- The collection of flowers, plants, fungal result changes in species composition
- Careless use of fire in tourist sports caused major setbacks to vegetation.
- Deliberate chopping of trees for tent poles and firewood remove many younger trees, which alters the age structure of plant community.
- Excessive dumping of garbage results changes in nutrient status of soil by blocking light and air.

- Pedestrian and vehicular traffic impact directly on vegetation and it becomes a problem when intensity of use is excessive and this depends upon the vulnerability and campaign of the Ecosystem. It results in the reduction of areas for gazing and recreational use.
- With the development of trails and picnic areas, damage caused to vegetation and also extended to surrounding areas.
- With continuous use of an area, there may be decline in the diversity of species and only the most tolerant species may appear.
- Some vegetation covers in grassland ecosystems vegetarian cover in grassland ecosystems vegetation little deterioration because of its high proportion of resilient spices. Hence, the effects of tourism will vary greatly from ecosystem to ecosystem.
- The reproduction rates are largely affected in trampled areas.

MEASURES TO DEVELOP TOURISM

This sector occupies an important place in the process of economic development of a poor country specifically. As this sector is having high potentiality to generate employment and income with less capital investment, it becomes more influential and vibrant. There is a need to take appropriate steps to develop this sector—

- Development and availability of facilities is highly needed in the tourist spots to attract more tourists.
- The development of basic existing infrastructure should be given highest priority.
- There should be positive attitude and hospitality towards the tourists.
- The government should take necessary steps, so that they can stay, move and enjoy safely.
- People living in the surrounding areas of the sport must cooperate and they should not try to exploit or criticize the tourists.
- The people dealing with the tourists must be respecting their culture and tradition so that the tourists will have the interest to know our culture and this can lead to more of tourists.
- Private and corporate investment in this sector should be encouraged hugely.

PROBLEMS OF TOURISM/NEGATIVE IMPACTS OF TOURISM

1. Opportunity Cost

It is highly essential to investigate the relative economic benefits of investing in tourism as opposed to invest in other industries.

2. Overdependence on Tourism

A perfect balance needs to maintain among other economic sources, as tourism is highly sensible to changes like price, policy, nature, etc.

3. Inflation and Land Values

Due to inflationary rise in the prices of goods, sometimes the commodities may not be accessible to the local people. Similarly, increase in land values creates habitat construction by reducing the agricultural and industrial lands that has a far-reaching social and economic impact.

4. Seasonality

This sector is highly sensitive towards seasonal fluctuations and it creates unsteady production during the off seasons. This ultimately leads to cut down of employment and income.

5. Leakage from the Economy

There is always a demand to import services especially to cater into the needs of tourists from developed nations which becomes a problem to manage during the off peak season.

6. External Costs

There is also the possibility of incurring external costs like garbage collection, maintenance of the spots during off peak, payment of salary to the employees, etc.

CONCLUSION AND RECOMMENDATIONS

The state of Odisha can be a successful tourist destination if the industry is encouraged. So far, tourism has been developed by the State in coastal regions and a few inland sites. The State Government has not given due and appropriate importance to develop and/or enrich tourism from an ecological and cultural point of view. For this reason, the following recommendations are made:

- Attempts should be made to conserve the physical ecology as well as the cultural ecology of the ethnic tribal communities by empowering them through a participatory protected area management approach.

- There should be a crackdown on illegal encroachments of the heritage sites. Permission for construction of structures within these zones should be denied.
- The ethnic communities should be encouraged to enrich their ethnic heritage and skills so as to make their traditions more attractive rather than less in the face of change. Eco-tourism (inclusive of its ethno-component) should provide an opportunity for these tribal communities to generate more income from the tourism business in a dignified manner.
- The whole approach of cultural integrity of the tribal communities and tourism should be honored on the basis of right perspective for the tribal communities, not merely on income generation perspective for the State. The fabric of native culture reflected through folklore, folk music, folk dance, and customs should be promoted through required value addition.
- Studies should be conducted to establish the possible effects of eco-tourism on the tribal people and their cultural life.
- Development of infrastructure and safety measures in the interior of the State should be of prime importance in the intervention to establish eco-tourism in the State.
- Greater investment should be made in the promotion and preservation of monuments with cultural, historical and mythological significance.
- Overall, the eco-tourism of the State should effectively protect indigenous people and cultures from external threats; recognize their traditional rights on land and water; recognize their rights to control and co-manage these resources; allow participation of traditional institutions in the management of natural resources; and recognize the rights of these people to determine their own development priorities.

References

Archer, B. (1996), *"Economic Impact Analysis, Annls of Tourism Research, 23(3)*, pp.704-7".

Briassoulis, Helen. (2002). Sustainable tourism and the question of the commons. *Annals of Tourism Research*, 29:1065-85.

Briguglio, L. Butler, R. Harrison, D. Filho, W. (eds.) (1996), *"Sustainable Tourism in Islands and Small States: Case Studies"*, Pinter.

Chopra, S. (1991). *Tourism and Development in India.* Ashish Publishing House, New Delhi.

Eadington, W.R., and Redman, M. (1991), *"Economics and Tourism"*, Annals of Tourism Research, 18, pp. 41-56.

Economic Intelligence Unit (1973). The Role of Tourism in the Economic Development a Benefit or Burden. *International Tourism Quarterly* (2): 53-68.

Economic Survey, Govt. of Odisha (Different Volumes).

Government of Odisha, (2002). *Odisha the Souls of India—Home of Art and Culture*, Department of Tourism and Culture, Bhubaneswar. Govt. of Odisha.

Henderson, D.M. (1975). The Economic Impact of Tourism: A Case study in Greater Toyside. *Tourism and Recreation Research Unit*, University of Edinburgh, Research Paper-13.

Mathieson, A. and Wall, G. (1992), *"Tourism: Economic, Physical and Social Impacts"*, Longman.

Oppermann, M. and Chon, K.S. (1997), *"Tourism in Developing Countries"*, ITBP.

Singh, P. (1992). Tourism Industry—Problem and Prospects, *Unpublished Ph.D* Work, APS University, Rewa.

T. John (1999), *"The Economics of Leisure and Tourism"*. 2nd Edition, Butterworth-Heinemann, Jordan Hill, Oxford.

Wiley, J, *"Travel Tourism and Hospitality Research: A Handbook for Managers"*.

APPENDIX

TABLE 1

Distribution of Tourists Flow to Orissa in Different Years

Year	*Domestic Tourists*	*Foreign Tourists*
1990-91	11,96,861	29,428
1991-92	12,42,746	30,445
1992-93	12,59,003	26,335
1993-94	13,09,330	25,824
1994-95	13,28,057	25,978
1995-96	13,71,642	30.218
1996-97	14.09,178	35,318
1997-98	14,33,246	35,390
1998-99	14,75,406	31,622
1999-2000	12,60,873	21,433
2000-1001	15,25,992	25,565

Source: Govt. of Odisha, Department of Tourism and Culture, Bhubaneswar.

TABLE 2

Distribution of Tourists Spending in Orissa in Different Years

(In Crore)

Year	*Domestic*	*Foreign*
1990	73.96	5.98
1991	86.37	6.33
1992	99.85	6.40
1993	108.75	6.43
1994	124.39	7.44
1995	139.76	8.82
1996	151.85	11.31
1997	479.32	41.61
1998	489.64	32.26
1999	443.56	30.55
2000	493.76	NA

Source: Govt. of Odisha, Department of Tourism and Culture, Bhubaneswar.

TABLE 3

Distribution of Foreign Tourists who Visited Museum of Men at SC and ST R and TI at Bhubaneswar

Sl. No.	*Foreign countries*	*No. of Foreign Tourists*	*Per cent of Foreign Tourists*
1.	France	16	5.90
2.	Italy	16	5.90
3.	Israel	7	2.58
4.	The Netherlands	32	11.75
5.	Belgium	21	7.74
6.	Switzerland	13	4.79
7.	Germany	26	9.59
8.	USA	23	8,48
9.	Japan	6	2.21
10.	Canada	6	2.21
11.	Spain	7	2.58
12.	Denmark	12	4.42
13.	UK	39	14.38
14.	Australia	18	6.64
15.	Other countries	27	9.96
	Total	**271**	

Source: Govt. of Orissa, SC and ST, R and TI, Museum Records, 1998.

CHAPTER

2

Eco-Tourism and Rural Development: A Case Study of Beaches in Goa

DR. BADRUDDIN

ABSTRACT

Goa, the tiniest state of India with its natural and long awaited beaches bordered by coconut groves, swinging wind, vast network of rivers and canals attract large number of tourists. A fun loving place, adventure, spirituality, even plain relaxation, wildlife sanctuaries, eco-plantation, and historical churches are bordered by beaches that add additional attraction. Goa Tourism Development Corporation with assistance of Beach Management Support and local NGOs continue to emphasize on emotional appeal through enlightening, participatory travel experience and welfare of villagers near the coast. Goa, presently having long queue of 32 beaches, acquired with the name of coastal villages, attract millions of tourists of diverse categories. During the peak season (October to March) the local bodies experience extraordinary rush in hotels, casinos, bars and places of tourist attractions. Emphasis is given "Keep Beaches Clean and Natural" so as to maintain the ethics of 'Green Principles'. Incentives are also given for sale of eco-friendly products, inhabitants in cottages built with bamboos and naturally filtered resources, collection of articles in natural museums, organization of open air yoga camps, hiking, trekking , birds watching, boat cruise water sports and host of more attractions. Maintenance of turtle nest sites, water springs, medicinal plants, unattended palm trees, coral reefs, hillocks, waterfalls, springs and naturally preserved animals are important for eco-tourism. The unexplored 106 kilometers coastal zones beaches in Goa have vast potentials for conservation and preservation, blessed with variety of flora and fauna, no have become integral part of rural Goa. Foreign tourists and their instance to enjoy on isolated rural beaches rather than crowded urban ones is more interesting to study the relevance of emerging rural tourism in Goa.

Keywords: Goa, Beach Tourism, Eco-Culture, Sustainability, Rural Development.

ECO-TOURISM AND RURAL DEVELOPMENT: A BACKGROUND

Eco-Tourism often associated with sustainable development, is responsible travel to fragile, pristine, and usually protected areas that strives to be low impact and (often) small in scale. It helps to educate the travellers, provide funds for *ecological conservation*, benefit the *community development*, enhances the political empowerment of locals, and foster respect for different cultures. Most serious studies of ecotourism including several university programmes now use this as the working definition. Ecotourism may appeal to ecologically and socially conscious individuals. Generally it focuses on volunteering, personal growth and environmental responsibility. Such process typically involves travel to destinations where *flora*, *fauna*, and *cultural heritage* are the primary attractions. One of the goals of ecotourism is to offer tourists insight into the impact of human beings on the environment, and to foster a greater appreciation of our natural habitats.[1]

Goa, a tiny emerald land with its natural scenic beauty, capital of rural beaches, feasts and festivals and above all hospitable people with a rich cultural milieu, has made the state as an ideal holiday destination with long queue beaches attached to fishermen rural community and development of local economy. Eco-lover beaches of Goa provide travellers to natural destinations by building environmental awareness and direct benefits to the villagers by respecting local indigenous culture and supporting human rights movements. Recent initiatives to promote Eco-Tourist Explores and eco-destinations are working with help of several NGOs in Goa. All are helping to understand cross cultural studies on tourism, emerging issues on educational training, community sensitization and participation activities. Goa's 106 Kilometer coastline is dotted with several world-class beaches which acquire their names after the nearby village or locality. Kissed by the passionate, azure sea and embraced by the land and shaded by swaying palms, the golden sand sparkling in the sun are the most desired places one can ever love on earth. The evergreen eco-friendly destination of Goa is a holidaymaker's paradise. Breathtakingly blue beaches, sensuous silvery sands, fabulous flora and fauna, rich cultural heritage, archeologically captivating churches, temples, feast and festivals, Goa is blessed by all. By any means, Goa continues to earn heavy exchequer while comparing the average population of any other Indian state. It has opened foreign revenue earnings, attraction of external/foreign investments for local infrastructure services, diversification of local employment and income, service sector employment opportunities, Agriculture, Fisheries, Handicrafts, Cottage industries, and Development of export potentials for local products. Besides, the Tourism has also benefited seasonality of income and employment, small number of stakeholders, exposure to new information, lifestyles and historical and cultural heritage. Most importantly, it has widened Incentives like creation of resource and management: e.g., waste removal and strong environmental regulatory mechanism. Unlike any other part of India, Goa has potential for higher safety, rich infrastructure, least possibility of bargaining in hotels for accommodation needs, and permanent venue of International Film Festival of India (IFFI) due to public awareness. A latest data shows that Goa tourism earns between Rs. 850 million to Rs. 1200 millions annual from tourist sectors only.[2] The state government is also negotiation and tie up with Spain, France, Indonesia, USA, Japan, etc. to promote cross regional tourism in Goa. [3]

GOA TOURISM: A HISTORICAL BACKGROUND

When Goa joined the Indian Union in December 1961, its' only large industry was the export of its rich ores of iron and manganese. Goans largely depend on indigenous fish and coconuts for their vital nutritional requirements. The popularity of milk and vegetables amongst local inhabitants was low, and most of their dietary and other consumer items were imported. Before 1961 during Portuguese law, Goa had very little .development in the modern sense of the term. With rich coastal biodiversity and abundant natural resources, the quality of life and the social fabric was comparatively good. Its main assets were its genial and peace loving people, a tranquil and unspoiled environment, and an absence of population pressure. Tourism was adopted as a key sector for Goa's development, not only for the well-established reasons of increasing income and employment but also for its potential to generate non-manual employment in a state with an increasingly educated work force and limited industrial growth.[4] Fearing industrial pollution, the planners and decision-makers opted for tourism as an avenue to earn income over increased industrial development in addition to mining. Very little awareness and understanding existed back then among planners about the processes of the life support systems of the coastal environment and the interactive roles played by each component.[4]Goa was under Portuguese rule for over 450 years and was liberated on 19th December 1961, when there was sudden demand of tourism due to reduction taxes on air services. The emergence of hippie culture from South East Asia, France, England and some other European countries, Goa became popular tourist destination. 'Hippie Trail' became the starting point to develop new ways of Tourism in Goa.[5] It made Several important areas including beach tourism, medical tourism, educational tourism, adventure tourism and eco-tourism though most of these remained incomplete for several years. During 1985. Goa then a union territory, came up with proposal to set-up tourism policy. According to official tourist statistics, while 1981 to 1991 saw an increase in domestic and international tourists. In 1991, the population of Goa was 1.2 million and the tourist inflow was as high as 8 Lakh. By 1998, this ratio had risen to 83. This increase incorporates the growth in both the numbers of international as well as domestic tourists, with the share of international tourists showing an increase from 6.7% to 22% between 1981 and 1997. During the period 1981 to 1983, the share was less than 10%. It touched 11% to 12% in 1985-86, declined in the following 3 years, and showed a continuous rise after 1991, reaching 22% in 1997 represents the pattern of growth in tourist arrivals to Goa from 1981 to 1997. By the end of 2011, it is projected that there will three million tourists in Goa. International tourism in Goa is directly connected with the number of Scheduled Flights operating from South East Asia, and Europe. Some of themes include: .Intasun (UK), Lufthansa (Germany), Club Mediterranean (France), JAL (Japan), Arke Fly (Netherlands), Aeroflot (Russia), SIL (Singapore), and Thai Airways (Bangkok) . During the year 2008-09, the charter flights comprised 40% of the total tourists who visited to Goa. [6] During the year 2009-10, over 637 chartered flights brought 155428 tourists. Goa Tourism Development Corporation (GTDC) agrees that the increase in the scheduled flights has given options to tourists to decide on their arrival and departure dates.

While accepting Goa as a major tourist destination, several NGOs have come forwards to "Keep Beaches Clean, Healthy, Wealthy-wise and natural" so as to maintain the ethics of 'Eco-Culture' where tourists are encouraged to purchase eco-friendly products built with bamboos, coconut wastes, and water shells. Goa's naturally filtered resources, collection of articles in natural museums, open air yoga camps, birds watching, boat cruise water sports, waterfalls and carnival. Maintenance of turtle nest points, medicinal springs and plants, palm trees, coral reefs, hillocks, naturally preserved animals are important source for rural tourism.[6]

RURAL COMMUNITY AND MOVEMENT FOR ECO-TOURISM IN GOA

Goa with immense kaleidoscopic ecosystem and the wealth of cultural heritage has better scope for eco-tourism. Efforts to activate NGOs by involving large number of environmentalists, teachers, students, professionals and host of other bodies that have already joined the mainstream to promote eco-tourism in Goa. Tourist cottages, sea food stuffs and beverages are served in a natural environment. Still there is a greater need to promote the unexplored avenues related to eco-tourism and better infrastructure on beaches, hotel managements, road safety, communication networks and recreational facilities. Goa Tourism Development Corporation (GTDC) and National Tourism Authorities (NTA) need to coordinate with local NGOs for sustainable developments. The emphasis on tourism with conservation of environment with support of Goa International Travel Mart (GITM) on responsible tourism is a step ahead for community confidence. In fact, tourism depends on nature which cannot exist independently. But Goa needs to work hard to ensure sustainability. GITM with support of other agencies is now engage to promote rural based tourism at a wider scale.[7] In several instance, Goa continues to earn dividends by taking up 'Beach Cleaning Abhiyan' with support of NGOs, and large number of Schools Children as welcome move for tourist arrivals.[8] In fact, 80% of Goan beaches are surrounded by rural Goa only. The initiative have come from large number of organization that strongly support rural tourism in Goa. It briefly include: Goa Foundation, Goa Desk Resource Centre, Goa Institute of Rural Development and Administration, Goa-Net, Goa-Can, Goa International, Goan World, Goan Link, and Global Goan Conventions.[9]

NEEDS OF ECO-TOURISM IN GOA

Beach culture in Goa has historically been associated with sustainable community, rural development and maintenance of fragile eco-system through their classical maintenance. Goa is blessed with world known beaches like Colva, Calangute, Anjuna, Vagator, Baga, Palolem, and Agonda and the indigenous population in the surrounding villages have openly criticize the local bodies to allow hotels, parlors, bars, clubs, relaxed centers, and dance centers where no equitable distribution of resources have been locally allocated. Rural lands have been either seized or sold while locals villagers, mostly fishermen have been deprived for their basic rights. Even large number of domestic and foreign tourists have been cheated by taxi operators and other transport services. Let transport operators be the friends of tourists to see Goa as a rising star. Directives have been taken up by Goa International Travel Mart (GITM) with support of several Asian and European nations to promote tourism in the best possible manners.[10]

TOURIST ARRIVALS IN GOA: A DEMOGRAPHIC PROFILE

Goa's population as per the 2011 census stands 1457723 though it does not include seasonal migration. In exceptional cases, tourist stay for more than six months and prefer to spend leisure time on beach side villages in huts, bamboo made shacks, open terrace and *Gaddo* (typical mini huts) truly enjoy the sun burns which are rare in Europe's cold climate. Goa receives the largest number of tourist from Russia followed by UK, Germany, Finland, France, Switzerland, USA and Thailand. More recently, the state has influx of tourists from Central Asian republics like Kazakhstan, Uzbekistan, and Kyrgyzstan. There is a large number of tourists coming from South East Asian with an average 7% growth. The domestic tourists come from north mainly Punjab and West Bengal. Table I shows the number of domestic and foreign tourist who have visited from 2000 to 2010.

TABLE I

Tourist Arrivals in Goa (Domestic and Foreign): 2001-10

Year		*Number of Tourists*		*Annual Growth (%age)*
	Domestic	*Foreign*	*Total*	
2001	1120242	260071	1380313	8.8
2002	1325296	271645	596941	15.7
2003	1725140	314357	2039497	27.7
2004	2085729	363230	2448959	20.1
2005	1965343	336803	2302146	-6.0
2006	2098654	380414	2479068	7.7
2007	2208968	388457	2597443	4.8
2008	2020416	351123	2371539	-8.7
2009	2127063	376640	2503703	5.6
2010*	2201020	441000	2642020	17.6
2011	25,0000	5,0000	30,0000**	20

*www.goacom.com/joel/news, Accessed on 16 March 2011. **Projected Data Rounded.

Source: Economic Survey, 2009-10: Government of Goa, Directorate of Planning, Statistics and Evaluation, Panaji.

GOA TOURISM: INFRASTRUCTURE DEVELOPMENT (2009-10)[11]

1. Government of India, Ministry of Tourism has sanctioned financial assistance to the tune of Rs. 430 million for the Development of Tourism Jetty and parking lot at Panaji. While Beach Safety Management System continues to work for the development of econ tourism and Capacity Building.
2. Union Government is serious in setting up Convention Centre at Capital city—Panaji. A site for setting a Golf Course has also been spotted and the work is already is on progress.
3. New Shack Policy has been promoted for smooth running of the temporary shacks on identified basis. During the year 2010, the state government has approved 37 shacks on different beaches of Goa.
4. Development of infrastructure faculties like beautification of important tourist destinations like unexplored sea beaches, wild life sanctuaries, improvement of roads, and several safely measures are on. The work is on full swing.
5. Several eco-friendly areas in and around Temples, historical mosques archeological excavations, and Churches of Goa with support of GTDC are already working.
6. Beach cleaning and regulation of the activities to maintain the free flow of tourists. Large number of NGOs are working to maintain the cleanness of Beaches and tourist destinations.
7. Extra Safety measures are being hunted to respects the sentiments of foreign tourists.

NEW TOURISM POLICY OF GOVERNMENT OF GOA[12]

The Tourism Policy and Tourism Master Plan, 2011, Goa is a step forward to promote private sectors in number of ways:

1. Diversification and value addition while emphasizing the basic infrastructure of the Government to upgrade the facilities without violating the rural and eco-culture.
2. Due incentives to the existing private tourism initiative through package of fiscal and friendly taxation measures on priority basis.
3. Regulatory measures to ensure social, cultural and environmental sustainability are already started.

4. Involvement of village community in the tourism sector for its own benefits and to create trained manpower primarily from the rural population at global measurement.
5. Life Safety Measure Beaches (LSMB) has been given a major thrust in order to reduce the drowning deaths near zero level with support of Beach Trained Life Guards with necessary equipments. Special Tourist Security Organizations are being provided. In recent years, LSMB is serving as an instrument of life security.

Goa being the best tourist destination in South West India, is being promoted through various communicative means to explore the possibility and avenues. It will provide an access by creating jobs for local community. It is being promoted through Tourism Marketing and Promotion participated in travel related overseas events at Beijing International Tourism Expo-2009, Leisure 2009—Moscow, Dubai Expo 2009, Love Goa Tourism Thailand 2009, WTM-2009 London.[13] Besides, the Road Shows at Finland, Denmark, Switzerland, Rome, London, Kuala Lumpur, and Jakarta have greater relevance. The domestic events at TF Calcutta, Hyderabad, Mumbai, Pune, IITM Bangalore, TTF Ahmadabad, Holiday Expo Travel and Tourism Exhibition at Visakhapatnam etc are instrumental for Goa tourism. In a recent move, the exhibit of 2nd Goa International Travel Mart (GITM), there was an extra ordinary Tourism Event of 1000 Exhibits at Goan capital Panaji where large number of tourists participated. Most exhibited exposed the relevance of rural handicrafts.[14] In UK itself there are large number of Goa Diaspora A Goan Tourism Company, Goan Hospital in UK, Holiday Destination Office, Goan Overseas Association, Goan Cultural Society, UK, Young London Goans, etc. are helping to promote English Tourists in Goa.[15]

BEACH-RELATED ECO-FRIENDLY ATTRACTIONS IN GOA

Goa offers wonderful climate, rich history and large number of unattended beaches. Away from the beaten track, traveler discover the romantic flora and fauna that catch the imagination of visitors that include prominently rural based elephant safari, open jungles, wildlife Sanctuaries and bird watch islands. But it is unfair to say that eco-tourism flourished only in jungles, wild lives, and natural plantation. A broad visionary eco-tourist finds better options in sea beach related activities like nature-friendly dolphin watch, fishing in deep sea beaches, the habit of sea turtles, bamboo made shacks as relaxed place, and several beach-related activities in Goa. Let us briefly examine some of these attractions in Goa.

1. Dolphin Watch

Dolphins belong to large group of small, gregarious, streamlined whales or one of two species of oceanic sport and food fishes. These dolphins are small toothed whales, usually with a well-defined, beaklike snout. They are loved ones friendly intelligent and playful mammals that delight visitors who want to enjoy the beauty of Arabian Sea. In Goa, dolphin tours include time for stretching out and relaxing, snorkeling and taking pictures. Don't forget your camera and some sunscreen. Since Key West guides for Dolphin adventures in Goa are environmentally aware, they prefer to take visitors to observe Dolphins in deep seas. The number of Dolphins are not known as these eco-lovers creatures freely move from one beach to another and across the sea. By a conservative estimate, over 2400 dolphins have been spotted, a combined strength all twenty seven eco-friendly beaches of Goa. Who can resist the thrill of watching dolphins frolic in their natural surroundings, cavorting in graceful curves above the sea in Goa, a really an adventurous experience, probably a rare example in India.

2. Crocodiles Watch in Goa—A Unique Experience

Crocodiles are long-snouted, lizard like carnivores whose feed depend on fishes, turtles, birds, and small mammals. Crocodiles swim and feed in the water, floating at the surface to wait for prey, but bask in the sun and breed on land. Goa offers wonderful experience. Stray crocodiles have been spotted at several prominent beaches and widely rural dominated like Morgim, Talpona, Galgibaga, Polem, Talpona, Agonda and Palolem while Water Lands of Galgibaga, Chorao, Siolim, Mayem and Carambolim are well read cases. Galgibaga highest on all beaches, has spotted over 39 while Agonda beach famous is home of Crocodiles of varied species. Goa Foundation, National Institute of Oceanography, Department of Maine Science, Goa University and several environmental related NGOs continue to hunt fresh data on such conservation essential for eco-tourism.

3. Sea Turtle Conservation

Sea turtles inhabit the land and the sea. They connect the shallow near shore waters to open sea, cold temperate to warm tropical waters. They migrate across ocean basins. Since thousands of years, they are connected us ecologically and culturally to the sea. Goa's three prominent beaches least known and rarely visited by the tourist—Morgim, Agonda and Galgibaga - are fortune to be home of turtles. These little marine-animals are being frequently found nesting throughout the coast. The Centre for Environment Education Goa (CEEG) has been working on 'Turtle Conservation' since 1984 while large number of environmentalists, teachers and students are being familiarize about such eco-friendly Sea Turtles and often visit the sites as part of educational tours. Cohabitation activities are being watched by marine researchers. 30th Annual Symposium on Sea Turtle Biology and Conservation, Goa, India, 27th-29th April, 2010 has come out with details about the tourist attractions in Goa.[16]

4. Beach Shacks: A Relaxed and Comfort Life

Beach Shacks have more recently come up as a part of relaxec and econ friendly approach. These are Conical roof made of eco-friendly coconut palms, the wooden filigree bordering the circular interior, the privately maintained spaces and the mouth watering dishes served hot by courteous smiling faces add to the lasting impression. Tourism shacks is essentially a family business since 1960s that came from 'hippies' or 'flower children' with a views to promote natural over for beaches against the western culture. Much before the advent of hotels, the shack culture attracted large number of tourists from Portugal once the Goan colony. These shacks also served the Goan Dishes a 'family spirit'—fish curry rice, coconut and most recently the dishes of varied interest . Several Goans used leafs to enjoy curry rice. [17] Several agencies like Benaulim Environment Trust, Colva Environment Trust, Vagator Beach Management, Anjuna Beach Trust are active in promoting shacks culture. During New Year celebrations the shacks are over crowded with domestic and foreign tourists. It is another means of entertaining by way of fast food, frolic and fun.[18]

There are working with support of Goan Traditional Shacks Owner Association (GTSOA) and Shacks Owner Welfare Society (SOWS) and Drishti Special Response Services (DSRS).[19]

Goa is has large number of famous beaches and islands are connected with mainland by various communications. It is a state of seven rivers and their estuaries, there are hills with lush green vegetation, and most of the tourists are needed to cover large distances to move from one place to other. A list of islands with very brief profile is mentioned in the following table.

5. Birds-Friendly Watching in Goa

Goa is a home for large number of migratory birds from Maharashtra and Karnataka, now an important source of eco-tourism in Goa. Humid forest of Anjediva sea island, Sao Jorge Island, Diwar and Chorao river island possess extraordinary collection of 300 species of birds. Some of them are sparrow, bulbuls, Maynas and house crow are important. Others are quails, courser, pintail ducks, winged Blackbird, Malabar Whistling Thrust Shams and Hill Maynas are increasingly attractive for foreign tourists. Several eco-friendly destinations are being promoted to attract birds watching in Goa.

Atlas: Selected Beaches and Other Tourist Attractions of Goa

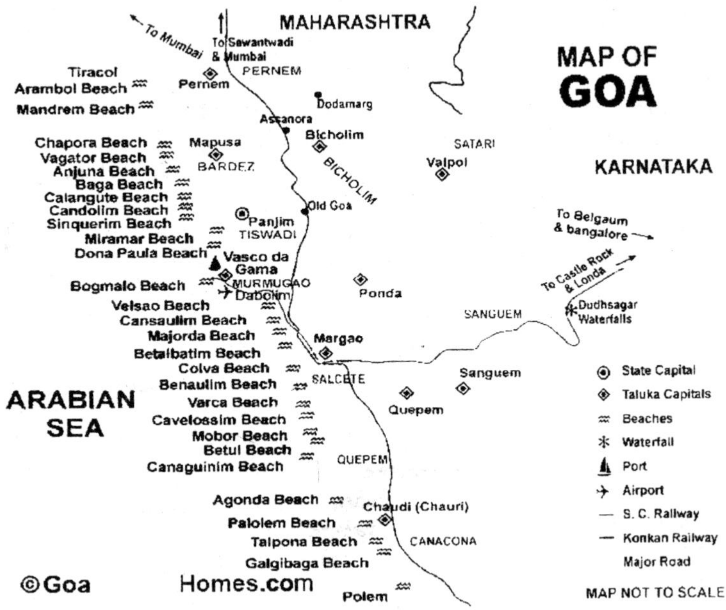

6. Uses of Mangroves-Range of Rural Services

The term 'Mangrove' applies to the thickets and forests of such plants deeply rooted beneath the mud. Mangrove fruits put out an embryonic root before they fall from the tree; the root may fix itself in the mud before the fruit separates from the parent. Likewise, branches and trunks put out adventitious roots which, once they are secure in the mud, send up new shoots. The common mangrove (*Rhizophora mangle*) grows to about 30 ft (9 m) tall and bears short, thick, leathery leaves on short stems and has pale-yellow flowers. Its fruit is sweet and wholesome and offer wide range of services and medicinal benefits to the mankind. Ecological and economical values of mangroves are recognized world-wide. They are instrumental in providing ecological, livelihood security to coastal regions and people. Large number of rural community in Goa are making use of mangroves by protecting them as a nursery ground for various fish and crab species, which form a part of their daily food. Large number of river islands are being promoted as major tourist destinations by making aware about the application of mangroves. In Goa, general services provided by the mangroves include prevention from soil erosion, stabilization of coasts and beaches, aqua culture, food beverage, boat making and most importantly reduction of the effect of Tsunamis.

7. Miscellaneous Eco-Tourist Attractions in Goa

Goa continues to attract foreign tourist who visit Goa not merely for fun and enjoyment but also to see the latent natural beauty where other sources and attractions like woodland Birds, Butterfly Diversity and Medicinal and (Ayurvedic) properties are not less significant to promote rural tourism. Ayurvedic medicine are essential hygiene, exercise, herbal preparations, and yoga cure ailments. Such treatment rely on herbal medicines, physiotherapy, and diet. Such medicines are still a popular form of health care in India, especially Kerala, Tamil Nadu, Karnataka and Goa. By a conservative estimate, over 25 to 30 percent Europeans visit Goa for Ayurvedic treatment in an eco-friendly peaceful environment and spend several months not known to commons. During post Monsoon, Goa attracts large number of eco-tourist packagers on several beach and river islands. With different eight location, Goa has collection of 257 butterfly species in various habitats. Such butterflies play an important role in various eco-system.[20]

Besides, several islands, Goa has been directly associated with rural-based wildlife eco-system where human intervention is least in exception of local communities. The recent move about environment brigade in Goa has declared these zones as 'Eco-friendly' where variety of migratory birds are found in astonishing diversity of endemic species, habitats and eco-system. Virtually 40 per cent land of Goa is covered by geographical diversity.[21] Goa's rural-based wildlife sanctuaries (Bondla, Cotigao and Salim Ali), sea islands, Lakes (Mayem) and natural springs (Kesarwal) are not less important for eco-tourists. About 1200 species of plants belonging to different families have been found in Goa.[22]

CONCLUSION

Goa, a resource rich centre yielded unimaginative foreign revenue, cultural assimilation and economic objectives. It's well defined rural supportive eco-tourist destinations all within the periphery of 50 Kilometers sea coast are of immensely important for the tourists of diverse interest. Though the present paper highlights mainly sea beaches in Goa apart from various beach-related activities, tourists are equally motivated by wild lives, historical churches, temples, historical mosques, natural springs, forts and plantations. Marine-related activities are incomplete without giving due consent to hinterland eco-friendly objects that help to learn new styles of living. Large number of tourists visit the plant herbs, trees, flowers, and local products. Several NGOs and Research organizations like International Centre, Peaceful Society,

GIRDA, GDRC, GTDC, and GPS increasingly stress the need for sustainability and environment-friendly Goa. More recently, Goa was awarded Indian's best beach Tourism destination at a conference in New Delhi. The award was given to Secretary of Tourism by Union Ministry. [23] It was in a views of keeping beaches in Goa, the most natural form unlike any other part of India. Thanks of several awareness campaigns at educational levels. There is greater need to clean the mind and thinking to promote "*Beach Cleaning Abhiyan* and *Goa Bachao Andolan*" as regular practice. More emphasis should be given on the diversification and implementation of strict law where large number of beach side located industries hotels cause serious environmental pollution. Usage of bio degradable products should be promoted among the tourists and maintenance of destinations should be under strict control of local NGOs. Education and training programme is the core of eco-tourism that needs vision and futuristic community based approach.[23]

Today, there is greater need to divert the crowd of tourists from the busiest beaches to rural based hinterland and lopsided beaches like Harmal, Kegadi, Bambolim, Sri Dao, Agassim, Arossim, Velsao, Talpona and Polem. These are most natural and are blessed with unlimited eco-friendly culture. Unfortunately, tourist are either little know or unfamiliar as the infrastructure is not developed and several agencies do not take tourists to these beaches due to public prohibition. Interaction between host and guests create new opportunities for peace and greater coordination. So it is imperative to understand pleasure and hospitality for better results. No doubt, Goa has been constructed to serve as one of the world's pleasure peripheries and cultural space for relaxation to preserve local custom. But eco-friendly culture cannot be guaranteed merely by the changes in the value system, traditional life styles, family relationship, individual behaviour and community structure. [24] Now the time has come to move hinterland tourism giving emphasis on 'Rural Tourism' that can better document the history, couture, flora and fauna and can open new avenue for eco-culture. In an nutshell, there is a greater need to establish people-friendly tourism with greater emphasis on eco-conservation, preservation and protection of land, water and other resources. Upgradation of infrastructure facilities without damaging the environmental culture and sustainable development is must. More emphasis should be given on hospitality industries by promoting the grass root level and complete ban on the sale of agricultural lands while the legislative bodies should be honest enough to hand over the ownership titles to local communities.

Notes and References

1. Prabhakar S. Angle, *Goa: An Economic Update* (Mumbai: Goa Hindu Association, 2001), p. 107.
2. "Incognito CM Quizzes Tourism on Safety", *The Times of India* (Goa), 22 October 2011, p. 4.
3. "Thrust on Responsibility, Quality Tourism", *Navhid Times* (Panaji), 22 October 2011, p. 2.
4. Visit www.http://siteresources.worldbank.org/WBI/Resources/wbi37134.pdf
5. Avinash V. Raikar, "Tourism in Goa: Is it Sustainable", *Atharva* (Goa), Vol. 4, No. 8, August 2009, pp. 4-5.
6. See, "A Bittersweet 09 for Tourism", *Times of India*, 30 December 2009.
7. "Experience for Tourists should not be at the cost of Goa's Environment, Say Expert", *The Times of India*, 23 October 2011, p. 68.
8. " Beach Cleaning Drive taken up at Colva", *Navhind Times*, 17 October 2011, p. 79.
9. For details, visit, http://worldgoaday.blogspot.com
10. "True Ambassador for Goa", *Navhind Times*, 21 October 2011, p. 10.
11. Economic Survey 2009-10: Government of Goa, Directorate of Planning and Statistics Evaluation, Panaji, pp. 43-44.
12. *Ibid.*
13. *Ibid.*
14. Joaquim Fernandez, "Most Goan Sellers Stay Away from Rs. 2 Crore Mart", *Navhind Times*, 24 October 2011, p. 3.
15. For details visit www.acroindia.com
16. http://www.seaturtle.org/iotn/pdfs/issue-9/iotn9_15.pdf
17. Sebastian D'Souza, 'Beach Shacks: Struggle for their Rightful Place of Beach", *Friday Balcao* (Mapusa), Goa Desk Resource Centre, Vol. 1, No. 1, August 1999 to February 2001.
18. *Times of India*, 10 October, 2010.
19. For the year 2010-11, *Navhind Times*, 30 September 2010, p. 3.
20. "Goa Tourism Bags Best Award", *Times of India*, 30 September 2010, p. 9.
21. Devidas V. Kotkar, "Butterfly Diversity: A Case Study of Mhadei Wildlife Sanctuary" *Atharva*, Vol. 6, No. 2, February 2010, pp. 11-12.
22. Sangrama Kesari Das, "Medicinal Plans of Goa", *Atharva*, Vol. 3, No. 6, June 2008, p. 9.
23. Devidas V.Kotkar, "Woodland Birds in Goa", *Atharva*, Vol. 5, No. 8, August 2010, pp. 2-3.
24. Ritz (2000) "Residents Perception of Socio-Cultural Impact of Tourism at Lake Balaton" in G. Richards and D. Hall, (ed.), *Tourism and Sustainable Community* (London: Rutledge, 2000).

CHAPTER

3

Rural Tourism

ASHISH K. MAKWANA

INTRODUCTION

India is an agricultural country. India is a fast developing economy. Despite this more than 65 percent of population still lives in villages. Livelihood of most of the person living in villages depend upon agriculture and allied activities. In India, still majority of agriculture output is dependent upon natural factors like rain. Due to natural calamities (drought as well as floods) villagers suffer loss. Thus, over the years returns from agricultural income has reduced. This forces the villagers to go to urban areas leading to the process of urbanization and its associated drawbacks. As the urban centres have a certain threshold limit to sustain the people and afford good quality life. This calls for urgent rethinking and stop the migration of villagers from rural to urban areas.

Rural tourism can play an important role in development of village by providing an excellent blueprint of development, carrier and quality of life. Rural tourism can be used as a means to: Improve the well-being of the rural poor, Empower the rural people, Empower the women, Enhance the rural infrastructure, Participate in decision-making and implementing tourism policies, Interaction with the outside world, Improve the social condition of lower sections of the society, Protection of culture, heritage, and nature.

DEFINITIONS AND CHARACTERISTICS

It is difficult to define the term rural tourism due to following problems: (1) It is difficult to arrive at a one single definition of rural area as rural tourism takes place due to different criteria developed by different nations. (2) Simply by occurrence of a tourist form in rural area, it cannot be categorized as rural tourism as it may be of urban form. (3) Emergence of different types of rural tourism forms in different geographic areas makes it difficult to have common characteristics throughout the world. (4) Due to impact of external forces like globalization, telecommunication, etc. rural areas are constantly changing and providing difficulty to define rural tourism, nevertheless the following definition have emerged.

Rajesh K. Aithal has defined rural tourism as a tourism form in which tourists/guest participate in various village activities and enjoy the typical culture of village life.

The World Tourism Organization defines Tourism as People who travel to and stay in places outside their usual environment for more then twenty-four hours and not more than one consecutive year for leisure, business and other purposes not related to the exercise of an activity remunerated from within the place visited.

The term 'Tourism' is defined as travel for recreational, leisure or business purposes. Tourism is the temporary, short-term movement of people to destination outside the places where they normally live and work and their activities during the stay at each destination.

Rural tourism is any form of tourism that showcases rural life, art, culture and heritage at rural locations, providing economic and social benefits and leading to interaction between local people and heuristic resulting in more enriching tourism experience.

Rural tourism is a multi-facilities activity targeted at countryside locations involving farming/agriculture tourism, cultural tourism, nature tourism, adventure tourism and eco-tourism.

Tourism can be categorized as rural, if the rural culture is the key component of the product on offer.

Thus, rural tourist place can be defined as a large area dominated by natural and/or farmed/forested environments where specific natural, economic and socio-cultural features such as tradition, local cooperation, trust and reciprocity are harmoniously embedded and as such generate a unique tourist product which is predominantly small scaled, nature-friendly, ethno cultured, i.e. sustainable.

The characteristic features of tourism products of rural tourism includes task of providing personalized contact to visitors, a task of physical and human environment of villages, allows the tourists to participate in the activities, traditions and lifestyles of local people.

Rural tourism occupies an important place in overall tourism market of any country. Many times rural tourism market plays critical role in development of specific rural areas.

CLASSIFICATION OF RURAL TOURISM

Agriculture rural tourism: Allows farmers to diversify their activities thereby enhancing value of their product and property. It reconciles farming interests and environmental protection through integrated land management wherein farmers play an important key role. Tourists who prefer and reside in farm accommodation are interested to experience rural atmosphere and share intimacy of house hold they reside in, enjoy home food drinks, etc. during their stay. Apart from enjoying the rural experience they also wish to learn traditional crafts and skills from rural people.

Heritage and cultural tourism: It is of many different forms. Each form is unique to an individual local place. It covers religious buildings (temples, mosque, church, etc.), historical monuments of literary artistic or scientific people, archeological style, traditional parkland, etc. India has a rich cultural history and reflection of its glorious past is still visible in its numerous places, forts, places of worship of various religions, ancient monuments, etc. So, tourism regarding above mentioned kinds is known as Heritage tourism. Heritage tourism itself can be further classified as colonial heritage, religious tourism, industrial, heritage, urban renewal and ethnicity. The Indian Government has started preserving the heritage sites from a tourism perspective.

Eco-tourism: Rural areas provide opportunity to watch birds and animals and obtain knowledge about different area specific flora and fauna. India has abundant Natural wealth. Eco-tourism belong to natural beauty and diversity. It is relatively new segment in India. India with it natural diversity is one of the pristine places in the world for eco-tourism. The Himalayans region, Kashmir valley, Kerala, Andaman and Nicobar Islands and Lakshadweep Islands, wild life in forest and Western as well as Eastern Ghats are some of the hot spots for eco-tourism in India.

FACTORS RESPONSIBLE FOR GROWTH OF RURAL TOURISM

Low cost gateway: The expenditure on food, travel, recreation and residence is less which widens the tourist base. In general the concept of travel and tourism is enjoyed by economically affluent urban population. But due to less overall cost of rural tourism, its scope is broadened to include larger population.

Curiosity of rural life: India is still considered as agricultural country. It is regarded as land of villages. People have affiliation to their native place and even after living in a city for one or two generations, people still refer to their ancestor village as our village. Rural people of India manifest a deep loyalty to their village and a family residing in a particular village does not easily change their residing location. With growing urbanization, the population in cities have increased. People move from rural to urban areas. All the people living in urban areas have their roots in rural areas. The growing generation of urban area who are born in urban areas have curiosity to learn about sources of food, plants, animals, raw materials like wood, handicrafts, languages, culture, tradition, dresses and rural lifestyles. Thus, there is curiosity to know about all this matters which is satisfied by rural tourism as it revolves around rural villages, farmers, agriculture and rural life style.

Demand for wholesome family-oriented recreational activities: Rural lifestyle encompassing rural games, festivals, food, dress and the nature provides recreational opportunity to persons of all age groups i.e children, young, middle and old age, male, female, etc. to the whole family at lower cost.

Health awareness: Modern lifestyle is completely different then old traditional lifestyle. In modern competitive world, in order to earn more money, people work hard. All this has made life more stressful. In order to increase happiness in life people are searching many peaceful ways. People are adopting ayurveda, organic foods. Thus, many health conscious urban people look to pro-nature village for peaceful life. Tourism is a means for searching peaceful location. Peace and tranquillity are inbuilt in rural tourism as it is away from urban areas and close to nature. Birds, animals, crops, mountains, water bodies, etc. in villages provide totally natural experience and help the visitors to relieve stress.

Educational purpose: Rural tourism also serves education purpose. Picnics, tours, educational visits/tours of varied durations are arranged by educational establishments (schools and colleges) so as to provide experience of rural lifestyle and gain knowledge.

SCOPE AND OPPORTUNITIES OF RURAL TOURISM DEVELOPMENT

There are some critical factors responsible in the evaluation and development of rural tourism as a product. Some of these changes are:

(i) Changes in need and preferences of visitors.
(ii) Destination of the natural and man-made environment.
(iii) Change or disappearance of those attractions which brought tourists to the area.

(iv) Identification of potential customer.

(v) Analyzing and responding to rural tourist destination must possess basic tourist requirements such as hygienic accommodation and catering, safety of visitors, medical facilities.

In rural areas, generally following types of products are available, e.g. historical, heritage, manmade and those based upon nature, medicine, religion, culture, etc. By undertaking necessary steps it is possible to use these products to promote rural tourism in India.

The rural tourism destinations are to be developed in a sustainable way which satisfies need of present as well as future generations. Thus, during development of a rural tourism site, along with economic, social and other factors are also to be considered.

There are many potential rural tourism sites which can be developed. But following are some of the limiting factors hindering the growth of this sites: poor infrastructure, lack of public and potential will, regional politics, lack of social awareness, illiteracy, traditional belief that only agriculture can help people, lack of tourism marketing, lack of management, poor people participation, absence of supporting industries such as hotels and restaurants, state and private transportation, etc.

Following are some of the specific rural tourism opportunities:

(1) Cultural and heritage tourism: India has a vast array of protected monuments with 22 world heritage sites, 16 of which are monuments. The integrated development of area around these monuments provides opportunities for the development of cultural tourism.

(2) India has a vast stretch of sea shore. There are number of sites providing opportunity for beach and coastal tourism. Apart from states having seashore, Andaman and Nicobar islands also provide great rural tourist opportunity.

(3) Wildlife sanctuaries and national parks of rural areas are good rural tourist attractions, e.g. Sasan Gir of Gujarat.

(4) India's fairs and festivals should as unique products also act as important rural tourism occasions, e.g. Kumbh Mela.

SUGGESTION AND RECOMMENDATIONS

Infrastructure facilities: Rural tourist destinations will be away from urban areas. The first requirement is to have good road network up to the rural tourist sites. In India there is excellent road network connecting major cities, which should be developed up to rural tourist sites.

Leadership: Successful tourism necessitates good leadership by open minded and enthusiastic people from different sectors viz community, local government, industries, trusts, etc.

Local government support and participation: Local governments' role in following areas is vital for growth of tourism: Creation and maintenance of necessary infrastructure, zoning and bifurcation of total area so as to make it appealing to tourist.

Necessary financial support: For any rural site to be developed and maintained as tourist destination, adequate financial support should be provided. Public funds may be insufficient. Therefore, it is essential to explore for other sources of financial assistance, e.g. private investments, donations, etc.

Strategic planning: This is fundamental to achieve the targets on time with efficient and effective use of resources and funds. Due to interdependence of the community and key aspects of tourism development and promotion, rural tourism planning of any site should be integrated into communities overall economic strategy. Thus, planning for rural tourism necessitates involvement of various stakeholders in the community.

Coordination and cooperation among rural tourism entrepreneurs: Tourism of any type necessitates working of different types of businesses together as by its nature itself tourism possess intertwined relation between different types of businesses such as shops, accommodation facilities, restaurants and tourist attractions. This creates different types of formal and informal networks. Many of such tourism networks require standardized, high quality business management which is lacked by small enterprises. A tourism network can substantially improve small tourism business performance by transforming their sporadically scattered products into a one stop shop selling wide verity of functionally interrelated tourism products.

Community support: Community support for tourism development and the attitudes and hospitality of local tourism workers is necessary for success of rural tourism.

Location: For rural tourism to be successful, it is necessary that the rural tourist destinations are located not more than two or three hours drive form airport or railway stations. They should also not be very near to the main city; otherwise the charm of rural environment will be lost. The rural tourist location should have inbuilt unique natural features which can be focussed and promoted e.g. unique arts and crafts, fairs and festivals, cuisines, etc.

Capacity building: For success of Rural tourism Project, regular capacity building of local people is essential. The training may be for creating souvenirs, promoting regional dance, customs and food which are based upon the local agricultural products. Instead of commercialization of rural tourism project, interest of local people should be given emphasis. The availability of trained manpower is essential to achieve excellence in the tourism sector. It is necessary to open more institutes of hotel management and catering technology and food craft institutes in the country. Certificate courses or training programmes should be organized by government to train service providers in the unorganized sectors such as small hotels, dhobis, restaurants, etc.

Necessity: Rural tourism project should match the rural environment and should not be based upon imported goods from urban areas. Therefore, it is necessary that rural tourism project should not be given urbanized look comprising latest furniture, good glass frame outside look, chandeliers, decorative center table, etc. There should be greenery and local plants be used. Eco-friendly aspects should be focussed for establishing rural tourist projects.

For a rural tourism project to succeed, it must have—

(i) *Something to see*: animals, birds, farms, culture of village, dress and festivals.

(ii) *Something to do*: Participating in rural operations, e.g. agricultural operations, riding on local animals such as camel, horse, etc., participating in rural games, e.g. kabaddi, gillidanda, gotti (marble), etc.

(iii) *Something to purchase*: Rural crafts, dress materials, farm gate fresh processed food items, etc.

Marketing aspect: For providing rural tourism experience, the natural, cultural, capital, human and other resources specific to the rural area must be attractive for the tourist. Important aspect includes attraction and its authenticity, quality of service and facilities, accessibility of the destination price, etc., it is necessary to create an environment in which tourist experiences reality. For marketing rural tourism products, the emphasis is on marketing an experience. This necessitates preparing attractive advertisement.

CONCLUSION

Rural tourism can help in shaping our society. It can have both positive and negative impacts on rural as well as urban communities. There is a scope of rural tourism in India. The government should encourage private enterprises to promote tourism in rural areas. For developing the rural tourism we need to understand the rural environment, demography, socio-culture, economic and political background of that place. How we can involve the rural people to enhance their socio-economic condition. To develop a strategic marketing plan for rural tourism we have to understand the target customer their needs and wants and how to match it with our rural infrastructure. Rural Tourism can develop a win-win situation for both the rural and urban communities.

References

Gopal, R., Verma, S., Gopinathan, R. (2008) Rural Tourism Development: Constraints and Possibilities with a special reference to agri tourism. Compendium on conference on Tourism in India—Challenges ahead 15-17 May 2008, IIMK.

Haldar, P. (2007) compendium on International Marketing Conference on Marketing and Society, 8-10 April, 2007, IIMK

Lesson-21. Emerging dimensions in Tourism: India Specific, www.ignou.ac.in

Muthe, P.R. (2010) 'Global Recession: Challenges and Opportunities for Indian Tourism and Hotel Industry' Research Analysis and Evaluation, Issue-8, RNI: RAJBIL/ 2009/30097, pp. 22-25

Trivedi, M.K. and Dixit, P.K. (2005) 'Rural Tourism in Globalization World' www.IndianMBA.com

www.planning commission.nic.in

http://gourkanjilal.com/new-articles/rural-tourism.htm

http://www.incredibleindia.org/newsite/pdf/Rural Tourism Sites.

Appendix

List of Commissioned Rural Tourism Sites of Ministry of Tourism (As on 30.6.2011)

Sr. No.	Name of Rural Site	District	State	USP of Site
1.	Pochampalli	Nalgonda	Andhra Pradesh	Cotton & Silk Sarees
2.	Konaseema	East Godavari	Andhra Pradesh	Eco-tourism (Coastal Development)
3.	Puttaparthi	Ananthpur	Andhra Pradesh	Culture (Spiritual life)
4.	Chinchinada	East Godavari	Andhra Pradesh	Eco-tourism (Coast development)
5.	Rengo	East Siang	Arunachal Pradesh	Culture and Bamboo Cane handicraft
6.	Ligu	Upper Subansiri	Arunachal Pradesh	Culture
7.	Ego-Nikte	West Siang	Arunachal Pradesh	Culture
8.	Dehing-Patakai Kshetra	Tinsukia	Assam	Culture and Eco-tourism
9.	Heritage village at Tera	Kachch	Gujarat	Heritage
10.	Hodka	Kachchh	Gujarat	Mirror work/ Embroidery
11.	Navagaon and Malegaon	Dang	Gujarat	Culture & Eco-tourism
12.	Nagar	Kullu	Himachal Pradesh	Topi and Shawl weaving
13.	Paragpur	Kangra Valley	Himachal Pradesh	Himachal Heritage
14.	Baroh	Kangra	Himachal Pradesh	Gurukul Culture
15.	Akingaam	Anantnag	Jammu & Kashmir	Culture (Folk Dance :Bhand Pathar)
16.	Kokkare Bellur	Bellur	Karnataka	Eco-tourism
17.	Attiveri Bird Sanctuary	Uttar Kannada	Karnataka	Eco-tourism
18.	Banavasi	Uttar Kannada	Karnataka	Stone machinery, Wood Carving and Musical instruments
19.	Anegundi	Koppal	Karnataka	Banana Fibre Craft
20.	Chaugan	Mandla	Madhya Pradesh	Lantana Craft
21.	Pranpur	Ashoknagar	Madhya Pradesh	Chanderi Sarees
22.	Orchha	Tikamgarh	Madhya Pradesh	Historical and Adventure (River rafting)
23.	Seondha	Datia	Madhya Pradesh	Wood and stone craft
24.	Budhni	Sehore	Madhya Pradesh	Historical, Spiritual and Wood craft
25.	Morachi Chincholi	Pune	Maharastra	Sufi tradition and Culture
26.	Mopunchuket	Mokokchung	Nagaland	Shawl weaving
27.	Avachekha	Zunheboto	Nagaland	Tribal Culture
28.	Changtongia	Mokokchung	Nagaland	Tribal Culture
29.	Leshumi	Phek	Nagaland	Tribal Culture and Adventure
30.	Thetsumi	Phek	Nagaland	Tribal Culture
31.	Longsa	Mokokchung	Nagaland	Tribal Culture
32.	Mitikhru	Phek	Nagaland	Art & Craft (Woodcraft), Handloom
33.	Chungli Yimti	Tuensang	Nagaland	Historical & Tribal Culture
34.	Raghurajpur	Puri	Orissa	Stone Craft and Pattachitra
35.	Samode	Jaipur	Rajasthan	Lac Work, Paper painting, Gems stone painting
36.	Lachen	North Sikkim	Sikkim	Rugs and Carpet
37.	Tingchim	West Sikkim	Sikkim	Trekking and bird watching
38.	Kazhugumalai	Thoothukudi	Tamil Nadu	Spiritual and Pottery making
39.	Theerthamalai	Dharmapuri	Tamil Nadu	Historical
40.	Devipattinam Navbhashnam	Ramnathpuram	Tamil Nadu	Stone Carving
41.	Thirukurungudi	Tirunelveli	Tamil Nadu	Historical
42.	Kombai (Kurangani)	Theni	Tamil Nadu	Spice
43.	Kamalasagar	West Tripura	Tripura	Historical
44.	Jageshwar	Almora	Uttarakhand	Spiritual
45.	Mana	Chamoli	Uttarakhand	Trekking Adventure
46.	Adi Kailash	Nainital	Uttarakhand	Adventure
47.	Padmapuri	Nainital	Uttarakhand	Adventure
48.	Nanakmatta	Udham Singh Nagar	Uttarakhand	Spiritual
49.	Mukhrai	Mathura	Uttar Pradesh	Folk Dance
50.	Ballabhpur Danga	Birbhum	West Bengal	Folk Dance
51.	Mukutmonipur	Bankura	West Bengal	Sari weaving
52.	Kamarpukur	Hooghly	West Bengal	Spiritual & Craft

CHAPTER

4

Present and Past Scenario of Tourism Industry in Jammu and Kashmir State

DR. AMRIK SINGH SUDAN AND DR. RAIS AHMED

OWING TO ITS RICH AND VARIED TOPOGRAPHY, VIBRANT CULTURE AND CAPTIVATING INTRODUCTION

Tourism in today's world is a very big economic and social activity, generating a large amount of income employment and foreign currency and investment opportunities. Though currently, the developed nation's are getting the lions share of the benefit of tourism, there is a promising future for the developing nation's also. The contribution of tourism towards socio-economic development and environmental conversation is immense.

The growth and unprecedented boom in travel industry has been incredible, but definitely not sudden. There has been a consistent growth pattern in the travel industry. World Travel and Tourism Council (WTTC) state that India will emerge as the second fastest growing tourism economy globally between 2005 and 2014 followed by China. Successful promotions such as Ministry of Tourism Incredible India's multimedia campaign and the budget travel boom are to have contributed to the tourism gold rush. India earned Rs. 296035 crore through tourism in

2006 and Rs. 5040320 crores in 2011. Economies of the world are changing at an alarming rate. Europe and America are growing way to shift of wealth power to China, India and for east.

Tourism along with travel, is now regarded as the world's fastest growing industry. Tourism and Hospitality sector is amongst the largest job producers around the globe. It is that investments sector which pushes the nations economically without any huge investments. According to the United Nations World Tourism Organization (UNWTO), 940 million international tourists spent US$ 919 billion in 2010 and the first two months of 2011 saw a healthy growth of 5% over the same months as compared with the previous year. It is, thus, widely recognized that tourism is globally one of the largest industries and the top generator of foreign exchange earnings for many of the world's economies. Recent statistics issued by the world Travel and Tourism Council (WTTC) clearly shows the importance and size of the tourism and hospitality industry with over 235 million jobs, 650000 alone at managerial levels, hospitality today definitely is one the fastest growing industries in the world with a Gross Domestic Product (GDP) of just over 9% in 2009 and an annual growth of 4.4%, the tourism and hospitality industry contributes billion of dollars to the global economy. Looking beyond the current economic outlook, the industry is expected to resume its leading, dynamic role in global growth. Talking this growth into consideration the world's leading tourism and hospitality. Companies are looking for skilled professional to become tomorrow's leaders.

Tourism sector provides a wide range of career opportunities both at home and overseas in many and varied industry sectors such as tour operators, Airlines, tourist attractions and hotel groups. On the other hand, hospitality industry is in continuous need of managerial Staff in hotel management, restaurant management, public house management, nightclub, management, and even includes working on cruise liners and theme parks. Both the industries are truly international with career opportunities at home and abroad. Furthermore, the Events industry is flourishing and the management of events has become increasingly important within the hospitality, tourism, leisure and Sports Sectors. Event Mangers can be fond working on a wide range of events from weddings to conferences, meetings to product launches, at festivals and major sporting occasions. Some high profile events include the Olympic Games, the Commonwealth Games, ICC cricket world cup, Grand prix, FIFA World Cup, etc.

Tourism has been traditionally viewed as a great force in promoting understanding among nations and within the national boundaries facilitating emotional integration. Tourism is the movement of people from their normal place of residence and work, for a period of not less than twenty four hours and not exceeding a year. Tourism is the most important industry in our economy without chimneys. Over the past three decades tourism has emerged as the world largest industry. Tourism is one of the most important items of exports. It is well said by Husker and Kraft, "Tourism is the sum of the phenomenon and relationship arising from the travel and residents in so far as it does lead to permanent residence and is connected to any earning activity. Today, we all know about the tourism but before the independence, in India tourism industry was not promoted by the Britishers, but soon after independence, the importance of tourism was felt vigorously by the Indian government. They set-up a committee under the committee under the chairmanship of Sir John Sergent. This committee submitted its report in 1948. It was thought at that time, it was beneficial for India to encourage and develop the tourist traffic by all means. A small tourist section was set-up in the ministry of transport in 1948. A separate department was set-up early in 1958 which is called as the department of Tourism. Tourism at that time was a new activity for India. Its importance was not realized by the people and government. This was so because the scope of tourism was low. There was not proper infrastructure such as the number of hotels was very less. There were no proper means of accommodation of the tourists. Same was the case of transport also. But now-a-days people have realized the importance and value of tourism and the economic benefits of the tourism. In India now tourism industry is developing a lot and this has increased the nation's income also.

The International nodal agency for tourism, i.e. United Nations World Tourism Organization (UNWTO) gives the definitions for tourism as "tourism comprises the activities of persons traveling to and staying in place outside their usual environment for not more than one consecutive year for leisure, business and other purposes not related to the exercise of an activity remunerated from within place visited. "As far as the tourism-international and national scenario is concerned, it is the sector which is growing at a rapid rate especially from the last two decades. As per the Travel and Tourism Competitiveness Report 2009 by the World Economic Forum, India is ranked 11th in the Asia Pacific region and 62nd overall on the list of the world's attractive destinations while the Indian Travel and Tourism Industry is ranked 5th in the long-term (10 years) growth and is expected to be the second largest employer in the world by 2019. In India, tourism is viewed and promoted as a "development paradigm" by the governments and it also considered as a major sector for economic as well as social growth.

No doubt that tourism plays its role as a backbone for most of the countries but one cannot ignore the adverse impacts of tourism on the environment and the socio-cultural. Moreover, several tourism destinations are facing increasing pressures on their social and economic structures as well s their natural environments. Although, tourism sector can generate high income and is known for its multiplier effect but inflation and economic leakages also come along with it. To earn the high economic yield from tourism, its infrastructural and super-structural impacts are underestimated. In our country, tourism heavily flourishes on environmental and natural resources, besides the biodiversity. The primary responsibility should be to protect these bio-diverse places.

Ecotourism is the vital type of tourism while the term "eco-tourism" was given by a marketing agency that was promoting Costa Rica as a rainforest destination way back in the early 1970s. Because of the ill-effects of tourism, the need to develop a certain form of tourism that was sensitive towards the environment and was felt to the local community. After 1970s, the United Nations World Tourism Organization (UNWTO) has seen ecotourism as a niche market. The several international and national organizations see ecotourism as sustainable development and it also helps to ensure the sustainable and responsible use of environmental and natural resources. Ecotourism has been formulated and studied as instrument for sustainable and equitable tourism by various international organizations like United Nations World Tourism Organization (UNWTO) and United Nations Environment Programme (UNEP). The International Ecotourism Society (TIES) defines ecotourism as "responsible travel to natural areas that conserves the environment and improves the welfare of local people." In India, eco-tourism also means to protect areas of ecological values like the national parks, wildlife sanctuaries, etc. and in short it assumes as "nature-based tourism" which include both sustainable and responsible tourism. In our country, there is a conflict between tourism, community and conservation because of the motive of high-income generation attitude of tourism service providers. They are less concerned with practicing the concept of sustainable and responsible tourism development. Basically tourism is linked to people and heavily rely on them for their overall survival and hence the role of local community or host community is vital. The immediate need to minimize the social impact and environment impact should be prioritized by the authorities so that there should be no hurdle between practicing the responsible tourism.

HUMAN CAPITAL AND TOURISM SECTOR

People are clearly fundamental to the sufficient operation and further development of the tourism industry. Many tourism products include people as a primary part of the proficiency offered whether as performers or as members of the cultural environment. There is a growing realization that labour should not be treated simply as variable costs but as human capital. A high quality skilled worksforce will ensure greater competitiveness and innovations, improve job prospects and ease the process of adjustment in changing markets. There is the commonly acknowledged proposition that economic prosperity depends on an educated workforce. It is also agreed that the increasing recognition of the economic importance of tourism itself has lent further prominence to the necessity for an expansion of tourism education. The competitive advantage of countries in a global economy increasingly depends on the availability of skilled labour. This is also true for the tourism-related industries. Therefore, the structure of and focus on (public and private) educational and vocational provisions are important issues.

A primary concern is that a lack of suitable staff will challenge the industry's capacity to meet the expectations of service quality that tourism promotions have created. The demand for tourism services to meet international standards has led to an increasing trend to employ a highly skilled labour force and although human resources are the most valuable asset of these industries paradoxically, the will to invest in education and training in some of the major sectors is comparatively low, compared with other industries. However, upgrading human resources and techniques, as well as improving management skills are essential requirement for the further development of the tourism industry.

Now let us talk about tourism and tourism industry in Jammu region as we know Jammu is popular in all over India as "City of Temples". The City is located on the bank of river Tawi. This land is known as "Dugger" and the people here are known as Dogras. We are aware that natural beauty of the vale of Jammu has deservedly won high praise from travelers. In a kind of historical spectrum in tourism, we have quoted enough in substance from accounts of travelers on the bewitching valley to show how the lovers of beauty not only adored the bounties that nature lavished on Jammu but found these a theme of Literature. Now, we will throw light upon some famous tourist places of Jammu region. In Jammu region, the most famous historical place is Bahu Fort Complex, originally built by Bahu Lochan, the rural of Bahu City on a hillock on the eastern bank of river Tawi. It was rebuilt by Maharaja Gulab Singh and served as a garrison for Dogra Army.

The famous temple of Bawey Wali Mata is inside the fort, where on every Tuesday and Sunday pilgrim went to worship to Goddess Kali. A little farther away, on a hill top, opposite the Bahu Fort, is a lovely spot overlooking the river Tawi, where a temple dedicated to Mahamaya has been constructed but people of Jammu know less about this temple. Now, there is one more attractive spot which is called as Amar Mahal. It is the most valuable monument of Jammu. Once the residential place of

Raja Amar Singh, it possess a 120 kg golden throne. Amar commands an excellent view of the city and its surroundings now houses, a rare collection of pahari paintings, a family portrait gallery and a rich library. It is the tallest structure of the city. There are famous Raghunath Mandir and Ranweshawar Mandir, in the city. Then there is also a popular site which is called as Mansar Lake. It is about 62 kilometer away from Jammu city. A beautiful lake covered by forest and hills. Boating facilities are available on the spot. Every year around Basaikhi, a foodcraft festival is organized here by Jammu and Kashmir Tourism Department. Then there is also another popular site know as Surinsar Lake which is 45 kilometer away from Jammu. Then comes Behr Deveta, who owes its origin to the great service rendered by him to the people of Jammu town and saved them Snake Bites and Scorpio during the region of Raja Jas Dev (1020 AD to 1053 AD).

Another important and famous religious and tourist places in Jammu region are:

1. Gurudawara, Nangali Sahib, Poonch.
2. Shiv Khori Mandir near Katra (Reasi).
3. Baba Dhansar Mandir near Reasi.
4. Gurudawara, Baba Banda Singh Bshadur near Reasi.
5. Ghar Jitto near Reasi.
6. Other tourist places are: Kud, Patni Top, Sanasar, Batote, Sudh Mahadev and Jager Kotli.

Apart from these 48 kilometers from Jammu, the Shrine of the Goddess Vaishno Devi is one of the most important tourist attractions in Jammu and Kashmir. It is considered one of the most sacred places of worship among the Hindus. A pilgrimage to the holy Shrine of Shir Mata Vaishni Deviji considered to be one of the holiest pilgrimages of our times popularly the world over as Moonh Maangi Muradein poori karne wali mata, which means the mother who fulfils whatever her children wish for. Shri Mata Vaishno Deviji resides in a holy cave located in the folds of the three peaked mountain named Trikuta Hills. The Holy Cave attracts lakhs of devotees every year. Inside the cave of Trikuta Bhagwati are the image of mata Vaishno Devi, Mata Saraswati, Mahakali and Maha Lakshmi. The following table shows the annual *yatra* to Shir Vaishno Devi Shrine figures over the years ranging from 2000 to 2011.

In Jammu, a two day workshop from April 6 to 7, 2011, on the topic, "Practicing Responsible Tourism" was organized jointly by the Ecotourism Society of India (ESOI) J&K Tourism Department and World Wide Fund (WWF) Green Hiker and it was supported by pacific Asia Tourism Association (PATA), India chapter on the first day of the workshop, the people from central and state tourism ministry highlighted

TABLE

Number of Inflow of Tourists/Pilgrimages Visting Shri Mata Vaishno Deviji Shrine, Katra

S. No.	*Year*	*No. of Yarties (in Lakhs)*
1.	2000	52.17
2.	2001	50.57
3.	2002	44.32
4.	2003	54.00
5.	2004	61.00
6.	2005	62.52
7.	2006	69.50
8.	2007	74.17
9.	2008	67.92
10.	2009	76.27
11.	2010	82.22
12.	2011	92.97

Source: Directorate of Tourism, Jammu.

the importance of responsible and sustainable tourism and also discussed the role of ecotourism in it. This was the step within our region to sensitize the people regarding the ill-effects of tourism and the ways to minimize these effects for better green full tomorrow. Several academicians, experts, scholars and students within and outside the state attended this workshop. The next two workshop will be organized in Srinagar/Kashmir and Leh region in the near future to aware the local people regarding the importance of conservation of biodiversity.

There is direct relationship between tourism and ecotourism and the survival of both depends each other. The tourism industry can highly contribute in sustaining the ecotourism and if ecotourism will flourish, the tourism will also flourishs: in short both compliment each other perfectly.

The need of the hour is to educate the local communities as well as the tourists. Service providers, consumers and the people on in decision-making seats. Students who study in the schools regarding the benefits of practicing the responsible environment practices. It is the high time to act now and do some constructive work and try to contribute in sustaining the environment.

TOURISM ACTION PLAN FOR JAMMU AND KASHMIR

To synergise the tempo of pilgrimage tourism development in Jammu and Kashmir (J&K), two set of measures are suggested—operational measures and operative measures:

(A) Operational Measures

The operational measures revolve around the problems. Among them the foremost is the dilapidated condition of the infrastructure. It is, therefore, suggested that the Government, the private sector, the local bodies, the Non-Governmental Organisations (NGO's) and the local people should make concerted efforts to create the required infrastructure keeping in view the projections for the future inflow of the tourists in the State. There is a need for the creation of infrastructure in the form of roads, public conveniences, transport, accommodation, etc. Since, both domestic and international Tourists/pilgrims have the potential for growth, efforts ought to made to create such infrastructure which is in conformity with the capacity to pay and the level of living standards of the tourists/pilgrims, coming from different parts of the world.

Simultaneously, while designing the infrastructure, the environment of the State should not be disturbed. In other words, the whole of the State should not look like a concrete jungle.

Equally important is the creation of services in the form of uninterrupted electricity, streetlights, safe drinking water, proper sanitation, etc. The condition of these essential services is deplorable at present in whole is deplorable at present in whole of Jammu and Kashmir (J&K). It becomes, therefore, imperative that sufficient provision in budget ought to be made so that the level of these services could be improved on was footing. Unless these steps are taken, the country may be deprived of the growing inflow of international tourists/pilgrims visiting the State.

The rush of tourists/pilgrims has encouraged irresponsible and greedy businessmen to create disorganized and shabby structures to accommodate the tourists/pilgrims, especially during the peak seasons. This has created the problems of congestion, insanitation and other problems of typical urban environment. These problems have taken away the very essence and spirit of the places like Katra, Jammu, etc.

The Government, the local bodies, the private sector, the Non-Governmental Organisations (NGOs) and the local people should make a united effort to improve the ecology and environment of the State.

They should start with the afforestation progrmme which ought to be followed by other efforts to ensure that the water bodies do not get polluted by the filth and the garbage which is coming from the drains.

It is suggested that a study of the environmental impact of the tourism in the State may be undertaken to known the carrying capacity of the places in terms of the capacity of the public amenities and utilities, the ecology of the locus, or the capacity of the people to live a life of comfort and dignity in their own cultural and social milieu.

Since tourism has been declared as an industry by the Jammu and Kashmir state government, it becomes necessary to take the requisite steps to regulate the functioning of this industry. For this purpose, the following steps are suggested:

- Restructuring of tourism related laws in the light of the present day needs so that the tourists could be provided with the required safety and other forms of protection during the course of their stay in the State. This will ensure dispensing with the activities of touts, middlemen and the irresponsible element among service providers who fleece tourists/pilgrims.
- In the operational stage of tourism development, there is a need for a systems approach in managing tourism development. For this purpose, it is the responsibility of the Government to ensure that any tourist who lands in State does not face any problems during the course of his stay in this region.
- It is suggested that some exposure, if not the training be provided to different groups of local people like the pujaris of temples, police personnel, drivers, dhaba-owners, etc., on selective basis regarding inter personal behaviour so that they could interact with the tourists in a healthy manner. The services of the Non-Governmental Organisations (NGOs) may be obtained to provide this exposure. The selective group so exposed will provide the spread effect among their colleagues.
- The paying guest accommodations should be encouraged. The locals may be provided finance to create better facilities of accommodation at their places for the tourists/pilgrims.
- The important component of the systems model is the law and order machinery of which police is the most important component. Apart from the deployment of police for maintaining law and order, there is need for strengthening the hands of the 'tourist police', to help the tourists in overcoming the problems faced by them and to curb

tourism related crimes. The duties of the tourist police will include guiding tourists politely and taking care of their security.

(B) Operative Measures

The operative measures include three important components:

- o Strategy for Tourism Development.
- o Creation of Jammu and Kashmir State Tourism Development Board.
- o Creation of three independent boards in all three regions of Jammu and Kashmir.

The strategy indicates the long-term goals and objectives, adoption of alternative course of action and the deployment of resources to achieve the desired objective.

The long-term goals and objectives of the strategy are to develop the whole of the Jammu and Kashmir State tourism. Creation of a Jammu and Kashmir State Tourism Development Board will be the first step.

The work of restructuring the whole of Jammu and Kashmir State should be entrusted to this board which would be a qualified body of town planners, architects and consultants. The Board will incorporate within its purview all areas of Jammu and Kashmir.

CONCLUSION AND FUTURE IMPLICATIONS

Tourism in today's world over is very big economic and social activity, generating a large/huge amount of income, employment and foreign exchange and investment opportunities. Though, currently, the developed nation's are getting the lion's share of the benefit of tourism, there is a promising future for the developing nations also. Despite, the immense tourism potential in Jammu and Kashmir, it has not got the right share in tourism. The political unstability and the ongoing militancy/disturbance is the main hurdle in the way of tourism promotion. Advertising, publicity, policy of the state government, lack of proper funds and above all lack of professionalism in Jammu and Kashmir state are the other main reasons for the slow growth of tourism. The State Government has created number of tourism development authorities for integrated development of tourist destination in the state, but none of these authorities is headed by professional trained in tourism management. Infact, the department of tourism and the J&K State Tourism Development Corporation Ltd., needs to formulate and implement a proper tourism policy and plan with an integrated tourism marketing efforts for the promotion of tourism.

However, if Jammu and Kashmir State has to achieve its objective to attract more and more tourists both domestic and international then the pre-requisition for its implementation of tourism policies with support an coordination from all segments like, central, State Government Department of tourism, Tourism Development Authorities and Travel and Trade Organisations both in public as well as private sector. With the growing recognition of the importance of responding to contemporary tourism and hospitality students needs, there is a growing body of education focused literature documenting the development and implementation of innovative pedagogical approaches for studying in tourism and hospitality sector. Empirical research highlight the commitment of educators to develop sound, academically rigorous, innovative and perhaps even entertaining lecturers and case studies. Moreover, a key feature of the tourism industry is that of extent and pace of change patterns of consumption, technological change and supply innovations in tourism as elsewhere are in constant state of change which means education must evolve with industry changes by incorporating a life long learning approach to tourism education with tourism now established as one of the principal global industries the need for an effective and industry relevant tourism education framework that captures and utilizes latest pedagogical as well as business trends to underpin the industry's development is of paramount importance. Education is considered by many as a vital determinant of success choosing what to study and where, is one of the most important decisions a person will make as it will steer and ensure a successful career that will last many years. If someone is looking for a career that offers travel, varied challengers and they enjoy working with people then gaining a degree in tourism or hotel management field could be beneficial for them.

The hoteliers are genuinely aggrieved over the apathetic attitude of the concerned authorities for neglecting tourists spots in Jammu. The Jammu landscape is as beautiful as any scenic spot in the state. Some of these scenic sports figure prominently on the tourist map only because they are being promoted by the government, while other have remained neglected and underdeveloped.

There are many beautiful places in Kishtwar, Bhaderwah, Rajouri, Poonch, Reasi, Ramban, Udhampur, Kathua, Basoli, Billawar, Akhnoor and Sunderbani in Jammu region which is given proper publicity will definitely turn out to be famous tourist spots. Not only that Jammu has great potential for heritage tourism. There are many places, forts and historical buildings which can attract tourists, historians and art lovers. These places need to be developed and given wide publicity through electronic and print media. Small exhibitions showcasing these aspects of life need to be held in different parts of the country in general and the Jammu and Kashmir State in particular.

Man has always shown keen interest in tourism from early times. In fact, man has always been able to spare some time to move to other places and gain satisfaction.

Tourism comes in many forms but interactive tourism has gained momentum in today's age. The conception of this form of tourism is quite broad. In this form of tourism, people do not normally visit the monuments and other scenic reports, but instead spend a lot of time in the outskirts of places. They generally visit the villages of that areas. They stay in the homes of villagers and live there as the members of the family. In this process, tourists interact with the members and learn a lot about their culture which includes languages, mode of dress, eating habits and many of the aspects. Thus, a interaction tourism they enrich their knowledge about the culture of other people.

In our state such an interactive tourism is flourishing in villages of Ladakh region. This has benefited many villages in respect of their economic condition. Another benefit of this form of tourism is that tourists are educating people in different spheres. Many tourists even attend classes in schools for months together and help in improving the standard of instructional process.

In the near future, interactive tourism has a lot of scope in our state and is proving to be very useful for the local populace. No wonder, that tourism in Jammu region because of the natural beauty, has become an important industry vital for the economy of the state.

CHAPTER

5

Awareness of Ecotourism in Tuticorin District: A Study

D. Amutha

ABSTRACT

Global tourism is increasing every year with the number of people travelling abroad reaching 903 million in 2007. In India alone, in 2007 over 4.97 million tourists arrived in the country, an increase of almost 1.06 million arrivals per year since 2005. Income from the tourist industry is also increasing with the total international receipts in India reaching US$ 10,729 million in 2007 representing a huge economic boost for the country. This paper highlights awareness of ecotourism, what class of people makes a visit, various factors that influences the tourist while arranging a tour and so on in Tuticorin District of Tamilnadu.

The present study was carried out with following objectives:

1. *To study the socio-economic conditions of tourists from various places of India.*
2. *To understand the places and purposes of visit in Tuticorin District of Tamilnadu.*
3. *To know the number of groups and awareness of ecotourism.*
4. *To identify the place and duration of stay.*
5. *To find out the various factors that influences the tourist while arranging a tour.*

The study is conducted with the help of both primary and secondary data. The study analyses the responses provided in the questionnaires by the tourists who have visited Tuticorin District. The study analyses perceptions, preferences, and satisfactions of respondents. The sample chosen consisted of 400 sample respondents representing the tourists. Primary data was collected from 400 respondents by administering them a detailed questionnaire. The major sources of data are primary, where a structured questionnaire is designed and administered to the select respondents covered only two months period (2011). The sampling is random. The secondary data was collected from different sources like searching net, traveling authorities, traveling guides. Percentage analysis, averages, t-tests, standard deviation, chi-square tests, Cramer's V and probability analysis were used.

INTRODUCTION

Global tourism is increasing every year with the number of people travelling abroad reaching 903 million in 2007 (WTO, 2008). The United Nations World Tourism Organisation (WTO) has predicted that the number of international arrivals will reach 1.6 billion by the year 2020, with 1.2 billion of those being intra-regional travelers and 378 million being long-haul travellers (WTO, 2008). In India alone, in 2007 over 4.97 million tourists arrived in the country, an increase of almost 1.06 million arrivals per year since 2005 (UNWTO, 2008). Income from the tourist industry is also increasing with the total international receipts in India reaching US$ 10,729 million in 2007 (WTO, 2008) representing a huge economic boost for the country.

Tourism is generally viewed by governments, especially in the third world, as a means of economic development and as a result, investment in the industry is usually encouraged (Glasson *et al.*, 1999 and Hall, 1995 as referenced in Mbaiwa, 2005). However, while tourism can boost a national economy, it is also possible that it can benefit only a small number of industries and not others (Stoeckl *et al.*, 2006) and that the majority of local people can live in poverty while the tourism industry around them thrives (Nbubano, 2000 as referenced in Mbaiwa, 2005). Furthermore, tourism can often have damaging impacts on the environment such as increased water use, increased pollution loadings through waste and emissions, and direct and indirect disturbance to wildlife and vegetation (Fennell, 2003; Green and Giese, 2004). This paper highlights awareness of ecotourism, what class of people makes a visit, various factors that influences the tourist while arranging a tour and so on in Tuticorin District of Tamilnadu.

Places of Tourist Attraction in Tuticorin District

Thiruchendur is said to be one of the six sacred abodes of Lord

Muruga. The annual income of this temple is more than 50 lakhs. It attracts large number of devotees not only from different parts of the state, but also outside the state every year other places of tourist interest in this district are: Thiruchendur-Seashore, Panchalankurichchi-Memorial of the freedom fighters, Ettaiyapuram, Ottapidaram-manimandapam for the national poet Subramaniabharathi, Tuticorin-port, Industries, Saltfields, the Wild-life sanctuary and Tiger sanctuary located at Kalakad and Mundanthurai are equally important tourist places. There are 7 tourist places located at 4 villages/towns in the historical heritage area. The tourist arrivals both domestic and foreign have increased during 1990-98 period.

Tourist Arrivals in Tuticorin District

Year	*In Lakhs*
1994	6.80
1995	7.29
1996	9.38
1997	10.29
1998	12.06

Objectives of the Study

1. To study the socio-economic conditions of tourists from various places of India.
2. To understand the places and purposes of visit in Tuticorin District of Tamilnadu.
3. To know the number of groups and awareness of ecotourism.
4. To identify the place and duration of stay and
5. To find out the various factors that influences the tourist while arranging a tour.

Methodology

The study is conducted with the help of both primary and secondary data. The study analyses the responses provided in the questionnaires by the tourists who have visited Tuticorin District. The study analyses perceptions, preferences, and satisfactions of respondents. The sample chosen consisted of 400 sample respondents representing the tourists. Sample respondents comprising both tourists from different socio-economic category of society. Primary data was collected from 400 respondents by administering them a detailed questionnaire. The major sources of data are primary, where a structured questionnaire is designed and administered to the select respondents covered only two months period (2011). The sampling is random. The secondary data was collected from different sources like searching net, traveling authorities, traveling guides. Percentage analysis, averages, t-tests, standard deviation, chi-square tests, Cramer's V and probability analysis were used.

RESULTS AND DISCUSSION

TABLE I

Age-wise Classification of the Respondents

Age group (years)	*Male*	*Female*	*Total*
Below 20	77 (35)	36 (20)	113 (28)
21-40	167 (76)	137 (76)	304 (76)
41-60	98 (45)	113 (63)	211 (53)
Above 61	54 (24)	21 (12)	75 (19)
Total respondents	220 (100)	180 (100)	400 (100)

Figures in parenthesis are percentages.
Source: Primary Data.

In the study, out of 400 samples, more than 70% of the respondents belonged to the age of 21-40 years and the other age group are low in numbers in the study area. The chi-square test value (÷2=22.52, P=<.0001) assured that there are marked significant differences in sex and age-wise classification of the respondents in the study area.

TABLE 2

Educational Qualifications of the Respondents

Education	*Male*	*Female*	*Total*
Higher secondary	22 (10.0)	21 (11.7)	43 (10.8)
Graduate	131 (59.5)	107 (59.4)	238 (59.5)
Postgraduate	64 (29.1)	44 (24.4)	108 (27.0)
Others	3 (1.4)	8 (4.4)	11 (2.8)
Total respondents	220 (100)	180 (100)	400 (100)

Figures in parenthesis are percentages.
Source: Primary Data.

In the study area, 60 percent males and 59 percent females were graduates. 29 percent male respondents were professional with post graduation, whereas, 24 percent amongst female respondents were qualified. The chi-square analysis result (÷2=4.46, P= 0.2159) shows no significant relationship between sex and educational qualification of the respondents in Tuticorin District.

TABLE 3

Size of Monthly Income of the Respondents

Size of Income (Rs.)	*Male*	*Female*	*Total*
Below 10000	29 (13)	27(15)	56 (14)
10001-20000	69 (31)	59 (33)	128 (32)
20001-30000	92 (42)	85 (47)	177 (44)
Above 30001	189 (86)	103 (57)	292 (73)
Total	220 (100)	180(100)	400 (100)

Figures in parenthesis are percentages to the total of the respective columns.
Source: Primary Data.

The study revealed that 14% of the respondents are earning an income up to Rs.10000, 76% of the respondents are earning Rs. 10001-Rs. 30000 monthly and 73% are earning above Rs. 30000 monthly. The result of chi-square test (÷2=9.83, P=0.0201) revealed that there is no significant difference between sex and earning of the respondents.

TABLE 4

Places of Visit

Places of visit	*Male*	*Female*	*Total*
Kattabomman Memorial Fort—Panchalankurichi	101 (46)	59(33)	160 (40)
Tiruchendur—Shore Temple	202(92)	108 (60)	310 (78)
Kulasekarapattinam—Sea shore	107(49)	76(42)	183 (46)
The Basilica of a Lady of Snows	156(71)	103(57)	259 (65)
Kattabomman Hanging memorial place—Kayathar	46 (21)	35(19)	81 (20)
Tuticorin Port	131 (59)	121 (67)	252 (63)
Ettaiyapuram Palace	42 (19)	30(17)	72 (18)
The Gulf of Mannar Marine National Park	46(21)	18 (10)	64 (16)
Total Respondents	220 (100)	180(100)	400(100)

Figure in parenthesis are percentages to the total of the respective columns.
Source: Primary Data.

t=1.4165, d.f=14, P=0.1785, Standard error of difference = 24.797
Mean: 103.88 68.75: SD: 58.10 39.29: SEM: 20.54 13.89.

The table reveals that the special significance of the Tuticorin District is Kattabomman palace, Bharathiyar and V.O. Chidambaram houses, Murugan temple, RC temple, and The Gulf of Mannar Marine National Park, Tuticorin Port, and Kulasekarapattinam—Sea shore showed significant difference (t=1.4165, P=0.1785, Standard error of difference = 24.797).

TABLE 5

Purpose of Visit

Purpose	*Male*	*Female*	*Total*
Holiday and Sight Seeing	104 (47)	91 (51)	195 (49)
Pilgrimage	53 (24)	45 (25)	98 (25)
Conference Study	27 (12)	22 (12)	49 (12)
Visiting Friends	17(8)	9 (5)	26 (7)
Others	19 (9)	13 (7)	32 (8)
Total Respondents	220 (100)	180 (100)	400 (100)

Figures in parenthesis are percentages to the total of the respective columns.
Source: Primary Data.

The study highlights that an overwhelming respondent say 49% arrange tour for sight seeking and to enjoy their holidays, 25% to pilgrimage, and only 7% for visiting their friends and relatives and the difference is statistically not significant (÷2=1.63, P=0.8034).

TABLE 6

Company and Awareness of Tourism

Company	*Male*	*Female*	*Total*
Family	53 (24)	121 (67)	174 (44)
Friends	104 (47)	15 (8)	119 (30)
Colleagues	27 (12)	22 (12)	49 (12)
Relatives	4(2)	9 (5)	13 (3)
Education Tour	32 (15)	13 (8)	45 (11)
Total Respondents	220 (100)	180 (100)	400 (100)

Figures in parenthesis are percentages to the total of the respective columns.
Source: Primary Data.

The table elucidates that an overwhelming respondents say 44% are accompanied with their family, 30% by their friends, 12% by their colleagues, and 11% by educational tour.

TABLE 7

Number of Groups and Awareness of Ecotourism

Number of Groups	*Male*	*Female*	*Total*
1-3	8 (3.6)	3 (1.7)	11 (2.8)
4-5	39 (17.7)	96 (53.3)	135 (33.8)
6-8	102 (46.4)	76 (42.2)	178 (44.5)
more than 8	71 (32.3)	5 (2.8)	76 (19.0)
Total Respondents	220 (100)	180 (100)	400 (100)

Source: Primary Data.

It also elucidates that a majority of respondents 45% arrange tour by 6-8 groups, and lowly 3% respondents form 1-3 groups while arranging tour and the difference was statistically significant (÷2= 84.3, P < 0.001).

TABLE 8

Travel Modes for Destination

Travel Mode for Destination	*Male*	*Female*	*Total*
Bus	39(18)	61(34)	100(25)
Train	120(54)	105(58)	225(56)
Own Car	61(28)	14(8)	75(19)
Total Respondents	220 (100)	180 (100)	400 (100)

Source: Primary Data.

It also highlights that a majority of the respondents say 56% of them travel through train for reaching their destinations, 25% of them through bus and only 19% of them by taxi and own car. The result of chi-square- test (÷2=31.61, P=<.0001,) revealed that there is significant difference between male and female tourists and travel mode for destination in Tuticorin District.

TABLE 9
Place of Stay

Place of Stay	*Male*	*Female*	*Total*
Guest House	72(33)	14(8)	86(21)
Hotel	130(59)	32(18)	162(41)
Relative House	18(8)	134(74)	152(38)
Total Respondents	220 (100)	180 (100)	400 (100)

Source: Primary Data.

The study revealed that an overwhelming respondents say 41% make their stay in hotel, 38% of them stay in their relative homes, and only 21% of the respondents stay in guest house in the study area.

TABLE 10
Duration of Stay

Duration of Stay	*Male*	*Female*	*Total*
up to 2 days	47 (50)	47 (50)	94 (23.5)
3-5 days	72 (55.4)	58(44.6)	130 (32.5)
6-8 days	68(58.1)	49(41.9)	117 (29.3)
above 8 days	33 (55.9)	26 (44.1)	59 (14.7)
Total Respondents	220 (100)	180 (100)	400 (100)

Source: Primary Data.

The table shows that a majority of the respondents say 33% make a stay for 3-5 days, 29% of the respondents stay for 6-8 days and only 15% tourists' stay more than 8 days. The result of chi-square analysis (÷2=1.44, P=0.6962) revealed that there is no significant relationship between sex and duration of stay of tourists in Tuticorin District of Tamilnadu at 5% level.

TABLE 11
Factors of Stay

Factors	*Male*	*Female*	*Total*
Cost related	13 (5.9)	28 (15.6)	41 (10.3)
Overall satisfaction	4 (1.8)	15 (8.3)	19 (4.8)
Infrastructure	22 (10.0)	17 (9.4)	39 (9.8)
Service and safety	81 (36.8)	26 (14.4)	107 (26.8)
Nature	32 (14.5)	63 (35.0)	95 (23.8)
Basic needs	68(30.9)	31 (17.2)	99 (24.8)
Total	220 (100)	180 (100)	400 (100)

Source: Primary Data.
t=0.4595, d.f=10, P=0.6557, Standard error of difference = 14.509
Mean: 36.67 30.00: SD 31.02 17.34: SEM 12.66 7.08.

Every respondent arrange a tour due to influence of many factors. The influencing factor may differ from one respondent to another. The present study is an attempt to find out the various factors that influence the tourist while arranging a tour. The study revealed that the most important factors adopted by the respondent in arranging tour are service and safety (27 per cent), nature (24 per cent), basic needs (25 per cent), cost-related (10 per cent), infrastructure (10 per cent) and overall satisfaction (5 per cent), showed no significant difference (t=0.4595, d.f=10, P=0.6557, Standard error of difference = 14.509).

CONCLUSION

The study investigated the awareness of ecotourism in Tuticorin District of Tamilnadu. If a proper marketing plan is done rural tourism, it could bring lots of benefit to our society. It could be a sustainable revenue generating project for rural development of our government. It can help inflow to resources from urban to the rural economy. It can prevent migration of rural people to urban. Shortterm and long-term planning, implementing and monitoring are vital in avoiding damage to rural areas. Environmental management, local involvement, sound legislation, sustainable marketing, and realistic planning are crucial for development of rural tourism.

Rural tourism will emerged as an important instrument for sustainable human development including poverty alleviation, employment generation, environmental regeneration and development of remote areas and advancement of women and other disadvantaged groups in the country apart from promoting social integration and international understanding. The government should promote rural tourism to ensure sustainable economic development and positive social change.

REFERENCES

Andrew Vladimir, Complete Travel Marketing Handbook, NTC Business Book, 1989

Coltman, Michael M., Introduction to Travel and Tourism, Van Nostrand Reinhold, New York, 2004.

Indian Tourism, Karan, P. Mazumdar, Prestige Books, 2004.

Karan Singh, Indian Tourism-Aspects of a great adventure, Department of Tourism, New Delhi, 1980

Tapan K. Panda, Marketing Management, Excel Books, 2006

Tourism as a Heritage—Text and Cases, New Age Publications, 1999

www.agnet.org

www.indiabusiness.nic.in

www.kerala.gov.in

www.nal.usda.gov

www.progya.org

www.tourismindia.com

www.wttc.org

CHAPTER

6

Improving Quality of Management Education in India: Changing Needs and Shifting Paradigms

DR. V.L. DHARURKAR AND DR. M.R. CHANDAWARKAR

I. INTRODUCTION

Management Education is a most challenging Area in the higher education of India. Management Education is on cross road's of change it Indranil Boss has been pointed that Management education in India is going through a metamorphosis. The economic structure is changing. New areas are now emerging and developing. Traditional specializations are also experiencing fundamental changes in applications.[1] The management education in India requires positive changes at this point when higher education in India is on the verge of change.

The emergence of new corporate culture is a greagt character of new millennium in India. M.V. Pylee observed that India is on the threshold of an industrial revolution. The progress that has been achieved during the fifteen years of planned development since 1950 has been impressive if not spectacular.[2] Pylee's observation is more significant because this industrial change has provided background for rise of management education in India.

The root cause of poor growth rate is lack of training it has been observed that the issues involved in India's development effort are of tremendous social significance. For, the basic fact that holds up not only a satisfactory rise in the productivity but even the overall rate of development is that India is by and large an untrained nation.[3] Hence, management education has tremendous scope in India.

II. PERSPECTIVE

Quality is need of the time in every field of education it has been observed that management education is the most sought after higher education option for a majority of the brightest and most talented graduates. However, unfortunately, many of these students aspiring to go for management education do not seem to understand what actually this means and how should they go about choosing the right institution. The right selection of management institute is require to get quality. It is trout that in quality management education we are basically referring to the shaping up of the student to develop the competency and capability either as an entrepreneur (to join and help an organization to grow) or as an entrepreneur (to establish and grow one's own organization).[4] Therefore, quality brought through the cooperation of students organization and corporate industries. In this paper practical approach has been adopted for improving management education in India.

III. THE CURRENT STATE OF CONDITION

Management Schools are also described has business Schools Indian Business Schools (B-Schools) such as the IIM's (Indian Institute of Management) at Ahmedabad, Bangalore, Calcutta, XLRI are world famous for the quality of students it churns out year on year. IIMs have historically the highest number of students who become enterpreneurs in their professional life. The demand for these B-School students has consistently grown over a period of last few years.[5] These are the schools that are generally regarded as the top B-Schools in India. It would be interesting to compare this business schools with other Indian the B-Schools in regional Universities. On Global level some studies have been conducted on global management education. It has been pointed that It is the pervasiveness of management practice, rather thanits exclusivity, that makes quality business schools so vital to individuals, organizations, and societies around the world.[6] In India there are following drawbacks in management education:

1. Indian Management Education is conceptual and it lacks practical base there is no proper linkage between B Schools and Industries.

2. The Management Schools in India lack infrastructure facility. Proper financial support is require for improving B Schools.
3. The B Schools in India should catter local needs such as rural development agriculture development and cooperative development.
4. Special management courses are required to strengthen social sector such as children women and weaker sections.
5. Management education in India must be changed to salve socio-economic and cultural needs of poor and marginalized people.[7]

IV. FIVE GLOBAL CHALLENGES IN MANAGEMENT EDUCATION

Due to the process of globalization every field of education is changing and management education is also not an exception to this. It has been observed that in management education, quality has become a necessity and the circumstances require total quality management. The future belongs to those who see possibilities before they become obvious. To make India an intellectual capital of the world, we have to rethink about the management education and effort should be made to create a dynamic environment, which can nurture superior quality technical education colleges.[8] The Five basic challenges of globalization can be noted below.

1. *Imbalanced Growth*: The Management education has not grown imbalanced manner it has been pointed that all the indicators point to continuing increases in the demand for management education. Driven by demographics, economic trends, business expectations, and initiatives that expand access to higher education, future demands will come not only from traditional college-age populations, but also from working professionals who need to retool and reinvigorate their careers.[9] In order to improve there career proper training is required.
2. *Balancing Global Aspirations and Local Needs*: After globalization people's hopes and aspirations have increased but there is a need to understand local problems it has been pointed that a recurring theme throughout this report has been tensions between global aspirations of countries, schools, faculties, and students—and pressing local needs. These tensions are revealed on many dimensions: curricula, strategy, and collaboration, for example. Further economic

integration calls for strengthening our curricula emphasis on global perspectives, but we cannot ignore unique histories, politics, and cultures.[10] By understanding this background Management education can solve local problems.

3. *Quality Assurance*: Quality has became mantra of new world, it has been pointed that:
   We have argued that expansion in management education has brought greater diversity among the programs and providers in management education. Schools have different missions and aspirations; vary in governance structures, faculty characteristics, and financial models; and are embedded in a wide array of cultures, histories, and governing systems. All of this diversity is to be nurtured and celebrated. Diversity means that students and employers have choices to meet their unique goals and accommodate their circumstances. It also fosters innovation among schools and programs.[11] Innovative ideas must be rooted in management courses.
4. *Sustaining Scholarship*: There is as need to sustain scholarship effectively in the world it has been pointed that Throughout this report, we have highlighted the difficulties that schools have had in recruiting and retaining qualified faculty. For many schools, the challenge is to recruit faculty with doctorates to support missions that include research and scholarly approaches to teaching. Clearly, the demand for doctoral faculty has been outstripping production, leading to concerns about the ability of some of these schools to introduce or sustain an emphasis on scholarship.[12] A proper reorganization to Scholarship must be given by proper awards and prizes.
5. *Aligning with the Future Needs of Organizations*: The management education should look at future needs of young generation it has been pointed that by examining global economic and business trends, we have attempted to isolate the emerging needs of organizations around the globe. For example, we argued that the integration of economies will require stronger emphasis on global perspectives, fracturing value chains will require graduates to master important skills rather than just apply knowledge, and emerging emphases on social responsibility and sustainability will require new ways of thinking about business strategy.[13] This can be possible if management

schools can become creative. It has been observed that We should caution that these are only examples and are rather subjective. The point here is not that the needs of organizations have changed over time; they have and always will. What's new is that the pace of change has been accelerating. How can business schools structure themselves and build systems to learn about, predict, and react quickly enough to emerging needs?[14] Developing Countries require such approach.

V. RECOMMENDATIONS

In order to improve B schools in India following recommendations can be made:

1. Advocate for quality assurance globally and locally : There is need to establish quality both in theory and practice.
2. Invest in mechanisms to engage business and government leaders in envisioning future organizational and societal needs: There is a need to remold management education in tune with organizational needs in third world.
3. Facilitate and encourage investments in doctoral degree education and other infrastructure development : We have to develop problem-oriented research in a systematic manner.
4. Create an international clearinghouse for data and information related to business schools and management education structures, trends, and practices : Such a clearing house can be more officiate for coordinating various agencies.
5. Facilitate multilateral collaboration among business schools: 14 All the business schools in the world must develop new strategy to face current problems in systematic manner.

VI. CARE AND CAUTION

It has been pointed that A nation or economy must be careful about Globalization. On the impact of globalization, there are two major concerns. These may be described as even fears. Under each major concern there are many related anxieties. The first major concern is that globalization leads to a more iniquitous distribution of income among countries and within countries. The second fear is that globalization leads to loss of national sovereignty and that countries are finding it increasingly difficult to follow independent domestic policies. These two issues have to be addressed both theoretically and empirically.[15]

Management education has to understand all these problems seriously. Economists have different views on Globalization. Some favor it while many oppose it. The argument that globalization leads to inequality is based on the premise that since globalization emphasizes efficiency, gains will accrue to countries which are favourably endowed with natural and human resources. Advanced countries have had a head start over the other countries by at least three centuries. The technological base of these countries is not only wide but highly sophisticated. While trade benefits all countries, greater gains accrue to the industrially advanced countries. Management Schools in third world countries have to shoulder this new responsibility in the new age. It has been pointed that this is the reason why even in the present trade agreements, a case has been built up for special and differential treatment in relation to developing countries However, there are two changes with respect to international trade which may work to the advantage of the developing countries. For a variety of reasons, the industrially advanced countries are vacating certain areas of production. These can be filled in by developing countries. B Schools in developing countries can play a vital role in this connection.

World has become a global village due to communication revolution International trade is no longer determined by the distribution of natural resources. By and large, this treatment provides for longer transition periods in relation to adjustment. With the advent of information technology, the role of human resources has emerged as more important. Specialized human skills will become the determining factor in the coming decades. Productive activities are becoming "knowledge intensive" rather than "resource intensive". While there is a divide between developing and the advanced countries even in this area —some people call it the digital divide—it is a gap which can be bridged. A globalized economy with increased specialization can lead to improved productivity and faster growth. What will be required is a balancing mechanism to ensure that the handicaps of the developing countries are overcome. The schools must try to bridge this gap.

VII. CHANGING PARADIGMS

Management Education in India is on cross road's of change. We have to shift from traditional management education to advance education. There are only five global level standard management school's in India. In every state there should be at least one global level management institution. We have to newly introduce management courses such as agricultural management rural development management, water management, energy management and communication management.

India is reach in cultural heritage we require hospitality management courses. Further India is an emerging country were growth rate of media is higher than that of GDP hence media management courses can also be much effective and relevant. Peter Drucker has written a book on non profit organizations management. As in India the number of NGO's is increasing we have to start courses on the management of non-profit organization's urgently.

Indian economic is developing economy and we are third country in the world having science and technology power. Further we have to begin courses in IT management also. Courses for the development of women and leadership can be a positive steps.

In the new millennium new courses are required for educational management, entertainment management as well as urban management.

Environmental management: All these courses can meet our changing socio-economic needs. Developmental management approach can be vitally significant in this regard.

A bold step ahead is required to change the profile of management education in India. We can develop a new man, a new society and new culture on the basis of management education.

VIII. SUMMARY

Management Education is under going tremendous change. Due to globalization competition has increased. In its reports knowledge commission has pointed that Management education must meet new challenges of quality education. The course content must be revised as per socio-economic needs. In country like India rural development, Agricultural and Industrial Development problems must be solved in management Perspective. The new human face of management education is required to strengthen it in the age of globalization.

Notes and References

1. Boss, Indranil, Intellectual Capital and Quality Management Education in India.
2. Pylee, M.V., Management Education in India, Management Science, Vol. 13, No. 10, June, 1967, Printed in U.S.A.
3. Prof. Arya Varun, Sustaining Quality in Management Education, p. 1.
4. *Ibid.*
5. Prabhudesai, Arun, The top 25 Business schools in India, 9 July, 2007.
6. GFME Report, 2008, Gfme.org.
7. *Ibid.*
8. Dr. Priya Ashutosh, Global World and Quality of Management Education in India p. 1.
9. *Op cit.*

10. *Ibid.*, p. 47.
11. *Ibid.*, p. 48.
12. *Ibid.*, p. 49.
13. *Ibid.*, p. 50.
14. *Ibid.*, pp. 53-57.
15. http://www.citehr.com Joshi Maheshchandra G http://www.citehr.com/75481-impact-globalization-management-education-india.

CHAPTER

7

Outsourcing Business to Cloud Computing Services: Opportunities and Challenges

ANU CHADHA, DR. AMRIK SINGH SUDAN, DR. VEENA RANI AND DR. RAIS AHMAD

ABSTRACT

There have been dramatic changes in information technology over the past 15 years. The alliance of information technology and globalization have boosted the export and import rates, capital flows, cross border mergers, acquisitions and foreign ownership of domestic firms. Information technology enabled services (ITES) is a new line of entrepreneurial activity catching up very fast and has created thousands of entrepreneurs at Global level. It has also created employment opportunity of more than 50 lakhs all over the world. The second half of 2003 shows a clear sign of improvements in Information technology sector. The overheating of North-American economies, Canada and USA has resulted in reduction in Information technology (IT) spending and was accompanied by another development: outsourcing. Business process outsourcing (BPO) is popularly referred as call centres is totally a new enterprise graving in a big way. Global competition and advancement in the field of Information Technology have made the overseas companies to outsource some of their business activities.

In the fast changing global business environment, outsourcing is viewed as an intelligent and inevitable option for many enterprises including financial service companies. Financial service companies are using third parties to carry out certain activities that the business themselves would normally have undertaken. Industry reports and regulatory surveys of industry practices indicate that outsourcing has the potential to transfer risk, management and compliance of third parties. To cut down the cost and to achieve strategic aims, several financial services companies throughout the world are increasingly using the services of third parties to outsource their activities. Today financial service industry is the second largest sectoral user of outsourcing services after manufacturing. But in recent times, the outsourcing arrangements are becoming more and more complex day-by-day.

Developing countries can play a significant role in the area of service marketing especially in the field of information processing industry. The world market of this industry is increasing very fast and is likely to group approximately to US $500 million within the next five years. It is also expected to provide employment to the largest number of people world-wide and more particularly in a developed economy. This industry needs sophisticated software which are manpower related in which many developing countries can have a major share. So software experts should have a free access to operate in any market internationally. The software-relared service such as advertising, accounting, legal services, management consulting and data processing are highly competitive and volatile in which developing countries can become major participants. Service activities now-a-days account for the largest share of employment and national production. Most of the job creation in the industrial countries in the past two decades has been in the services which are tradeable across national boundaries and charges in the organizations of production of goods involve the corporation of ever more services.

Advances in service oriented architecture (SOA) have brought us close to the once imaginary vision of establishing and running a virtual business, as business in which most or all of its business functions are outsourced to online services. Could computing offers a realization of Service Oriented Architecture (SOA) in which Information Technology (IT) resources are offered as services that are more affordable, flexible and attractive to businesses. In this paper, we briefly study advances in cloud computing, and discuss the benefits of using cloud services for businesses and trade-offs that they to consider. We then present (1) a layered architecture for the virtual business, and (2) a conceptual architecture for a virtual business operating environment. We discuss the opportunities and research challenges that are ahead of us in realizing the technical components of this conceptual architecture. We conclude by giving the outlook and impact of cloud services on both large and small businesses.

Keywords: Could Computing, Virtual Business, Business Outsourcing, Service Oriented Computing, Software, Data Processing.

I. INTRODUCTION

There have been dramatic changes in information technology over the past 15 years. The alliance of information technology and globalization have boosted the export and import rates, capital flows, cross border mergers, acquisitions and foreign ownership of domestic firms. Information technology enabled services (ITES) is a new line of entrepreneurial activity catching up very fast and has created thousands of entrepreneurs at Global level. It has also created employment opportunity of more than 50 lakhs all over the world. The second half of 2003 shows a clear sign of improvements in Information technology sector. The overheating of North-American economies, Canada and USA has resulted in reduction in Information Technology (IT) spending and was accompanied by another development: outsourcing. Business Process Outsourcing (BPO) is popularly referred as call centres is totally a new enterprise graving in a big way. Global competition and advancement in the field of Information Technology have made the overseas companies to outsource some of their business activities.

In the fast changing global business environment, outsourcing is viewed as an intelligent and inevitable option for many enterprises including financial service companies. Financial service companies are using third parties to carry out certain activities that the business themselves would normally have undertaken. Industry reports and regulatory surveys of industry practices indicate that outsourcing has the potential to transfer risk, management and compliance of third parties. To cut down the cost and to achieve strategic aims, several financial services companies throughout the world are increasingly using the services of third parties to outsource their activities. Today, financial service industry is the second largest sectoral user of outsourcing services after manufacturing. But in recent times, the outsourcing arrangements are becoming more and more complex day-by-day.

Developing countries can play a significant role in the area of service marketing especially in the field of information processing industry. The world market of this industry is increasing very fast and is likely to group approximately to US $500 million within the next five years. It is also expected to provide employment to the largest number of people world-wide and more particularly in a developed economy. This industry needs sophisticated software which are manpower-related in which many developing countries can have a major share. So software experts should have a free access to operate in any market internationally. The software-relared service such as advertising, accounting, Legal services, management consulting and data processing are highly competitive and volatile in which developing countries can become major participants. Service activities now-a-days account for the largest share of employment and national production. Most of the job creation in the industrial countries in the past two decades has been in the services which are tradeable across national boundaries and charges in the organizations of production of goods involve the corporation of ever more services.

Advances in service-oriented architecture (SOA) have brought us close to the once imaginary vision of establishing and running a virtual business, as business in which most or all of its business function are outsourced to online services. Could computing offers a realization of Service Oriented Architecture (SOA) in which Information Technology (IT) resources are offered as services that are more affordable, flexible and attractive to businesses. In this paper, we briefly study advances in cloud computing, and discuss the benefits of using cloud services for businesses and trade-offs that they to consider. We then present 1) a layered architecture for the virtual business, and (2) a conceptual architecture for a virtual business operating environment. We discuss the opportunities and research challenges that are ahead of us in realizing the technical components of this conceptual architecture. We conclude by giving the outlook and impact of cloud services on both large and small businesses.

The idea of creating and running a business over the internet is not new. Banks and large manufacturers are among the first in exploiting the electronic network capabilities to conduct business-to-business (B2B) interactions through technological such as EDI[1]. With the introduction of the Web and the rapid increase of internet users in the early 1990s, companies such as Amazon and eBay were among the early entrants to the business-to-consumer (B2C) model of e-commerce. As the internet is a fast, easy-to-use and cheap medium which attracts millions of users online at any time, today there are very few businesses that do not have a Web presence, and there are many small and medium businesses (SMBs) such as retail shops that solely offer their services and products online.

Looking at the enabling technologies, B2B and B2C e-commerce have benefited from many innovations in the Internet and Web. Moving from static content delivery to dynamic update of page content and the introduction of XML created the first evolution in the path to more efficient and interoperable running of electronic businesses. A main characteristic of using technologies of the Web 1.0 era is that almost all the backend Information Technology (IT) systems are created, operated and maintained by the business owners. Motivated by business agility, operational efficiency, cost reduction and improved competitiveness, during the last decade, businesses have taken advantage of business process outsourcing (BPO)[2]. In Business Process Outsourcing (BPO), businesses delegate some of the company's non-core business functionality such as Information Technology (IT) operations to third – party external entities that specialize in those functions. It is estimated that by 2011 the world-wide market for Business Process Outsourcing (BPO) will reach $677 billion[3].

Up until recently, outsourced services were not necessary fulfilled online. Business Process Outsourcing (BPO) has become attractive to both large and small businesses with the advent of service oriented computing[15] and specifically Web services and Web 2.0[5] technologies. This has enabled offering of business process functions as online Web services and actively engaging customers via the Web[4]. It is estimated that Business Process Outsourcing (BPO) represents around 25% of the overall services market[3].

The next evolutionary wave in this space is cloud computing. Cloud computing refers to the offering of hardware and software resources as services across (distributed) Information Technology (IT) resources[6]. As a relatively new concept, cloud computing and related technologies have rapidly gained momentum in the Information Technology (IT) world. In this article, we study how advances in cloud computing impact the processes of creating and running businesses over the internet. In particular, we investigate the question of whether the technology is ready to allow business owners to create and run a business using services over the internet. We refer to this as a "virtual business" in which most or all of its functions are outsourced to online services. It should be contrasted to the concept of "virtual enterprise"[7] which often refers to creating a temporary alliance or consortium of companies to address certain needs with an emphasis on integration technologies, knowledge sharing, and distribution of responsibilities and capabilities.

In the following, we first give a short survey of advances in cloud computing, and through an example scenario, highlight trade-offs that businesses have to consider in moving to cloud cervices. Then, we discuss the requirement of an environment for creating and running virtual businesses, and present a conceptual architecture for such an environment. We study to what extent they can be realized. We also present challenges that are ahead of us in offering such an environment and provide an outlook of the impact of cloud services on businesses.

2. CLOUD COMPUTING: A SHORT SURVEY

Cloud computing has emerged as the natural evolution and integration of advances in several fields including utility computing, distributed computing, gird computing, web services, and service-oriented architecture[6]. The value and originality of cloud computing comes from packaging and offering resources in an economical, scalable and flexible manner that is affordable and attractive to Information Technology (IT) customers and technology investors. We introduce a framework to study and survey advances in cloud computing. It consists of four dimension: cloud cervices, public *Vs.* private clouds, cloud service customers, and multi-tenancy as an enabler.

2.1. Cloud Services

As promoted by the vision of "everything as a service"[8] many products are now offered as services under the umbrella of cloud computing.

We summarize them in the following:

Infrastructure as a service (IaaS): Hardware resources (such as storage) and computing power (CPU and memory) are offered as services to customers. This enables businesses to rent these resources rather than spending money to buy dedicated servers and networking equipment. Often companies are billed for their usage following a utility computing model, where usage of resources is metered. As examples in this category, Amazon[1] offers 53 for storage, EC_2 for computing power, and SQS for network communication for small businesses and individual consumers. HP FCS[2] (Flexible Computing Services) provides computing and storage infrastructure as services for businesses.

IaaS providers can allocate more computing power and hardware resources of applications on an as-needed basis, and allow applications to scale in a horizontal fashion (several machines running the same application with load balancers distributing the workload). This enables flexibly scaling up or down the amount of required resources on-demand. This is a valuable feature for companies with occasional large computation needs or sudden peaks in demand such as flash crowds. On the other hand, IaaS providers, e.g., Amazon EC_2, offer virtual machines to customers where they can deploy and run their programs. Typically several virtual machines run on the same physical machine to address the problem of under utilization of physical resources. Statistics show that 80% of computing power and 65% of storage capacity is not efficiently utilized, where a single company privately owns dedicated machines[9].

Database as a service (DaaS): A more specialized type of storage is offering database capability as a service. Examples of service providers are Amazon SimpleDB, Google BigTable[3], Force, COM database platform and Microsoft SSDS[4]. DaaS on the cloud often adopts a multi-tenant architecture, where the data of many users is kept in the same physical table. Also, in most cases, the database structure is not relational. For instance, Microsoft SSDS adopts a hierarchical data model, and data items are stored as property-values or binary objects (Blobs). Google Big Table, Apache HBase[5] and Apache Pig[5] also provide abstraction, which allow users to save data in a key-value pair fashion. Each DaaS provider also supplies a query language to retrieve and manipulate data. However, not all support operations such as joins on table (such as Apache HBase and Amazon SimpleDB).

Software a as service (SaaS): In this model, software applications are offered as services on the Internet rather than as software packages to be purchased by individual customers. One of the pioneering providers in this category is Salesforce.com offering its CRM application as a service. Other examples include Google web-based office application (word processors, spreadsheets, etc.), Microsoft online CRM and SharePoint, or Adobe Photoshop and Adobe Premiere on the Web. Commercial applications in this category may need a monthly subscription per user (salesforce.com) or can be billed per use, both of which are considerably cheaper than owing and maintaining the software as an in-house solution. In this approach, there is no official software release cycle, and the customer is free from applying patches or updates as this is handled by the service provider. Customer data is kept in this cloud, potentially based on DaaS.

Platform as a service (PaaS): This refers to providing facilities to support the entire application development lifecycle including design, implementation, debugging, testing, development, operation and support of rich Web applications and services on the Internet. Most often Internet browsers are used as the development environment. Examples of platform in this category are Microsoft Azure Services platform[6], Google App Engine[7], Slaesforce.com Internet Application Development platform[8] and Bungee Cannot platform[9]. PaaS enables SaaS users to develop add-ons, and also develop standalone Web-based applications, reuse other services and develop collaboratively in a team. However, vendor lock-in, limited platform interoperability and limitations of programming platforms in supporting some language features or capabilities are major concerns of using current platforms.

There are other types of capabilities that are offered as services in the cloud. Offering management, monitoring and integration as services are examples. In monitoring as a service, a third-party provider (e.g., Red Hat Command Center) observes SaaS applications or the Information Technology (IT) network of an enterprise on behalf of a customer with respect to SLAs and reports performance metrics to the customer. Management as service generally includes monitoring but adds responding to events rather than just reporting them. Provider of integration as a service such as Bungee Connect[9] and Boomi[10] aim at providing facilities (PaaS) to allow user to develop adapters and mappings that enable the interoperation of heterogeneous services or data.

Another important type of service that is offered on the cloud is people as services. Offering of services by people, e.g., their programming skills per hour on the net, is possibly as old as the Web itself. However, what is new in the could is there are people specializing in SaaS or PaaS platforms and offering consolation for businesses that need to use or customize SaaS solution or integrate solutions from multiple SaaS providers. For example, Salesforce.com AppExchange opens up an opportunity for such people to offer their services.

2.2. Public *Vs.* Private Clouds

It can be argued that the cloud is the result of natural transformation of the Information Technology (IT) infrastructure of enterprises over the last decade. The traditional Information Technology (IT) architecture was based on having dedicated resources for each business unit in an enterprise. This model leads to under-utilization and waste of Information Technology (IT) resources due to resource fragmentation and unequal distribution of workload. To overcome this, enterprises have implemented adaptive infrastructure techniques[10]. These include employing virtualization to address the under-utilization problem complemented with automation techniques to reduce the significant labour costs of IT operations. This type of cloud is called a "private" cloud as it is privately owned by enterprises. Examples of this category are clouds maintained by manufacturers such as Boeing or GM.

On the other hand, there are other cloud offerings (e.g., those provided by Amazon, Google, Microsoft and Salesforce.com) for public use. Some of these offered by Amazon and Google, are indeed extensions of their private clouds that are offered to the public. There are also cloud providers such as Salesforce.com that have created and offered cloud services solely for public use. It is interesting to notice that enterprises and large businesses are mainly the owners and users of private clouds, while public clouds are used by smaller businesses and millions of individual consumers.

In addition to cloud vendors, who own and operate cloud services, there are other providers called out-clouders (re-sellers). Out-clouders are gaining popularity and trying to acquire and re-sell unused computing resources of enterprises with private clouds[11]. Outclouding is also a source of income for enterprises who rent out part of their Information Technology (IT) resources which they are not utilizing efficiently.

2.3. Cloud Service Consumers

In addition to the coarse-grained categorization of cloud users as enterprises, SMBs and individual consumers, it is useful to identify and study various types of customers of cloud services. Understanding the target customers of cloud services and their requirements allows determining what type of services can be used by which customers. In general, cloud customers are of the following types: Information Technology (IT) administrators, software developers, managers and business owners, and finally individual (business) users. Table 1 shows the distribution of various cloud customers for various cloud services.

TABLE I

Cloud Customers *Vs.* Cloud Services

Customer Type	*IaaS*	*DaaS*	*SaaS*	*PaaS*	*Others*
IT administrators	Use to deploy images of existing software	Configure to store data	Only for administration use	N/A	Monitoring as a Service (to Set-up and monitor SLAs)
Software development	May use to deploy software	Store data	Mainly to browse and find existing services to reuse and add-ons	Main users of PaaS	Integration as a Service
Managers and business owners	N/A	N/A	Occasional users to mange their business	N/A	Monitoring as a service (dashboards), may employ people as services
Business users	N/A	N/A	Main users of SaaS, may perform simple configuration tasks and use add-ons	N/A	

2.4. Multi-tenancy as an Enabler

Multi-tenancy refers to sharing resources among users and/or applications. It is preferred over single-tenancy in cloud services due to higher utilization leading to cost reduction. Enterprises often have thousands of users but typically operate a variety of software environments an applications. Thus, in private clouds multi-tenancy is often about having multiple applications and environments deployed on shared resources. In contrast, in public clouds there are millions of users and service providers try to minimize the number of software applications and environments. Therefore, multi-tenancy is about sharing resources among millions of users (e.g., keeping various users' data in the same table and secured). If public cloud providers offer PaaS, then a variety of application environments are also supported. In this case, multi-tenancy techniques need to enable sharing resources among volumes of applications and users.

3. CLOUDRETAIL AS A VIRTUAL BUSINESS

Exemplary scenario: As an example scenario, let us consider a small fictional company called CloudRetail, from the category of SMBs with a few hundred employees across the country. CloudRetail designs and sells fashionable and eco-friendly clothing and accessories. They use contract manufacturers but sell directly to their customers via their catalog and Website. Their core competency is eco-friendly product design quickly capitalizing on trends in the marketplace. CloudRetail runs software in-house for some functions, such as human resources, customer relations management (CRM), and their customer-facing web site. They have Information Technology (IT) department which maintains the Information Technology (IT) infrastructure inside the company. This Information Technology (IT) infrastructure has grown more complex and expensive to maintain as it has grown with the company. It now includes dozens of servers, specialized storage and network equipment, and ever-growing list of software, much of it to ensure smooth and secure operation of the company and the website.

Cloud Retail invested heavily last year in website hardware and network bandwidth to be prepared for the rush of orders during the holiday shopping season. The vision o CloudRetail is to reduce operational costs and enhance focus on their core competencies by benefiting from the advances in technology, and if possible, to transfer all non-core business operation, including support functions, to external companies.

Evolving CloudRetail into a virtual business using cloud service: CloudRetail can take advantage of many existing cloud services including CRM, HR. IT infrastructure and the hosting and operation of their website. Using cloud service provides the following benefits: (1) avoiding huge initial investments in hardware resources and software, (2) reducing ongoing operational, upgrade and maintenance costs, (3) scaling up and down hardware, network capacity and cost based on demand, (4) higher availability compared to in-house solutions for small businesses and individual-consumer maintained resources, and (5) access to a variety of software applications and features offered as SaaS that otherwise CloudRetail would have to purchase separately.

However, the potential risks of using cloud services include: (1) loss of direct control of resources and software, e.g, website infrastructure and operations staff for CloudRetail, (2) increased liability risk due to security breaches and data leaks as a result of using shared external resource, (3) decreased reliability since the service providers may go out of business, causing business continuity and data recovery issues, and (4) SaaS solutions are mainly built as one-size-fits-all customers, although there are sometimes add-ons to complement the functionality. CloudRetail is limited to the functionality offered by the SaaS providers and it is hard to customize solutions based on its needs.

Beside considering the above trade-offs, the difficult questions to answer for CloudRetail are (1) which functions to move to the cloud in what order, (2) how to ensure a smooth migration process given legacy applications in their environment, and (3) how to find and select service offerings that meet their requirements and establish seamless interoperation between services. For instance, assume they would like to more their website operation, Customer Relationship Management (CRM), accounting, and Human Resource (HR) systems to cloud services. Customer behaviour information from the Website has to be sent to Customer Relationship Management (CRM) systems and the accounting function needs information from the Website on sales and taxes. There is a need also for data integration solutions to migrate date from CloudRetail's legacy applications to cloud services.

Currently there is no environment to help CouldRetail in addressing the second and third concerns above, i.e., locating services, facilitating the process of using them and managing the whole lifecycle of engagement with cloud services. We discuss issues related to offering of such an environment in the next section.

4. VIRTUAL BUSINESS OPERATING ENVIRONMENT

A large and increasing number of services are available most of which target small businesses and individual consumers (the long tail of service customers). The wide variety and low cost of cloud services provides an unprecedented opportunity and financial motivation for businesses to more their Information Technology (IT) infrastructure to services in the cloud. There is a pressing need for an environment that allows SMBs and individual consumers to create and run a virtual business using cloud services. We call this a virtual business operating environment. A natural question whether and how it is possible to provide it as a service in the cloud?

4.1. Requirements of a Virtual Business Operating Environment

A virtual business operating environment provides facilities that allow business owners to build their business in a holistic way: define their business, express their requirements, find and engage cloud services that match their needs, compose services if needed, and monitor their business operations over outsourced services. In particular, it should provide the following sub-environments:

Business definition environment: This should allow the business owners to define the business, its structure (e.g., organization chart), Goals and strategies in some form that can be tracked down to the service execution level and managed.

Information Technology (IT) services marketplace: This refers to an environment where the requirements of users in terms of Information Technology (IT) services can be captured, existing services could be advertised, and facilities to browse and find available services are provided. The services marketplace may support various business models of offering services, e.g., bidding for user tasks, pay-per-use or subscription-based payments.

Business service design, integration and composition environment: This refers to an environment where selected services from the marketplace can be configured, composed, new services developed if needed and deployed to IaaS platforms. Interpretation among services for both functional and non-functional aspects can be checked and integration solutions developed.

Business services management environment: Once the business services are designed, the management environment provides facilities to initiate the interaction with services, monitor and manage the interactions and provide reports to the business owners using business dashboards.

In the following, we present a conceptual architecture for a virtual business environment, and discuss how it can be realized.

4.2. Virtual Business Operating Environment: Conceptual Architecture

Business architectures have been extensively studied during the last thirty years. Frameworks such as Zachman[11][12] and industry standards such as TOGAF[12][13] describe enterprise architecture. In particular, the Zachman framework identifies a number of orthogonal (horizontal and vertical) aspects. The horizontal layers include contextual (goals and strategies of business), conceptual (high-level design), logical (systems-level design) and physical (technology model) definitions for an enterprise. The vertical dimensions identify aspects that characterize each horizontal dimension. Other recent work also shows how a service oriented design and implementation of systems in an enterprise can fit in the Zachman framework[14]. However, these frameworks consider cases where the Information Technology (IT) infrastructure and services are developed and deployed as in-house solutions. We present the business architecture in an outsourced service environment as depicted in Figure 1 and described in the following.

FIGURE 1

Business goals Business structure Business policy Business performance	Business Context	Business owners, executives	Business definition environment
Business functionalities Business conceptual design Non-functional aspects	Business Services	Business/IT architects LoB managers	Business service management environment
Business behaviour Service integration and composition	Business Processes	Business users Developers	Business services design, integration and composition
IT services from public and private clouds	IT Services	Business users Development IT Administration	Services market place

We define the architecture of business in an outsourced services environment in four layers: *business context, business services, business processes and Information Technology (IT) services.* Business context layer provides for the definition of business goals, strategies, structure, policies and performance metrics and indicators. The main users of facilities at this level are business owners and executives who are rarely Information Technology (IT) experts. The major functions (supporting or core) of a business such as human resources, payroll, accounting, etc. are defined as coarse-grained services, referred to as "business services" in the business services layers. Users such as business/It architects may define or select the required business services from out-of-box business services blueprints. The Information Technology (IT) services layer represents the services that are offered in the cloud. Finally, the business processes layer is the representation of selection, design, integration and composition of Information Technology (IT) services in the form of workflows that fulfil the requirements of outlined business services. Figure 1 also shows the correspondence of sub-environments of a virtual business operating environment with the virtual business architecture, and the users of various layers/sub-environments.

4.3. Realizing the Virtual Business Operating Environment: Opportunities and Challenges

Let us review how the current advances in SOA, cloud computing and existing standards and methodologies help in realizing a virtual business operating environment, and identify the limitations and challenges. Note that besides the new and unique challenges posed by offering and using services in the cloud, some of which we review in the following, many challenges of realizing a virtual business operating environment are related to locating, composing, integrating and managing services. Most of these are the same as those identified for general services in SOA[15]. In the following, we highlight why fresh solutions for tacking these problems are needed in the cloud services environment.

Business context layer: The Object Management Group (OMG, www.omg.org) has proposed a set of complementary business modeling (BMM) specification (v1.0)[13] can be considered as a baseline for the business context layer. It models a business having elements including "end" (vision, goals, and objectives of the business), "means" to realize the end (mission, strategy, tactics, and directives including business policies and business rules) and assessment elements to define and evaluate the performance of the business. It should be noted that for an SMB not all these components may be necessary, however, these provide guidelines that can be customized to define a business in a virtual business scenario.

Business service layer: Business services can be divided into three categories: common (found in most businesses such as HR or CRM), industry-specific (found in vertical industries of the same type of business) and company-specific (unique to the given business). The environment has to provide blueprints of business services in the above categories, and also allow businesses to define company-specific business services. These descriptions can be used to find services from the marketplace that may fulfil the requirements. There is some initial work on the definition of business services[16], however, a more through study is needed to capture various business- level properties and policies into this definition.

Information Technology (IT) Service Layer: While there is a large body of work in SOA on Information Technology (IT) service description, search and management based on both functional and non-functional aspects[15], the following challenges remain:

Service description and search: A first challenge is that not all services that are available on the Internet are described using Web services interfaces (e.g. WSDL) nor are they actually offered online. Some of these services only have textual descriptions with some form-based data entry for service request. Existing service search techniques are mainly focused on the interfaces (functional aspects) of services and only support Web services, e.g., UDDI[14] and Woogle[17], or are merely catalogues with keyword search, e.g., seekda.com. Innovative approaches in service search technology are required to combine techniques to consider Web services, REST services as well as services with non-structured and non-standard descriptions. These approaches also need to be highly scalable to index millions of services that will be available in the could and allow service seekers to pose potentially diverse constraints on service functionality as well as cost, performance, ratings, usage controls, regulatory requirement, and policies for data retention, transfer and protection.

SLA, data privacy and security concerns: A consequence of using services in the cloud in that the location where customer data is kept is out of the customers control. Currently, there is no support for mandating specific data protection policies to service providers, e.g., where, how long and how data is kept. Another more serious providers. Information is routinely leaked from sub-contractors with poor data management practices[18]. Indeed, there is a need for approaches to tag directly the data with security and privacy policies that would travel with sensitive data from one provider to another so that the roper technical controls can be employed by the various cloud providers to protect the data. As another issue monitoring service quality experienced by the customers of businesses that outsource functions to could services requires new techniques that ensure service providers follow the agreed upon service level agreements.

Business processes and integration layer: Although there are significant advances in service and data integration[19,20] and service composition[21,22] in SOA, problems such as how to automatically discover various Web services (including services with text-based interfaces, peoples services, etc.) that collectively fulfil a business service, how to automatically compose services, and integrate data and services are among the hard challenge yet to be addressed[15]. Given that most SaaS users are business users (not expert IT people), there is a need for lightweight solutions for composition and integration, e.g., based on advances in mashups[22,23], to address these problems rather than only relying on heavyweight solutions of conventional SOA that are targets for enterprise settings and require expert users' intervention.

As an additional factor, there are millions of users in the cloud but fewer business cases in which services need to be integrated or composed. We anticipate that the integration and composition of services become recurring problems and their solutions are packaged as services that can be deployed in the cloud environment. Therefore, a prevailing approach is to exploit the power of crowd (service users), by exploiting advances in Web 2.0[5], to allow reusing solutions that are ready-to-use with minor configuration or customization for new settings, or approaches based on capturing solutions as part of integration and composition patterns. Emerging cloud services for integration such as

Boomi and Bungee Cannot have taken the first step by providing pre-defined connectors between various SaaS and existing applications and also basic data mapping tools, however, they are inadequate for un-supported applications or complex or unstructured data models. To enable scaling such approaches to the diversity of applications on the Internet, a service and data integration development platform and marketplace has to be built that allows users (individuals, SMBa as well as enterprises) to easily develop, share, locate, customize and consume integration and composition solutions.

5. DISCUSSION AND OUTLOOK

Small businesses such as CloudRetail have already seen the benefits of using services in the cloud for most non-core functionally. Customers benefit form the economies of scale and the highly optimized Information Technology (IT) operations of cloud service providers. The opportunity to avoid capital costs and incur predictable expenses which scale up down with the current needs of the business is very attractive. Customers with occasional or bursty usage see tremendous benefits, as they only pay for resources when they are using them. Customers with stable usage patterns also benefits due to the lower cost of purchasing services than building them in-house. Unless Information Technology (IT) is a core competency of the business, most customers will not be able to attain the same capabilities cheaper by doing it themselves. As one example, Google's corporate email solution is, on average, ten times less expensive than in-house email solutions.

We envision that the low cost of using cloud computing is a key driver of its wide acceptance by individual consumers, SMBs as well as large enterprises. However, large enterprises will employ a hybrid cloud model in which both private and public clouds are present. Many enterprises will run mission—critical applications and store business-sensitive data in the private clouds, while outsourcing their supporting services to the public cloud.

The role that Information Technology (IT) staff is playing in the enterprises and businesses is also undergoing a significant transformation due to the push for further Business process outsourcing (BPO) based on cloud service. As most Information Technology (IT) staff more from business Information Technology (IT) departments to service providers, the role of remaining staff converts into support and consultation of service selection, engagement and management.

In terms of usage of services in the cloud, small SMBs and individual consumers will be the main users of: (1) Infrastructure as a service, (2) Database as a service, (3) Software as a service, and (4) Platform as a service. Enterprises may demand customization of services as the APIs provided by service provider may not offer the flexibility and features they require. In addition, they may demand instances of services to be deployed in their private clouds for the sake of keeping data onsite and retaining control. This can be seen as a transformation of how enterprises use commercial software as services in the cloud.

The virtual business operating environment for creating and conducting virtual businesses using cloud-based services is a missing piece and the current article lays the foundation o architecture for an environment that addresses this pressing need for businesses that intend to use cloud services.

Notes and References

1. Leyland, V.A., Electronic Data Interchange. Prentice-Hall, Inc., 1993.
2. Halvey, J.K, Melby, B.M., Business Process Outsourcing: Process, Strategies, and Contracts, John Wiley & Sons, Inc., 2007.
3. Anderson, C., *et al.* Worldwide and US Business Process Outsourcing, 2007-11 Forecast: Market Opportunities by Horizontal Process, IDC Market Analysis 208290, Sep. 2007.
4. HP, The benefits of combining business-process outsourcing and service-oriented architecture http://h20195.www2.hp.com/4AA0-4316ENW.pdf.
5. Murugesan, S., Understanding Web 2.0./EEEIT Professional 9(4):34-41, Jul. 2007.
6. Weiss, A., Computing in the clouds. ACM netWorker 11(4): 16-25, Dec. 2007.
7. Petire, C., Bussler. C., A service Agents and Virtual Enterprises: A Survey. *IEEE Internet Computing*, 7(4): 68-78, Jul. 2003.
8. Robison Shane. The next wave: Everything as a service Executive Viewpoint, http://ww.hp.com/hpinfo/execteam/articles/robison/08eaas.html, 2007.
9. Carr, Nicholas. The Big Switch: Rewiring the World, from Edison to Google, W.W. Norton Publishing, Jan. 2008.
10 HP, HP Adaptive Infrastructure, http://h20195.www2.hp.com/PDF/4AA1-0799ENW.pdf
11. Yarmis, J, *et. al.* Outcloduing: New Ways of Capitalizing on the Economics of Cloud Computing and Outsourcing, AMR Research, Nov. 2008.
12. Zachman, J.A., A Framework for information systems architecture. *IBM Syst. J.*. 26(3): 276-92, Sep. 1987.
13. TOGAF, The Open Group Architecture Framework Version 8.1.1, www.toggaf.og
14. Ibrahim, M., Long, G., Service-Oriented Architecture and Enterprise Architecture http://www.ibm.com/developerworks/webservices/library/ws-soa-enterprise1/?STACT=105AGX04&S CMP=ART
15. Papazoglou, M.P., Traverso, P., Dustdar, S., Leymann, F., Service-Oriented Computing: State of the Art and Research Challenges. *IEEE Computer*, 40(11), Nov, 2001.
16. Scheithauer, G., *et al.* Describing Services for Service Ecosystems, International Workshop on Enabling Service Business Ecosystems (ESBE'08), Dec. 2008.
17. Dong, X., *et al.* Similarly search for web services. In Proceedings of VLDB 2004, 372-83.

18. The Breach Blog, "BNY Mellon Shareowner Services loses backup tape," http://breachblog.com/2008/03/27/bny.aspx.
19. Hamid, R. Motahari Nezhad, *et al.*, Web Services Interoperability Specifications, IEEE Computer (3(05:24-32, May 2006.
20. Halevy, *et al.* Data integration: the teenage years. In proceedings of VLDB 2006.9-16.
21. Dustdar Schahrram, Wolfgang Schreiner: A survey on web services composition. *Int. J. Web and. Grid Services*, 1(1):1-30, 2005.
22. Brogi, A., Corfinio, S. and Popescu, R., 2008. Semantics-based composition-oriented discovery of Web services. *ACM Trans. Internet Technol.* 8(4): 1-39, Sep. 2008.
23. Benslimane Djamal, Dustdar Schahram, and Sheth Amit, "Services Mashups: The New Generation of Web Applications", *IEEE Internet Computing*, 12(5): 13-15, 2008.

CHAPTER

8

Small and Medium Enterprises: Support for Women—Barriers and Challenges

DR. AMRIK SINGH SUDAN AND ROHINI GUPTA SURI

INTRODUCTION

Women at present have broken the monopoly of men and taken rapid strides and proved themselves that they are not inferior when compared to men. Over the past few decades, the sex role of women has been breaking the barriers of four walls. Today, the role of women is not confined to the traditional one as a mother or house wife. The role of modern women is much more than what it was previously. However, even after the 64 years of India's independence, the bias against women continues because of the cultural and traditional roles which has affected lot on the female status and role.

In modern era, when women are showing their capabilities in various walks of life, the entrepreneurial class of women are playing an increasingly prominent role in the various sectors of industrial growth and economic development. Entrepreneurship amongst women is a relatively recent phenomenon, which is gradually changing the growing sensitivity to the roles, responsibility and economic status of women in the society in general and family in particular. At the same time, it is also recognized that their challenges are immense and complex. For women entrepreneurs, starting and operating a business involves considerable risk and difficulties, because in the Indian Social environment, women have always lived as subordinate to man. There has been noticeable changes in the socio-psycho Cultural and economic norms of our society due to liberalized polices of the government of India, increase in education level of women and increased social awareness in respect of the role women play. It has now recognized that to promote self-employment and to reduce the incidence of poverty. Some drastic efforts have to be made to accelerate self employment in women in various sectors.

Women, it is generally agreed, make a significant contribution to local and national economies all over the world but this fact is largely ignored or not given due recognition in national income and other statistics. Women are also largely excluded from economic decision-making at family, community, local and national levels.

The Potential impact of small and medium enterprises (SMEs) on women's economic and social development has been studied by academics, development agencies and policy-makers over the past two decades. The basis of this interest has been a growing belief that through support for women's small enterprise development and entrepreneurship, wider developmental goals and targets can be met. These include poverty reduction, improved health of families and communities, improved literacy and educational achievements, women's empowerment and wider economic development.

Given these arguments and beliefs, many donors and Non-Governmental Organisations (NGOs) have focused their effort on supporting the development of women's enterprise, although this has largely been at a micro level. There have been a very large number of projects throughout the world aimed at increasing women's participation in enterprise with funding made available to provide support through training, advice and credit.

It must be recognized, however, that women engaged in small-scale enterprise do not operate in a vacuum and must survive in a market place that is changing. They will inevitably find themselves having to compete in a market that is becoming increasingly global. Local women engaged in tailoring will have to compete with internationally based clothing manufacturers, those making cosmetics and soap will have to compete with the giants in the cosmetics industry, and those selling goods in the market place will have to compete with the wider availability of products made and sold around the world. The consumer is demanding access to these international products and will make choices between those that are locally made and those he/she sees advertised and promoted on a huge scale.

As yet there have been few studies that examine the impact of globalisation on women's small scale enterprise. Commonwealth Ministers responsible for Women's Affairs, at their meeting in April 2000, made a recommendation that this issue be addressed. They also noted the impact of globalization on the 'weak and vulnerable economies and especially on women and children'. In particular, the Ministers commented on the negative effect that globalization has had on the informal sector that offers employment and provides livelihoods for many women.

To date, much of the focus of women's small and medium enterprises (SMEs) programmes has been on the poor and marginalized, with support targeted to assist women develop an income stream from a trading activity.

Given the added difficulties of competition and the changing demands of the customer, the expectations of increased wealth from small and medium enterprises (SMEs) activity may not met. Without strategic support to help them compete, many women may continue to struggle on the margins.

Despite the potential challenges for women's small and medium enterprises (SMEs) development in the wider context of trade liberalization, there are still enormous benefits to be gained from women's participation in enterprise. There is potential economic benefit to be gained by families, communities and countries from women's entrepreneurship.

Gender and Small and Medium Enterprises (SMEs) Support for Women—Barriers and Challenges

Recognition of women's potential contribution has resulted in a major shift over the past twenty five years towards women as a key target group for programmes using small and medium enterprises (SMEs) development as a way to achieve wider poverty reduction and economic development targets. Despite this strong focus, many programmes have failed to adequately address gender imbalances and achieve real gender equality, with women and men able to enjoy equal opportunities and access to small and medium enterprises (SMEs) development support. Well-meaning policy statements will not bring about real change unless they are backed up with appropriate strategies and operational practices. Gender equality is not simply about increasing the number of women who access support for small and medium enterprises (SMEs) development. Unfortunately, there appears to have been a wide reliance on quantity as a key success indicator. Many programmes use a simple measure of the number of women involved to judge the performance and success of gender equity are often ignored or not understood. For some, women being helped out of poverty through small-scale enterprise is a legitimate strategy, but to effect gender change where women can achieve real economic power is possibly still too worrying to some governments and organizations.

Over-Reliance on Micro-Credit as a Mechanism for Achieving Gender Equity

Whilst the availability of credit has clearly been a critical factor in the development of women's enterprises, there is a growing recognition amongst support agencies that the availability of micro-credit on its own does not necessarily lead to successful enterprise and increased income.

Small Scale entrepreneurs need more than money to start up an enterprise and survive in the market place. They need access to strategic support that takes account of their needs and the complexity of small and medium enterprises (SMEs) development, as well as the markets within which they operate.

The date, most of the focus of support for women's enterprise development has been on the start-up of new enterprises through the provision of small-scale loans or short training programmes. There has been significantly less emphasis on supporting these enterprises and helping them to survive beyond basic start-up, and to compete in changing markets or achieve planned business growth.

Enterprise development is a complex issues and demands long-term and highly individual and flexible support if it is to achieve the impact that is the aim of many small and medium enterprises (SMEs) programmes. Women's small and medium enterprises (SMEs) development is still largely concentrated in low margin, competitive markets where women struggle to achieve profits.

One Government run programme that provides support for new and developing small- and medium-enterprises (SMEs) recognized that its services were largely taken up by men and that women were not being targeted in strategic plans. The number of women accessing support was extremely small and it was agreed that this must be addressed.

With support from a donor agency, a training programme was designed and delivered that sought to build the capacity of small and medium enterprises (SMEs) support staff to develop and deliver women-friendly and gender-aware support that understood, and took account of, the different circumstances of women as compared o their male counterparts that resulted from the additional social and household/ family responsibilities women had to meet.

Whilst this was clearly a major step forward, the plan failed to address some of the core issues of institutional gender awareness. Only the relatively small number of staff who received that training were aware of the wider issues involved in delivering support for women's enterprise. In addition, all the senior officers and approximately 98 percent of the staff who were actively engaged in support delivery were men, who were very unused to working with women as clients. In addition, no extra funds were made available to develop the range of services for supporting women's enterprise. Thus, the institutional reform that would have been necessary to achieve a gender-sensitive strategy never took place.

The female clients lost out in that only a small percentage were able to receive support from trained staff, and the trained officers were themselves struggling to operate without the necessary institutional support. The key lesson from this experience, which is by no means unusual, is that development organizations cannot achieve gender equity in their service delivery unless they also make fundamental changes in their own working practices, structures and culture.

A key programme of support for women's micro-enterprise was developed by one Commonwealth government and supported by a range of donors who assisted in the programme design. The programme sought to encourage 4,500 women living in rural areas to set-up and run small scale income enterprise with assistance from community workers who had received some training in enterprise development and credit control.

The prgramme placed strong emphasis on the credit aspects of the support and there was very little enterprise development support other than a short training course linked to the access to credit. The support workers were each required to provide support for up to 80 women who were engaged in individual income generating activities, as well as sometimes working in small groups.

Overall, the reliance on credit, while allowing women to operate at the margins and develop some income generation, did not offer a comprehensive range of support and therefore women were not expected to do more than generate income for survival.

The women involved largely chose to engage in similar activities and therefore created more tensions in an already saturated market. The overstretched development workers did their best to support these women but their ability in terms of time and skills could not offer the more detailed support required to take any of these small scale activities beyond a subsistence level and to help women to compete successfully.

For many women, small and medium enterprises (SMEs) development is less about enterprise, life choices and empowerment, and more about survival on a day-to-day basis. The dream of being a woman owning and running her own small scale enterprise can rapidly change to a daily fight for economic survival.

For many women, micro-finance and small scale enterprise is not empowering. It can be marginal and can increase tensions in the family, and create contradictions between the need to generate an income, repay loans and fulfil wider family and community responsibilities.

Overall, women have yet to be accepted as being competent to own and manage successful businesses. By focusing on women's enterprise as an anti-poverty measure, it is possible to overlook how women can access mainstream support with policies and measures designed to achieve equitable opportunity.

Successful Support Strategies

Over the past twenty-five years, and despite the difficulties, there has been a growing understanding of the role that finance plays in supporting the development of women's enterprise. A range of funds has been established throughout the world to widen access for poor people, particularly women, to micro-credit and thereby the income generation through enterprise.

The pioneering funds of the Grameen Bank, and Bangladesh Rural Advancement Committee (BRAC), K-Rep in Kenya and Bancosol in Bolivia have all demonstrated that poor women can, and will, borrow money, make interest payments and repay loans in a responsible manner. Indeed, it is the view of many funders that women are more responsible and conscientious borrowers than men and will make every effort to honour their commitments to pay interest and repay capital borrowed. The experience of very low default rates for funds that make loans to women has made the certain of micro-credit funds a very attractive option to donors who see their finance revolving and reaching a larger number of women.

In turn, the wider availability of micro-finance has been a significant factor in the development of small-scale enterprises run by women, and poor women in particular. Without access to credit on terms that are appropriate to their needs, it is unlikely that the growth in this sector would have had the same impact.

These funding organizations have learned lessons and developed practices for delivering credit to suit women's specific needs, as well as working towards the overall sustainability of the fund itself. Access to credit and women's enterprise development have become so inextricably linked that, for many Non-Governmental Organisations (NGOs) and other support organizations, women's enterprise development is all about setting up and running credit programmes and funds.

Continued access to credit is key to the development of a strong small and medium enterprises (SMEs) sector run by women. However, women must to able to access mainstream banking and finance, and barriers to this access must be challenged. Women need flexible finance that meets the developmental needs of the enterprise. For women to achieve wider goals of empowerment and equality of opportunity, they must ultimately be able to access mainstream services.

Business Development

Women engaged in setting up an enterprise have a range of complex needs; the development of Business Development Services that can run alongside micro-finance is now being recognized by an increasing number of Non-Governmental Organisations (NGOs), do nor agencies and governments as an essential part of the overall package of support. Business Development Services are particularly important as the market place becomes more complex, particularly if women are to be assisted to develop marketing strategies that can help them to compete more effectively.

However, to be successful, Business Development Services must develop their own models of good practice that are based on meting the needs of the client, in this case women. These services can be a critical factor in the development of sustainable and successful businesses run by, and for, women. This can, and should, go beyond the specific support for start-up and tackle the wider issues of enterprise development and the creation of enabling environments.

Business Development Services can and should assist women to develop their products and effective marketing strategies. If enterprise development is to be more than an income generating strategy, women must be supported to move beyond the traditional saturated markets and product lines that many now choose. This in itself demands high levels of competence from the enterprise support staff.

One example of a successful programme developed to provide a wide range of business development support for men and women was designed and implemented by an organization which is business-like in its approach, and sets targets and plans that include provision of support on a gender equitable basis. The organization ensures that women are represented at the Board level, and that this representation is both meaningful and contributes to the planning processes. The Board as a whole has addressed the gender considerations of its work, and every effort is made to ensure that the Board and senior staff operate within a gender aware culture and working environment.

The gender focus of Business Development Services is an essential component of supporting women's Small and Medium Enterprises (SME) development that can tackle some of the wider impacts of globalisation of women's enterprises. Effective gender-aware support that recognizes and attempts to redress/compensate for the different circumstances and needs of women, as compared to their make counterparts, is required to develop products and services that take account of the marketplace. Without it, women's enterprises will continue to operate at the margins.

Good Practice for Women's Small and Medium Enterprises (SME) Development

The Enabling Environment: Supporting women's enterprise development may seem a desirable strategy for those supporting economic development, but it is a strategy that can and does impact upon much wider issues and demands long term and strong commitment by those offering support. Goetz and Gupta (1996) point out that 'improvements in women's productivity, mobility, social status and control of household decisions take time, require considerable commitment by development workers, a long-term investment is local-level processes of social changes, as well as a willingness to cope with the sometimes violent and disruptive consequences of challenging class and gender privilege'. The enabling environment has several key components illustrated on next page.

Women's Small and Medium Enterprises (SME) Development—Programme Design Checklist

Successful enterprise support depends, to a very large extent, upon the design of support programmes and the ways in which these are delivered. There are key some factors that should be considered when planning and designing any support programme.

Fiscal and Legislative: For small business to succeed and thrive in an increasingly competitive market, the state must ensure that barriers are reduced and that an enabling fiscal and legislative framework is in place. In addition, women must have legal rights to ownership of property and control over their business decisions. Women must be able to take legal

Good Practice for Women's Small and Medium Enterprises (SME) Development

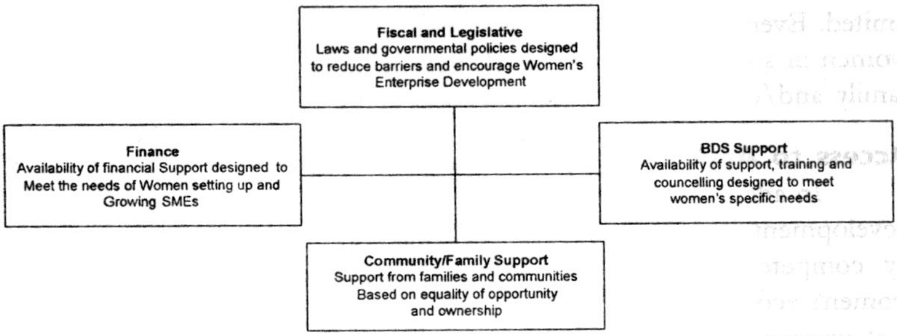

Does the programme take account of women's practical needs as well setting targets for Small and Medium Enterprises (SMEs) support?	The programes design should take account of women's roles and of responsibilities within the family and as wider community. This may mean that programes of support must plan around the time demands of those responsibilities.
Has the programme design taken account of any fiscal or legal barriers that may prohibit women from achieving the prgramme targets that have been set?	It is essential that, as part of the overall design, consideration should be given to any fiscal or legal barriers that may exist and that every effort is made to ensure that women can overcome the challenges these pose when setting up and running an Small and Medium Enterprises (SMEs).
Who will deliver the support, and do they have the overall competence and awareness of the different circumstances and needs of women and men to deliver a programme targeting women's small enterprise development?	It is essential that those delivering the support not only have the full competence to deliver the specific support, for example training in business development, advisory Small and Medium Enterprises (SMEs) counseling or small scale credit, but that they are also fully aware of the gender aspects of enterprise development and operate from within an organizational culture that understands and takes account of the different circumstances and concerns of women and men.
Does the programme take account of the wider enterprise development needs for Skills training and development, Small and Medium Enterprises (SMEs) advice and counselling and access to resources?	A programme can focus on a single intervention but should consider how the women as clients will access wider support. It is essential that any SMEs can have access to wider support and draw upon advice and councelling, and that it can have access to skills and enterprise management training, as well as to a broad range of resources, including equipment, premises, transport and credit.

responsibility for property and loans, and enter into contractual agreements in their own right with a range of bodies that form the support framework for small business, including banks. If women are denied these rights, then their ability to develop their enterprises will be limited. Even if legal rights of ownership exists, it is not uncommon for women in some countries to be denied ownership at a practical level by family and/or community customs and culture.

Access to Appropriate Business Development Support (BDS)

In addition, women must have access to a wide range of business development support, including training, advice and counseling delivered by competent advisers who themselves have a strong commitment women's economic development and an understanding of the constraints that women often have to overcome when setting up and running an enterprise, due to the different 'traditional' structural and social expectations and responsibilities of women and men. The Committee of Donor Agencies for Small Enterprise Development (CDSED) in the Business Development Support Study Report (2001) identify eleven principles of good practice for donor funded Business Development Support (BDS) interventions. Amongst these is the recognition that Business Development Services must themselves be business like the best Development Support (BDS) organizations are those that are like Small and Medium Enterprises (SMEs) in terms of people, systems and values. This in itself demands high levels of competence from the enterprise support staff.

Access to Finance and Financial Services

Continued access to finance is key to the development of a strong Small and Medium Enterprise (SME) sector run by women. However, women must be able to access mainstream banking and finance, and barriers to this access must be withdrawn. Women, like anyone else running an enterprise, need flexible finance that meets the enterprise's developmental needs. For women to achieve wider goals of empowerment and equality of opportunity they must ultimately be able to access mainstream financial services.

Community and Family Support

For a woman's small enterprise to succeed, she must have the support of her family and community. This includes both the practical support required to make the venture possible, such as time and access to resources, and the psychological support that will give her the confidence and freedom to try things out and move forward. If this support is denied, then it is difficult for the woman to have the necessary time and strength to develop and run the enterprise, raise the resources required and to control the decision-making.

Schemes for the Development and Promotion of Women Entrepreneurs in India

In order to encourage more and more women enterprises in the Micro and small Enterprises (SMEs) sector, several schemes have been formulated by the ministry of Micro and Small Enterprises (MSEs) government of India and some more are in the process of being finalized, targeted only at the development of women enterprises in India. These schemes are:

(i) *Trade Related Entrepreneurship Assistance and Development Scheme for Women (TREAD)* with a view to encourage women in setting up their own ventures, government implements a scheme, namely, "Trade Related Entrepreneurship Assistance and Development (TREAD) during the 11th Plan. The Scheme envisages economic empowerment of women through the development of their entrepreneurial skill in non-form activities.

(ii) *Micro and Small Enterprises Cluster Development Porgramme (MSE-CDP)*: The cluster Development programme (CDP) implemented envisages diagnostic study of identified clusters of traditional skill based MSEs to identify appropriate technologies and their providers and to facilitate adoption of available technology meeting the specific needs of the end users. The cluster Development aims at enhanced competitiveness, technology, improvement adoption of best manufacturing practices marketing of products, employment generation, etc. The Scheme provides assistance for capacity building common facilities, marketing, etc. and the delivery assimilation and diffusion of the identified technology from its producers to the recipient user/cluster of small enterprises.

(iii) *Credit Guarantee Fund Scheme for Micro and Small Enterprises*: The schemes was launched in August 2000 to ensure better flow of credit to micro and small enterprises by minimizing the risk perception of financial institutions and banks in lending without collateral security Under the scheme guarantee cover is provided to collateral free credit facility extended by member lending (MLIS) institutions to the new as well as existing micro and small enterprises on loan upto Rs. 50 lakhs. The guarantee cover available as up to 75% of the loans extended. The extent of guarantee cover is 80% for (i) Micro enterprises for loans up to Rs. 5 Lakhs; (ii) MSEs operated and/or owned by women; and (iii) all loans in the North-East Regions.

(iv) *Support for Entrepreneurial and Managerial Development*: MSME-DIs regularly organise a number of Entrepreneurship skill Development programme (ESDPs)/Entrepreneurship Development progamme (EDPs)/Management Development programmes (MDPs) to train the potential entrepreneurs in improving their techno/managerial knowledge and skill with a view to facilitating tham to start MSEs in various fields. Many of the programmes are tailor made for the target group for SC, ST, OBC Women, Minorities and other weaker sections and exclusively for women also. These programmes are also called "out Reach-programmes" as they are conducted in rural/less developed areas.

(v) *Exhibitions for women under promotional package for Micro and small Enterprises Approved by CCEA under Marketing Support*: Under this scheme, participation of women entrepreneurs in 25 international exhibitions is envisaged during the 11th Plan with a view to encourage women entrepreneurs to participate in the International exhibitions under MDA scheme it has been decided to:

1. provide rent free space (6/9 square meters) in the exhibitions.
2. reimburse 100% economy class air fare for one representative. The overall ceiling shall however be Rs. 1.25 lakh. During the period April 2008 to September 2008, Office of Development Commissioner (MSMF), New Delhi, participated in the following exhibitions with 36 women out of total 75 participants.

S. No.	*Name of Fair*	*Date & Duration*	*Number of women participants*	*Total participants*
1.	Fujian China	May 18-22, 2008	7	17
2.	Imported goods Fair Seoul, S. Korea	May 2008	11	13
3.	Exporive Garda Italy	June 14-17, 2008	2	7
4.	SAARC Fair Colombo, Sri Lanka	August 28 -31, 2008	4	7
5.	Budapest International Trade Fair, Budapest, Hungry	September 10 -14, 2008	3	16
6.	5th China, Small & Medium Enterprises Fair, Guangzhon, China	September 22 -25, 2008	9	15
7.	Total	-	36	75

CONCLUSION

Indian woman was like a well defined predictable masterplan. They were to act as per the liking and dislikings of their male counterparts. Slowly, changes started creeping in. Women migrated to cites and towns, got education and economic independence. Her role as a earning women was no more looked down upon. Women entrepreneurship is the process whereby woman organize the business and provide employment to others. A women entrepreneur/enterprise is termed as an SSI unit/industry-related service or business enterprise, managed by one or more women entrepreneurs in proprietary concerns, in which she/they individually or jointly have a share capital or not less than 51 percent as shareholders/partners/directors of a private limited company/members of a co-operative society. Unlike other countries Indian women run just 8 percent of SSIs while in USA, their number is as high as 50 percent. Indian women have to face a number of problems beginning right from the commencement of the enterprise. Male dominated society, heavy competition, non-cooperative attitude of financial institutions, marketing problems, family conflicts, etc. are just some of the problems faced by minority entrepreneurs. However, we can say that the transitional phase has set in. Since 1990s, India has become a major centre for business outsourcing. Besides, world of fashion has opened and education has developed rapidly as an industry. Government has come forward with umpteen facilities, concessions and incentives exclusively for women entrepreneurs. Various schemes like PMRY, TRYSEM, DWCRA, IRDP, TREAD, SGSY, SIDBI development schemes, etc. have been launched for women where by they are provided various relaxations, as compared to their make counterparts. Inspite of all these incentives, women have to go a long way fulfilling their multiple role of a mother, wife and a business women as well.

Women, which constituted nearly 49% of the countries total population can effectively and actively activate their latent human potentials through entrepreneurship. This initiative will lead to creation of self employment and wage employment avenues for large number of people. Women can engage themselves in small and tiny sector industries, both in manufacturing and service sectors and play a significant role in generation of employment.

Small and Medium Enterprises (SMEs) significantly contribute to industrial, economic technological and regional developments across the developed and developing economies. Small and Medium Enterprises (SMEs) make a substantial contribution in terms of creation of employment, output and exports. They are capable of producing quality goods and contribute around 35% to total exports.

Small and Medium Enterprises (SMEs) have enormous potential for innovations and incremental development, which need to be nurtured for production of new goods and services at competitive costs. The critical issues which entrepreneurs are likely to face are recruiting talent and developing potential leaders. Considering these challenges, looking ahead to have clear vision is no longer an option if Small and Medium Enterprises (SMEs) want to remain in business for the next ten years.

CHAPTER

9

Status of Women Entrepreneurship in India

DR. RADHA GUPTA, DR. AMRIK SINGH SUDAN AND DR. RAIS AHMAD

ABSTRACT

Women is often described as better half of man, but ground reality is different because women participation in social and economic activities is very low. Handful of successful women entrepreneur cannot fill this gap, but efforts have been started as women have started realising their potential and they are now taking active participation in all commercial activities by making proper balance between their personal and professional life. Approach of modern women is different from the traditional one as they have broken the barrier and domain set by male. Earlier women used to act according to choice of their counterpart that is why their concentration was in low paid and low status occupation which indicates that women were marginalized in the labour force. Though global evidence shows that women have started showing their gray mettle by starting and handling industries activities successfully. But "ALL IS STILL NOT WELL", because women entrepreneurship is considered as 'opportunity entrepreneurship' rather than 'necessity entrepreneurship'. In the present paper an attempt has been made to identify and understand skill that this gender bias population is already possessing and what new skill they should imbibe for running their enterprise. Paper further highlights how education, media exposure and, present successful women entrepreneur are creating urge in the women to perform economic activities independently. It further discuss the difficulties before women towards entrepreneur and in the end suggestion have also been made to foster women entrepreneurship for empowering women on the one hand and socio-economic development of the country on the other.

1. INTRODUCTION:

The Indian sociological set-up has been traditionally a male dominated one, where women are considered as weaker section and always to depend on men folk in their family and outside, throughout their life. Though the situation has not much improved but the traditional set-up is changing in the modern era where approach of modern women is different from the traditional one as they have become aware of their status, right and their work situation broken the barrier and domain set by male. Earlier women used to act according to choice of their counterpart that is why their concentration was in low paid and low status occupation which indicates that women were marginalized in the labour force. Though global evidence shows that women have started showing their gray mettle by starting and handling industries activities successfully. While women entrepreneurs have demonstrated their potential, fact remains that they are capable of contributing much more than what they already are. In order to harness their potential and for their continued growth and development, it is necessary to formulate appropriate strategies for stimulating, supporting and sustaining their efforts in this direction. Such a strategy needs to be in congruence with field realities, and should especially take cognizance of the problems women entrepreneurs face within the current system. This paper seeks to discuss in detail the following issues relating to status of Women Entrepreneurship in India:

1. Introduction;
2. Journey of women entrepreneur;
3. Recent Trends;
4. Factors impede the growth of women entrepreneurs; and
5. Conclusion and suggestions for developing women entrepreneurship.

Women is often described as better half of man, but ground reality is different because women participation in social and economic activities is very low. Handful of successful women entrepreneur cannot fill this gap, but efforts have been started as women have started realising their potential and they are now taking active participation in all commercial activities by making proper balance between their personal and professional life. Approach of modern women is different from the traditional one as they have broken the barrier and domain set by male.

Earlier women used to act according to choice of their counterpart that is why their concentration was in low paid and low status occupation which indicates that women were marginalized in the labour force. Though global evidence shows that women have started showing their gray mettle by starting and handling industries activities successfully. But "ALL IS STILL NOT WELL", because women entrepreneurship is considered as 'opportunity entrepreneurship' rather than 'necessity entrepreneurship'. In the present paper an attempt has been made to identify and understand skill that this gender bias population is already possessing and what new skill they should imbibe for running their enterprise. Paper further highlights how education, media exposure and, present successful women entrepreneur are creating urge in the women to perform economic activities independently. It further discuss as the difficulties before women to be entrepreneur and in the end suggestions have also been made to foster women entrepreneurship for empowering women on the one hand and socio-economic development of the country on the other.

Women Entrepreneurs may be defined as the women or a group of women who initiate, organize and operate a business enterprise. 'Women Entrepreneurs' is a person who accepts challenging role to meet her personal needs and become economically independent. A strong desire to do something positive is an inbuilt quality of entrepreneurial women, who is capable of contributing values in both family and social life. Women Entrepreneurs may be defined as the women or a group who initiate, organize and operate a business enterprise to women. Like a male entrepreneurs a women entrepreneur performs many functions. She undertakes the prospects of starting new enterprise, manages risk and introduces new innovations. Moreover, coordination, administration, control of business and providing effective leadership in all aspects of business are other key business function which she does perform.

FIGURE 1

Function of Women Entrepreneurship

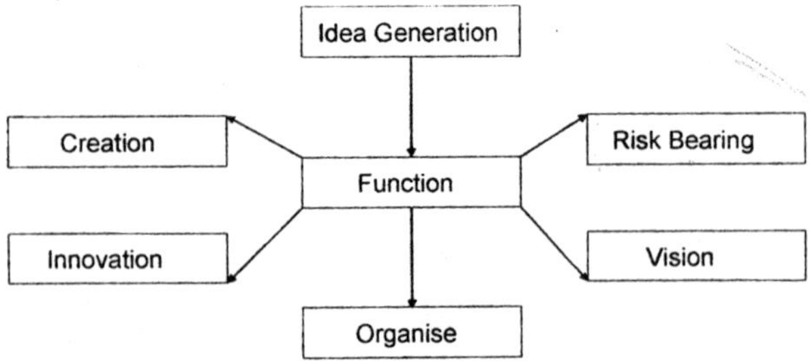

Women set enterprises due to certain reasons. These to reason be compelling or inspirational factors. They may be deriving, motivational or facilitating forces.

FIGURE 2

Force of Women Entrepreneurship

Motivational Forces	*Facilitating forces*
1. Economic necessity	1. Availability of capital
2. Independence	2. Innovative ideas
3. Family occupation	3. Family cooperation
4. Education	4. Self-satisfaction

2. JOURNEY OF WOMEN ENTREPRENEUR

The fact that women want to carve their own nitch in the society, they gradually started coming out from their shell for initiating economic activities. The journey depicting their status in different time period can be analysed in the following stages:

FIGURE 3

Journey of Women Entrepreneur

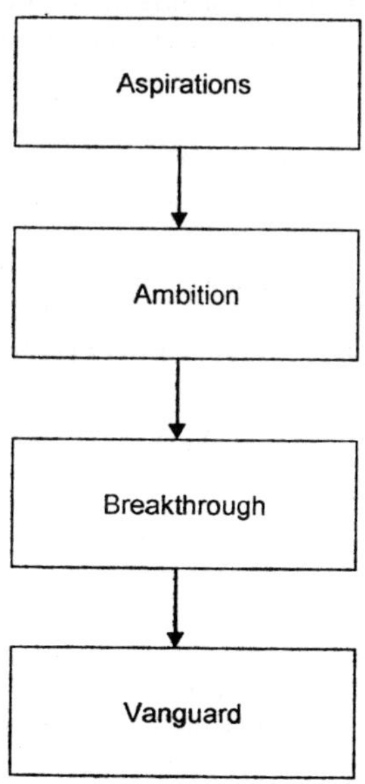

As Virgina Woolf has treated women as 'Angle of the house' just confined within the four walls of the houses. But the concept of the angle was changed by the women in the sixties when on the one hand they accepted the code of socio-cultural influence and on the other hand aspired to become educated. In the decade of 70's, women unlike their mothers become self-employed not because of the financial compulsions but it was the result of the choice. Next decade in the history of women entrepreneurship was considered as period of break through in which women took professions like medicine, engineering and moreover they joined the race of their fathers and husbands. In 19th century and 20th century different breed of women has emerged having their women role models who are an epitome of liberty and strength. The concept of women as heir to father's property emerged. 21st century which is the era of telecom revolution, women got their expertise in different walks of personal and professional life. Though women are competing and handling all the situation very efficiently but still there are so many hurdles like gender biasness which are creating problems for women to flourish.

3. RECENT TRENDS

Women entrepreneurs have been making a significant impact in all segments of the economy not only in India but also in other part of the globe like Canada, Great Britain, Germany, Australia and the United States. Apart from spreading their wings in Engineering, Electronics, and Energy women have also shown their grey mettle in the field of retail trade, restaurants, hotels, education, cultural, cleaning, insurance and manufacturing, etc. Women owned businesses are highly increasing in the economies of almost all countries. The hidden entrepreneurial potentials of women have gradually been changing with the growing sensitivity to the role and economic status in the society. Skill, knowledge and adaptability in business remain reasons for women to emerge into business ventures. Today women want to be financial independent. They want to make money to support family and improve their standard of living.

In the present scenario, women are more inclined to business as successful women entrepreneur are role model for them and they want to follow their foot step and touch the zenith of success.

With the advent of media, women are aware of their own traits, rights and also the work situations. The glass ceilings are shattered and women are found indulged in every line of business from papped to power cables. The challenges and opportunities provided to the women of digital era are growing rapidly that the job seekers are turning into job creators. They are flourishing as designers, interior decorators, exporters, publishers, garment manufacturers and still exploring new avenues of economic participation. In India, although women constitute the majority of the total population, the entrepreneurial world is still a male dominated one. Women in advanced nations are recognized and are more prominent in the business world.

Women are Not Only Starting Businesses, They are Staying in Business

Between 1997 and 2006, businesses fully women-owned, or majority-owned by women, grew at nearly twice the rate of all U.S firms (42.3% *vs.* 23.3%). During this same time period, employment among women-owned firms grew 0.4%, and annual grew 4.4%.

In 2006, reports on women-owned (or majority owned by women) in the United States returned the following impressive statistics:

- There were an estimated 10.4 million privately-held firms;
- This accounted for two in five (40.2) of all businesses in the country; and
- These firms generated $1.9 trillion in annual sales and employed 12.8 million people nationwide (Lahle Wolfe, 2011).

4. FACTORS IMPEDE THE GROWTH OF WOMEN ENTREPRENEURS

Though women want to get ahead their life in business as they have the skill for handling a business but they virtually find it hard to do. However Women face following challenges.

Lack of awareness about the financial assistance: Various institutions in the financial sector extend their maximum support in the form of incentives, loans, schemes, etc. Even then every woman entrepreneur may not be aware of all the assistance provided by the institutions. So the sincere efforts taken towards women entrepreneurs may not reach the entrepreneurs in rural areas.

Exposed to the training programs: Training programs and workshops for every type of entrepreneur is available through the social and welfare associations, based on duration, skill and the purpose of the training program. Such programs are really useful to new, rural and young entrepreneurs who want to set-up a small and medium scale unit on their own.

Identifying the available resources: Women are hesitant to find out the access to cater their needs in the financial and marketing areas. In spite of the mushrooming growth of associations, institutions, and the schemes from the government side, women are not enterprising and dynamic to optimize the resources in the form of reserves, assets mankind or business volunteers.

Concern for family: Women give more emphasis to family ties and relationship. Married women have to make a fine balance between business and home. Moreover the business success depends on the support the family members extended to women in the business process and management. The interest of the family is a determinant factor in the realization business aspirations.

Low level management skill: Another argument is that women entrepreneurs have low-level management skills. They have to depend on office staffs and intermediaries, to get things done, especially, the marketing and sales side of business. Here there is more probability for business fallacies like the intermediaries take major part of the surplus or profit. Marketing means mobility and confidence in dealing with the external world, both of which women have been discouraged from developing by social conditioning. Even when they are otherwise in control of an enterprise, they often depend on males.

Competition: The male-female competition is another factor, which develops hurdles to women entrepreneurs in the business management process. Despite the fact that women entrepreneurs are good in keeping their service prompt and delivery in time, due to lack of organizational skills compared to male entrepreneurs women have to face constraints from competition. The confidence to travel across day and night and even different regions and state are less found in women compared to male entrepreneurs. This shows the low level freedom of expression and freedom of mobility of the women entrepreneurs.

Lack of awareness: Knowledge of alternative source of raw materials availability and high negotiation skills are the basic requirement to run a business. Getting the raw materials from different source with discount prices is the factor that determines the profit margin. Lack of knowledge of availability of the raw materials and low-level negotiation and bargaining skills are the factors, which affect women.

Lack of interest: Low-level risk taking attitude is another factor affecting women folk decision to get into business. Low-level education provides low-level self-confidence and self-reliance to the women folk to engage in business, which is continuous risk-taking and strategic cession making profession. Investing money, maintaining the operations and ploughing back money for surplus generation requires high risk-taking attitude.

5. CONCLUSION AND SUGGESTIONS

While women entrepreneurs have demonstrated their potential, the fact remains that they are capable of contributing much than what they already are. In order to harness their potential and for their continued growth and development, it is necessary to formulate appropriate strategies for stimulating, supporting and sustaining their efforts in this direction. Such a strategy needs to be in congruence with field realities, and should especially take cognizance of the problems women entrepreneurs face within the current system. "When women move forward, the family moves, the village moves and the nation move." These were the words of Pandit Jawahar Lal Nehru which is fact. Until and unless women section of society is given the empowerment family can not grow, with the result society cannot move upword and socio-economic status of the country can not be developed because women and men are two pillars of the country. In present scenario due to modernization, globalisiation and development of education women are very much awarded about getting gainful employment and heading business units despites they are facing socio-cultural economic, technical, financial and managerial difficulties.

6. HOW TO DEVELOP WOMEN ENTREPRENEURS?

Right efforts from all areas are required in the development of women entrepreneurs and their greater participation in the entrepreneurial activities. Following efforts can be taken into account for effective development of women entrepreneurs.

Promoting women entrepreneurs in micro-enterprises: An increasingly large number of women in India are engaged in the informal economy and operating in the small scale sector, cottage industries, and micro-enterprises. With proper governmental and societal support, these women can create sustainable, organized and growth-oriented enterprises with a vision.

Institutional and policy-based support: Government policies and regulations regarding business and industry can be formulated with the intent to encourage women entrepreneurs. The first part of the approach must be to review the existing regulatory framework, and make necessary modifications. The second part should be to take vital steps to reduce the administrative hurdles especially for women entrepreneurs. The third part must be to include easy and subsidized especially for women entrepreneurs. The third part must be to include easy and subsidized financing for women's entrepreneurial projects.

Directing women entrepreneurs into high growth areas: Another key part of the strategy must be to facilitate the entry of women entrepreneurs into areas where high growth is expected. Special incentives, tax rebates, duty cuts and subsidized land and machinery can be provided to encourage women in emerging sectors. Special recognitions and award can be instituted for women participating in such targeted industries.

References

Ahlawat, T., 1999. Impact of Financial Assistance Schemes on Economic Empowerment of Women. M.Sc. Thesis, CCS Haryana Agricultural University, Hisar.

Aswathappa, K. (2000). "Human Resource and Personnel Management". Tata McGraw Hill, New Delhi.

Desai, V. (1999). "Dynamic of Entrepreneurial Development and Management", Himalaya Publishing House, Mumbai.

Gupta, C.B. and Khanka, S.S. (2000), "Entrepreneurship and Small Business Management", Sultan, Chand and Sons, New Delhi.

Gupta, M.C., 1987. "Entrepreneurship in Small Scale Industries", Anmol Publications Pvt. Ltd., New Delhi.

Jain, Kumar Arun, "The Making of an Entrepreneurial Manager". Vikas Publishing House, *Indian Journal of Industrial Regulation*, Vol. 39, No. 3, Jan. 2004, p. 495.

Kamulun, N. and A. Kumar, 1992. "Entrepreneurship Development in Orissa: Some Issues", *Yojna*.

Kapoor, K., 1998. "Entrepreneurial Behaviour: A Study of Selected Women Entrepreneurs". M.Sc. Thesis, Punjab Agricultural University, Ludhiana.

Reddi, P.N., 1991. "Problems of Women Entrepreneurs in Goa: A pilot study", *Khadi Gramodyog*.

Sharma, K and Basotia, G.R. (2003). "Entrepreneurship and Small Business". Mangaldeep Publications, New Delhi.

Shukla, M.B. and Gupta, Ranjita (2003). "Entrepreneurship Development in India" Bokland India, Varanasi, U.P.

CHAPTER

10

Growth and Performance of Milk Cooperatives in Punjab

PARMINDER KAUR

The importance of dairying in a country like India hardly needs introduction. India has vast resources of livestock, which play an important role in the national economy and also in the socio-economic development of millions of rural households. Currently, the livestock sector in India contributes about 27 percent to the Agricultural Gross Domestic Product (GDP) and provides employment to 20 million people, particularly women, in principle and subsidiary status. It possesses the largest livestock population in the world (520.6 million heads) and accounts for the largest number of cattle (16.1% of the world population) and buffaloes (57.9%) in the world. During the past three decades, milk production in the country has increased from about 22 million tons in 1970-71 to 112.5 million tons in 2009-10. The per capita availability of milk, which had decreased during the pre-Operation Flood period, not only kept pace with the growing population but increased from 107 grams per day in 1970 to 263 grams per day in 2009-10. Operation Flood, launched in 1970, has been instrumental in helping the farmers mould their own development. As a result of the cooperative structure, the whole exercise of production and distribution of milk and milk products has become economically viable for farmers to undertake on their own.[1] In this manner the farmer himself can enjoy the fruits of his own labour, instead of surrendering a majority of the profit to corrupt middlemen. The scheme sought to establish milk producers' cooperatives in the villages and make modern technology available to them. The broad objectives are to increase milk production ("a flood of milk"), augment rural incomes and transfer to milk producers the profits of milk marketing which are hitherto enjoyed by well-to-do-middlemen. In India, about 46 percent of the produced milk is retained for home consumption while 54 percent is disposed of to various agencies. About 80 percent of milk produced in the country is handled by the unorganized sector and the remaining 20 percent is shared equally by cooperatives and private sectors for processing with varying degree of product mix. The organized sector (large scale dairy plants) processes about 13 million tonnes milk annually while the unorganized sector (halwais and vendors) processes about 22 Million tonnes per annum.

The Punjab State Cooperative Milk Producers' Federation Limited popularly known as MILKFED Punjab, came into existence in 1973 with a twin objective of providing remunerative milk market to the milk producers in the State, value addition and marketing of milk as its derivatives on one hand and to provide technical inputs to the milk producers for enhancement of milk production on the other hand. The milk cooperatives as such act as a catalysts in promoting economic progress of the rural folk by providing assured market for milk and supply of dairy inputs including compounded foods, veterinary and artificial insemination services in rural areas.

As a result, the milk production in Punjab has shown a quantum jump from 32.21 lakh tonnes in 1980-81 to 93.89 lakh tonnes in 2009-10. This has enhanced per capita per day availability of milk from 541 to 915 gm in above mentioned period. Dairy development has, thus, assumed a position of paramount importance in the rural economy of the state. These cooperatives have an immense potential for creating self-employment and supplementing the incomes of the dairy farmers, especially of the small and marginal cases, and landless wage earners. Lucrative alternate employment opportunities are often not available in rural areas of the state, making dairying an attractive option for many villagers. A majority of milk producers have one or two milch animals, and these small producers account for some 70 per cent of the milk production. On average, 22.5 percent of the income of rural households is contributed by milk. In the backdrop of this, the present study has been undertaken to examine the role of milk cooperatives in the rural development of Punjab.

DATA AND METHODOLOGY

In order to assess the impact of milk cooperatives on the rural development in Punjab, the data on different parameters was compiled from various sources. The data pertaining to the number of cooperative societies, their membership, value of milk and milk products sold, numbers of milch animals, share of livestock in total income of Punjab, etc. for the period 1980-81 through 2009-10 were collected from Statistical Abstracts of Punjab. The compound growth rates for number of societies, membership, milk production and per capita availability of milk were estimated with following model.

$Y_t = AB^t$

where,

Y_t = Number of societies/membership/milk production/per capita availability of milk for the year 't'.

t = Time variable (1, 2...n) for each period.

A = Constant

Log transformation of the above function is:

$\ln Y_t = \ln A + t (\ln B)$.

where,

$\ln B = \ln (1+r)$, and

$r = [\text{antilog} (\ln B) -1]$

$CGR (\%) = [\text{antilog} (\ln B) -1] \times 100$.

In addition to this simple statistical tool such as averages, percentage, etc were also used to analyze the remaining data. The results presented and discussed with the help of simple tabular technique.

RESULTS AND DISCUSSION

The results of analyses of these data have been discussed under the following sub-heads as follows:

Growth of Milk Cooperative Societies

The dairy farmers are commonly exploited by milk traders due to low marketable surplus of milk, lack of transportation facilities and so on so forth. To overcome these problems the Planning Commission of India felt the need to organize the small dairy farmers in the form of cooperatives so that the milk production enterprise became a profitable venture. This would not only ensure adequate supplies of fresh milk and its derivatives to the urban consumer but also bring handsome returns to the milk producers.

The pattern of growth of milk cooperative societies and their membership has been depicted in Table 1. A perusal of the Table 1 shows that there were 2593 Milk Producers' Cooperative Societies (MPCS) in 1980-81 which increased to 5070 in 2008-09. This has exhibited a compound growth rate of 2.06 per cent per annum. The

TABLE 1

The growth of Milk Cooperative Societies and Membership in Punjab, 1980-81 to 2004-05

Year	*No. of societies*	*Membership (000)*	*Per cent of Villages running cooperative societies*	*Membership per society*
1980-81	2593	96.75	21.01	37.31
1985-86	4315	170.95	34.96	39.61
1990-91	5263	252.80	42.64	48.03
1995-96	5061	286.50	41.01	56.60
2000-01	5660	320.33	45.86	56.60
2005-06	5905	330.68	47.85	56.00
2008-9	5070	319.04	41.29	62.92
CGR(%)	2.06*	3.92*		

* Significant at one percent level.
Source: Statistical Abstract of Punjab.

pattern of membership reveals that it increased by 3.92 percent annually during 1980-81 through 2008-09. Thus, the rate of growth of membership was more than the number of societies established. The split up of data reveals that the percent of villages owning a cooperative society has increased from 21.01 in 1980-81 to 41.29 per cent in 2008-09. The membership per society also reflected an increasing trend as it increased from 37.31 to 62.92 in the above said periods. This clearly shows that the milk cooperative movement in the state is picking up with every passing year. It has been reported in the studies conducted earlier that in most of the MPCS were constituted by landless labourers, small and marginal farmers as its membeRs.

The studies conducted by Sidhu and Sidhu (1986) and Chahal (1991) brought out another interesting aspect that in the case of successful cooperative societies a major chunk of membership of these societies was constituted by weaker section of the rural population. This has created a sense of equality among masses in rural areas. As all the members, irrespective of their milk production. This has encouraged the weaker section to take up dairying extensively, which in turn helped to improve their economic status. These findings are in consonance with the study conducted by Devaria (1986).

Volume of Business and Pattern of Profit

The perusal of Table 2 shows that volume of business per society depicted an increasing trend during the study period. The volume of

TABLE 2

Performance of Milk Cooperative Societies in Punjab, 1980-81 to 2008-09

(Rupees)

Year	*Owned funds per society*	*Volume of business per society*	*Volume of business per member*	*Per cent of societies earning profit*	*Profit per society*	*Profit per member*
1980-81	3978	154740	4147	45.20	439	11.77
1985-86	5283	86452	2182	63.50	1966	49.63
1990-91	9304	127200	2648	59.22	2595	54.01
1995-96	22689	302466	5343	71.47	5703	100.73
2000-01	43961	383276	6772	74.12	9356	165.31
2005-06	70391	557863	9990	76.82	11290	195.47
2008-09	92966	883520	14040	81.92	13712	217.90

Source: Statistical Abstract of Punjab.

business per society has increased from Rs. 1,54,740 to Rs. 8,83,520 during 1980-81 to 2008-09. As such, volume of business has increased by six times over the period. The volume of business per member also depicted a similar pattern. This became possible with increasing number of societies earning profit during the study period. This has demonstrative effect on the societies which were in red and try to improve their working by overcoming their weaknesses when compared with successful MPCS. The results further reveal that the percentages of societies earning profit have increased from 45.20 in 1980-81 to 81.92 percent in 2008-09. The profit per society registered an increase by about three fold during the period of study. As such, the profit per society increased from Rs. 439 in 1980-81 to Rs. 13,712 in 2008-09.

Similarly the owned fund per society has increased to 92,966 in 2008-09 from Rs. 3978 in 1980-81. The volume of business per member also increased to Rs. 14,040 in 2008-09 from Rs. 4147 in the base year. It may be pertinent to mention here that the profit which accrues to a cooperative society is distributed to members in the form of bonus. This facility is not available with any other milk marketing system. Whereas this bonus helps to boost rural development, it is also attracts dairy farmers to adopt cooperatives for the disposal of their milk. The milk cooperatives, therefore, ensure better deal of milk-sellers, which in turn helps them to supplement their income by removing unemployment/ underemployment and gave them a much more edible way of life. The cooperatives also provide competition to private milk vendors/traders who in turn pay competitive prices to milk producers if not at par with the cooperatives. This has helped to adopt dairying on the commercial lines.

Milch Animal Scenario

With the strengthening of cooperatives not only resulted into assured market for milk but also helped to improve the quality of herd in the rural areas. This has encouraged the masses to take up dairying enterprise thereby resulted in increase in the number of milch animals in the state.

The perusal of Table 3 revealed that bovine population in Punjab increased from 7421.80 thousand in 1977 to 8810.5 thousand in 1997 and then declined to 8033 thousand and 6796.55 thousand in the year 2003 and 2007 respectively. The number of buffaloes increased from 4110 thousand to 6170.70 and then declined to 5994.50 thousand in 2003 and 5035.63 thousand in 2007. The sharpest decline was witnessed in the population of indigenous cows since 1977. The number of crossbred cows increased from 1579.50 thousand in 1990 to 1829.30 thousand and then declined to 1530.50 thousand in 2003 and 1258.39 thousand in 2007. But this negative trend should not be taken alarmingly as the low yielding desi cow breeds are being replaced by the high yielding exotic/cross bred cows. Thus, rearing of high yielding cross bred cows helped to remove unnecessary burden on economy by way of saving costly feed and fodder fed to desi cows. The dairy farmers responded positively to animal breeding programme for improvement of their stock. The animals inseminated artificially increased from 1269 thousand in 1994-95 to 2285.71 thousand in 2006-07. The Technical

TABLE 3

Total Bovine Population during Census Periods in Punjab

('000)

Category	*1977*	*1990*	*1997*	*2003*	*2007*
1. Cattle	3311.80 (44.62)	2832.50 (33.67)	2639.80 (29.96)	2038.50 (25.38)	1760.92 (25.90)
(i) Indigenous	3311.80 (44.62)	1253.00 (14.89)	810.50 (9.19)	508.00 (6.33)	502.53 (7.39)
(ii) Crossbred	---	1579.50 (18.78)	1829.30 (20.77)	1530.50 (19.05)	1258.39 (18.51)
2. Buffaloes	4110.00 (55.38)	5577.00 (66.33)	6170.70 (70.04)	5994.50 (74.62)	5035.63 (74.09)
Total Bovine Population	7421.80 (100.00)	8410.50 (100.00)	8810.50 (100.00)	8033.00 (100.00)	6796.55 (100.00)

Source: Statistical Abstract of Punjab.

Input Programs have enhanced milk production. Computerized feed formulation has led to improvement in the quality of feed which lead to better quality milk. Import of good breeds under our Breed Improvement Programmes as well as our Artificial Insemination Programmes have led to an improvement in the quality of the cattle. They are today healthier, stronger and of excellent breeds.

Milk Production Performance

India's milk production increased from 21.2 million metric tonnes in 1968–69 to 112.50 million metric tonnes in 2009-10. The per capita availability of milk increased from 112 gm per day in 1968-69 to 263 gm per day during the same period. In India milk production is increasing at the rate of 4 percent per annum whereas world's milk production is increasing at the rate of 1-1.5 percent per annum. The results presented in Table 4 reveal that with the adoption of animal breeding programme and with better management by dairy farmers, the milk production has increased from 3221 thousand tonnes in 1980-81 to 9389 thousand tonnes in 2008-09 in Punjab. This increase in milk production not only enhanced the availability of milk per capita, but also proved as a means to augment the income of the dairy farmers and created employment on the farm itself.

The perusal of Table 4 further reveals that per capita availability of milk has increased from 541 to 915 gm per capita per day during the study period. The marketable surplus in corresponding period has

TABLE 4

Growth and Performance of Dairying in Punjab, 1980-81 to 2008-09

(000' tonnes)

Year	*Milk Production*	*Per capita availability (gm)*	*Marketable surplus*
1980-81	3221	541	2029
1985-86	4035	597	2542
1990-91	5142	682	3239
1995-96	6424	798	4047
2000-01	7774	870	4897
2005-06	8554	901	5588
2008-09	9389	915	5822
CGR (%)	4.33*	2.41*	

Source: Statistical Abstract of Punjab.

increased from 2029 to 5822 thousand tonnes in the state during the period of 1980-81 to 2004-05. This is an enough evidence to show that milk cooperatives helped the milk producers supplement their incomes with adoption of dairy enterprise by providing competitive prices and assured market.

Milk procurement network: Working on Anand Pattern, the process of organizing socirties at village level started in Punjab as early as 1978. Fresh milk is procured from the milk producers twice a day through village level societies direct without the assistance of any middlemen. Milk procurement by milk plants in cooperative sector in Punjab has been depicted in Table 5.

TABLE 5

Milk Procurement by Milk Plants in Cooperative Sector in Punjab

(lakh litres per day)

Year	*No. of milk plants in Cooperative sector*	*Chilling centre*	*Milk procurement*
1980-81	11	35	4.17
1995-96	11	32	6.99
2000-01	11	36	7.22
2005-06	11	35	7.88
2009-10	11	34	9.49

Source: Statistical Abstract of Punjab.

The total milk procurement by cooperative sector milk plants has been increased from 4.17 lakh litres per day in 1990-91 to 9.49 lakh litres per day. This increase in milk procurement depicts the interest of milk producers in milk cooperative societies. These cooperative societies procure milk from milk producers without intermediaries in the rural areas. Besides they offer better prices as it is based on fat percentage in milk. This has also reduced the chances of adulteration of milk to some extent as the higher fat in SNF in milk assure better prices of milk as compared to the milk vendors..

Input Services

It is one of the fundamental objectives of Milkfed to carry out activities for promoting milk production in the State. In this view, various technical input services like veterinary health care, artificial insemination services, vaccination, supply of Verka balanced cattle feed and quality fodder seed are being provided for enhancing milk production and economic development of farming community.

Employment Creation

The dairy cooperative have brought perceptible change by way of creating more employment opportunities for dairy farmers in the rural areas. MILKFED and its units have a work force of about 5000 employees. Every morning and evening milk is lifted from the villages through private transport vehicles which are providing a regular employment to about 600 transporters. Most of them are self-employed people. Similarly, lay seminators and first aid workers are also finding employment and at the same time providing input services to milk producers' right in their villages. More than 5000 village cooperative societies are having 10,000 workers to man their milk procurement, technical input, etc. Many new milk plants, seed processing units and cooling units have sprung up throughout Punjab. As a result, there has been an overall increase in the supply of milk and cattle feed production. Today the farmers are getting better prices because of increased turnover and thereby making increased profits. At the national level milk/dairy cooperatives generate employment opportunities for about 12 million farm families. Women accounted for 93 percent of total employment in dairy production (World Bank, 1991).

CONCLUSION

With the launching of Operation Flood Programme the milk cooperatives have been strengthened through time. This has resulted into assured market for milk in the rural areas. It helps to provide competitive market for milk, thereby, freeing the milk seller from the exploitation of private milk traders in the state. India's milk production increased from 21.2 million metric tonnes in 1968-69 to 112.5 million tonnes in 2009-10. This increase in milk production not only enhanced the availability of milk per capita, but also proved a means to augment the income of the dairy farmers. The milk cooperative movement has helped to breakdown barriers between caste, as the members belonging to diverse social and economic status have to stand in same queue to sell the milk to the milk producers' cooperative society. The milk cooperatives provide a competitive, assured and remunerative market for milk in the producing areas itself. Along with this the milk cooperatives are promoting employment for the other skills as it involved a big infrastructure starting from the milk collection to the final disposal of milk and milk products to the consumers. It can be concluded safely that the milk cooperatives are acting as a catalyst in the augmentation of milk production, income and creation of employment in the rural areas of the state. Both dairy farming and crop husbandry are complementary to each other as crop husbandry provides fodder to the dairy farmers and dairy farming provides the farm yard manure to crop husbandry. This also helps to improve the eco-agri system as the residue generated by them is utilized and not realized in environment.

References

Chahal, S.S. (1991), A case study into factors determining the success or failure of milk producers' cooperative societies in Punjab, *Journal of Agricultural Development and Policy*, 3(1 and 2): 34-42.

Deveria, Leelamma (1986), Cooperatives and rural development, *Kurukshetra*, 34(6): 31-34.

Sidhu, D.S. and Sidhu, J.S. (1966), *Case studies on farmers' organizations in India* (mimeograph), Department of Economics and Sociology, PAU, Ludhiana, pp. 106-11.

Sidhu, D.S. and Sidhu, R.S. (1989), Case studies of successful and unsuccessful cooperative service societies and milk producers' cooperative society in Punjab, *Indian Journal of Agricultural Economics*, 45(3): 367-73.

World Bank, 1991, *A World Bank Country Report: Gender and Poverty in India*, World Bank, Washington, D.C.

CHAPTER

11

Women Entrepreneurship: Problems and Prospects—An Empirical Study

LEKHA NAGAR AND DR. AMRIK SINGH SUDAN

ABSTRACT

The study was conducted to have an overview of the existing problems and prospects of women entrepreneurs in Jammu city. It was also conducted to come up with recommendations and suggestions to address existing problems to promote a gender-friendly business environment. The analysis was based on primary data collected through personal interviews spread over a period of two months. Number of total samples is (100) hundred. Samples were taken from Jammu city. A set of questionnaire was prepared to conduct the field survey. The questionnaire consisted both closed and open-ended questions . Women entrepreneurship in Jammu is more than just income generation. It is about working to build a more prosperous region, to lift the many people out of poverty. It is about achieving economic independence and women's empowerment. And it is about striving for emancipation of women's gender roles, to create a better future for generations of women to come. The study finds that women are engaged in different types of enterprises. Most of the cases show that women entrepreneurs selected their enterprises with what they were familiar. The study clearly indicates that lack of capital is still the main problem. The other major problems include absence of marketing facilities and proper sales centers. Another common barrier faced by the women entrepreneurs is balancing time between the enterprise and the family. Others constants are machinery, equipment, technology and power supply. This study conveys the message that if the women are provided with appropriate training and need-based financial and related assistance, they will enter into the entrepreneurial occupation in a large number and will prove their worth to contribute to the economy of India . Both the government organizations and private sector have a major responsibility to promote entrepreneurship development for women. Without their interventions the advancement of women and women entrepreneurship cannot be achieved.

INTRODUCTION

The Indian economy has been witnessing a drastic change since mid-1991, with new policies of economic liberalization, globalization and privatization initiated by the Indian government. India has great entrepreneurial potential. At present, women involvement in economic activities is marked by a low work participation rate, excessive concentration in the unorganized sector and employment in less skilled jobs.

Any strategy aimed at economic development will be lop-sided without involving women who constitute half of the world population. Evidence has unequivocally established that entrepreneurial spirit is not a male prerogative. Women entrepreneurship has gained momentum in the last three decades with the increase in the number of women enterprises and their substantive contribution to economic growth.

In this dynamic world, women entrepreneurs are an important part of the global quest for sustained economic development and social progress. In India, though women have played a key role in the society, their entrepreneurial ability has not been properly tapped due to the lower status of women in the society. It is only from the Fifth Five Year Plan (1974-78) onwards that their role has been explicitly recognized with a marked shift in the approach from women welfare to women development and empowerment. The development of women entrepreneurship has become an important aspect of our plan priorities. Several policies and programmes are being implemented for the development of women entrepreneurship in India.

There is a need for changing the mindset towards women so as to give equal rights as enshrined in the constitution. The progress towards gender equality is slow and is partly due to the failure to attach money to policy commitments. In the words of president APJ Abdul Kalam "empowering women is a prerequisite for creating a good nation, when women are empowered, society with stability is assured. Empowerment of women is essential as their thoughts and their value systems lead to the development of a good family, good society and ultimately a good nation."

When a woman is empowered it does not mean that another individual becomes powerless or is having less power. On the contrary, if a women is empowered her competencies towards decision- making will surely influence her family's behaviour.

In advanced countries, there is a phenomenon of increase in the number of self- employed women after the World War-II. In USA, women own 25% of all business, even though their sales on an average are less than two-fifths of those of other small business. In Canada, women own one-third of small business and in France it is one-fifth.

Concept of Entrepreneur

The word 'entrepreneur' derives from the French word "Entrepreneur" (to undertake). In the early 16th Century it was applied to persons engaged in military expeditions, and extend to cover construction and civil engineering activities in the 17th century, but during the 18th century , the word 'entrepreneur' was used to refer to economic activities. Many authors have defined 'entrepreneur' differently. Generally, an entrepreneur is a person who combines capital and labour for production. According to Cantillion "entrepreneur is the agent who buys means of production at certain prices, in order to sell at prices that are certain at the moment at which he commits himself to his cost". According to P.F. Drucker "he is one who always (1) searches for change, (2) responds to it, and (3) exploits it as an opportunity."

Concept of Women Entrepreneur Enterprise

"A small scale industrial unit or industry-related service or business enterprise, managed by one or more women entrepreneurs in a concern, in which they will individually or jointly have a share capital of not less than 51% as shareholders of the private limited company, members of co-operative society".

Categories of Women Entrepreneurs

- Women in organised and unorganised sector.
- Women in traditional and modern industries.
- Women in urban and traditional areas.
- Women in large scale and small scale industries.
- Single women and joint venture.

The study was conducted to have an overview of the existing problems and prospects of women entrepreneurs in Jammu city. It was also conducted to come up with recommendations and suggestions to address existing problems to promote a gender-friendly business environment. The analysis was based on primary data collected through personal interviews spread over a period of two months. Number of total sample is (100) hundred. Samples were taken from Jammu city. A set of questionnaire was prepared to conduct the field survey. The questionnaire consisted both closed and open-ended questions . Women entrepreneurship in Jammu is more than just income generation. It is about working to build a more prosperous region, to lift the many people out of poverty. It is about achieving economic independence and women's empowerment. And it is about striving for emancipation of women's gender roles, to create a better future for generations of women to come. The study finds that women are engaged in different types of enterprises. Most of the cases show that women entrepreneurs selected their enterprises with what they were familiar. The study clearly indicates that lack of capital is still the main problem. The other major problems include absence of marketing facilities and proper sales centers. Another common barrier faced by the women entrepreneurs is balancing time between the enterprise and the family. Other constants are machinery, equipment, technology and power supply. This study conveys the message that if the women are provided with appropriate training and need-based financial and related assistance, they will enter into the entrepreneurial occupation in a large number and will prove their worth to contribute to the economy of India. Both the government organizations and private sector have a major responsibility to promote entrepreneurship development for women. Without their interventions the advancement of women and women entrepreneurship cannot be achieved.

Field Study

As mentioned before, the qualitative information for this study was collected in the month of June 2011. Quantitative data were collècted during the months of June, July and August 2011.

Data Collection Tools

Different types of data collection tools were applied in this study considering types of the respondents and methods of the study. Major data collection tools applied in this study is as follow:

Semi-Structured Interview Schedule (SSIS): A semi-structured interview schedules was used to collect data from women entrepreneurs.

Limitation of the Study

All possible cautions were maintained during the planning of study, selection of methods and techniques, data processing and analysis. In spite of this, the study is not free from Limitations. Some limitations related to the study could be cited here as follows:

- Considering the volume of the study, the timeline was limited, particularly for questionnaire survey. Therefore, it was a great challenge for us to reach all respondents within the timeframe. To get it happened, we had to pay extra efforts. Time limitation created some barriers to collect some qualitative data.
- To get response from women entrepreneurs for all questions was really a great challenge as women entrepreneurs were found to be very busy. As a result, some important information was missed.
- Some qualitative findings were related to the perception of respondents. These findings may not be representative to predict the overall situation of woman entrepreneurs.
- The women entrepreneurs, involved in the large scale of business, were not included in this study, Therefore, this study dose not tell anything about their situation.

DATA ANALYSIS AND INTERPRETATION

1. It is generally believed that Indian women give more emphasis to their motherly role and family/home responsibilities rather than venturing for a career/growth through 'entrepreneurship'. If they have to plan and settle for a career, they would at best prefer a safe risk free job than setting up a business venture of their own.

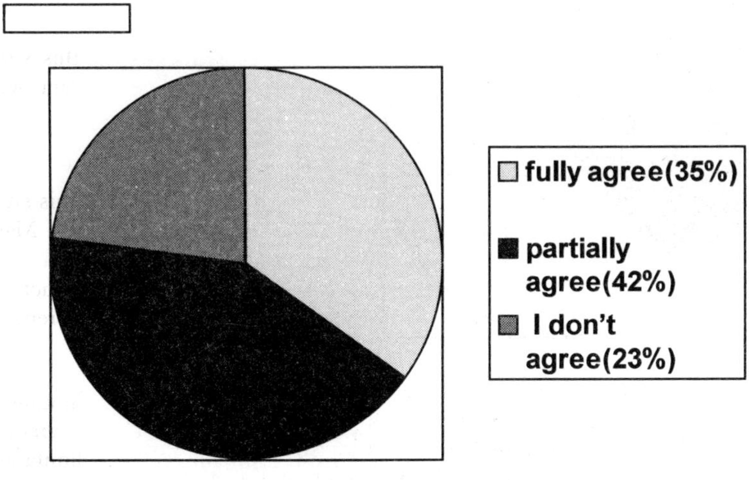

2. Women's family obligations also bar them from initiating and becoming successful entrepreneurs.

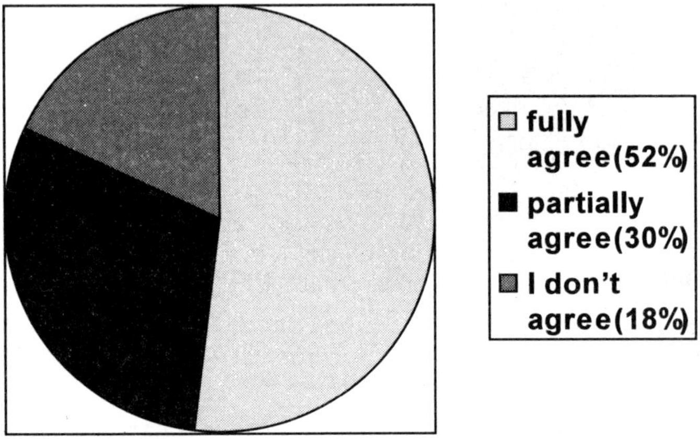

3. Financial institutions are generally skeptical about the entrepreneurial seriousness and abilities of women entrepreneurs. As a result, the women entrepreneurs are suffering from inadequate financial resources and working capital.

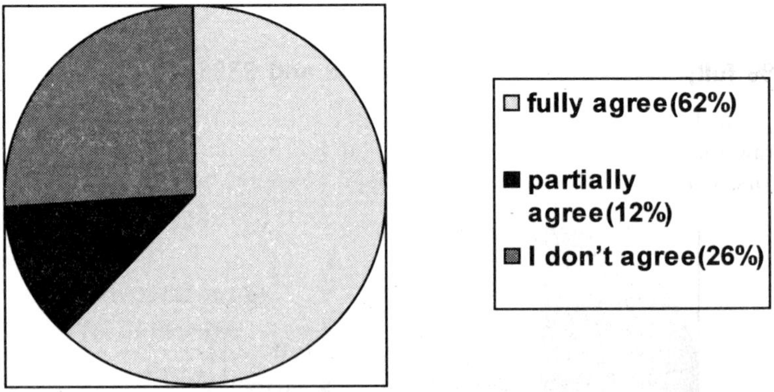

4. Although women managers and executives are proving to be more and more effective at their work places, they are still not considered efficient enough to handle labour and logistic issues as efficiently in self-owned business enterprises. They would prefer men to handle such issues.

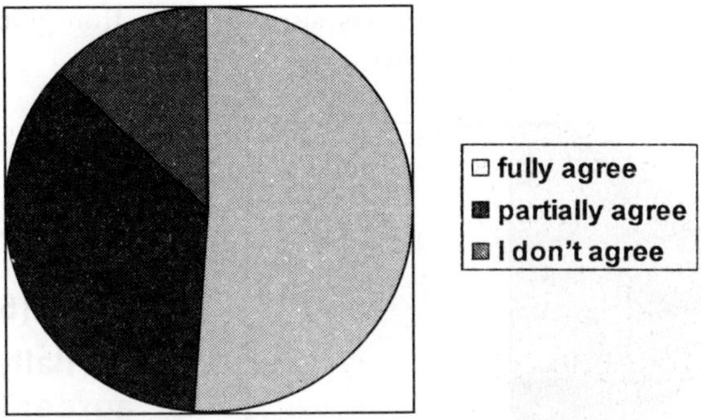

51% fully agree, 36% partially agree and 13% I don t agree

5. The overall male domination in the entire business and entrepreneurship area also works as a hurdle for women entrepreneurs.

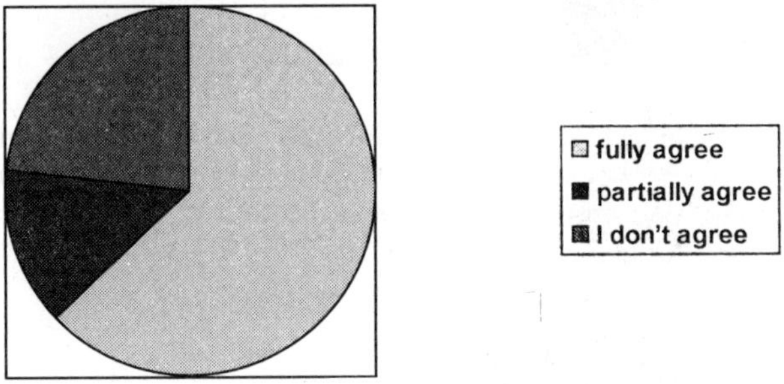

63% fully agree, 14% partially agree and 23% I don t agree

6. It is believed that lack of proper knowledge about availability of raw materials and low-level negotiation and bargaining skills are also factors, which affect women entrepreneurs business adventures.

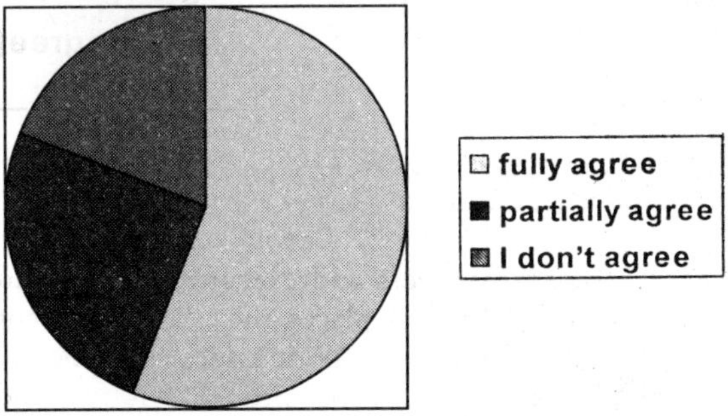

56% fully agree, 25% partially agree and 19% I don t agree

7. Low-level risk-taking attitude is another factor affecting womens decision to get into business.

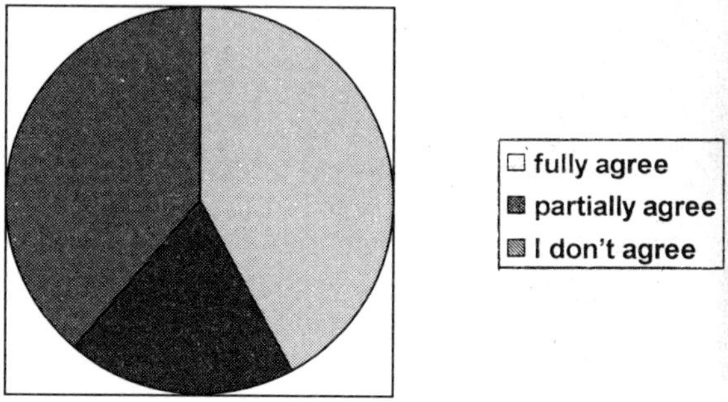

42% fully agree, 19% partially agree and 39% I don t agree

8. Educational Institutions and even Government agencies are doing little to promote entrepreneurship amongst women particularly.

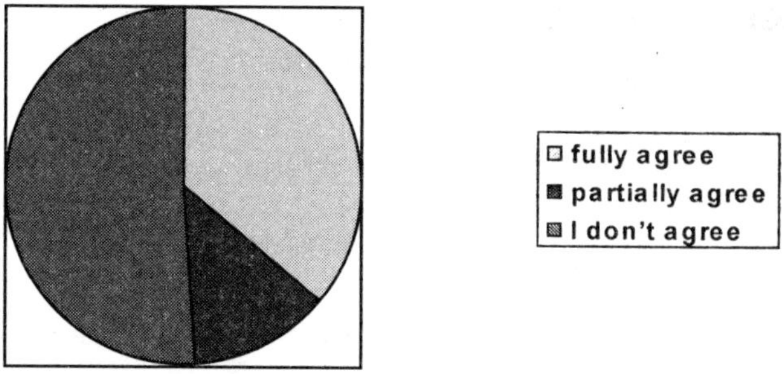

36% fully agree, 13% partially agree and 51% I don t agree

9. Women would perhaps prefer to be active partner entrepreneurs with their spouses and provide necessary support rather than initiating and running a business enterprise entirely on their own.

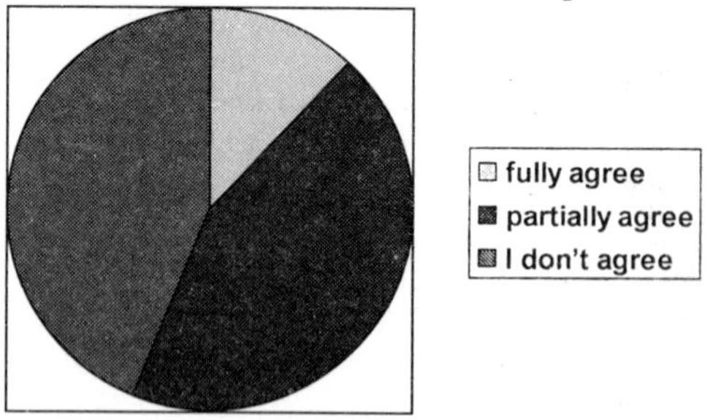

12% fully agree, 44% partially agree and 44% I don t agree

10. Larger number of women do not have much needed financial independence that would enable them to conceive, plan and execute a business plan independently.

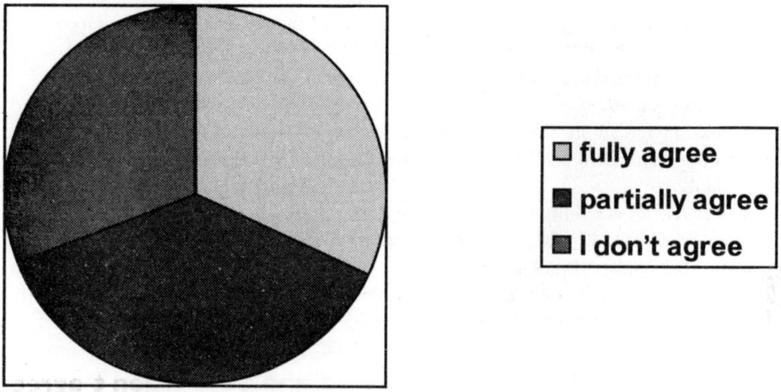

32% fully agree, 37% partially agree and 31% I don't agree

SUGGESTIONS TO IMPROVE PROBLEMS OF WOMEN ENTREPRENEURS

(1) *Keeping your life in balance is a constant struggle for many women entrepreneurs*: Here are some personal and organizational tips for both business and personal parts of your life that can help put you on the path to success.

(2) *Select a business that you love and know something about*: Being passionate about your business will help you make it during the challenges that all entrepreneurs face.

(3) *Assess the market*: Is there a need for your product/service? What type of customer base do you need to stay in business? You should decide on location.

(4) *Write a business plan*: This includes a description of the business, financial projections, clearly defines the product or service, and discusses the legal entity, market, location, competition, management and personnel. There are many "How to write a business plan" guides out there for you to use.

(5) *Network*: Join professional organizations, attend industry-related conferences and meetings, and go to other meetings that are of interest to you. You will be surprised where you meet clients!

References

Acharya, M. (2001). "Efforts at Promotion of Women in Nepal", paper, Kathmandu: Friedrich Ebert Stiftung.

Asian Development Bank (1999a). "Women in Nepal", Country Briefing Paper, December, Programs Department, West Division 1, Manila: Asian Development Bank.

Asian Development Bank (1999b). Women in Sri Lanka, Manila.

Asian Development Bank (2001a). SME Development Program, Sri Lanka, Report and Recommendation, Manila: Asian Development Bank.

Asian Development Bank (2001b). Women in Bangladesh, Manila: Asian Development Bank.

Aftab, K. (1991). "Growth of Agriculture-related Engineering Industry in Pakistan Punjab", an unpublished Ph.D. thesis, University of Strathclyde, UK.

Charumathi, B. (1998). "Women entrepreneur's challenges and prospects", in C. Swarajya Lakshmi (ed.), *Development of Women Entrepreneurship in India: Problems and Prospects*, New Delhi: Discovery Publishing House.

Cuong, T.T., Le Xuan Sang, Nguyen, K.A. (2007). "Small and medium enterprises development in Vietnam: The case of electronics and motorcycle", Draft report for the ERIA Related Joint Research of SME Project, IDE-JETRO, Hanoi: Institute for Economic Management, Vietnam.

Cuong, T.T., Le Xuan Sang, Nguyen, K.A. (2008). "Vietnam's small-and medium-sized enterprises development: Characteristics, constraints and policy recommendations", Paper presented at the fourth workshop, the ERIA Related Joint Research of SME Project, IDEJETRO, 3-5 January, Bangkok.

Das, D.J. (2000). "Problems faced by women entrepreneurs", in K. Sasikumar (ed.), Women Entrepreneurship, New Delhi, Vikas Publishing House.

Das, K. (2007). "SMEs in India: Issues and Possibilities in Times of Globalisation", in Hank Lim (ed.), *Asian SMEs and Globalization*, ERIA Research Project Report, 2007, No. 5, March, ERIA, Bangkok.

Finnegan, G., Katrine, D. (1997). Promoting the Development of Women Entrepreneurs—Means and Ends in Entrepreneurship Development, Geneva: International Labour Organisation (ILO).

Ganesan, S. (2003). Status of Women Entrepreneurs in India, New Delhi: Kanishka Publications.

Giovannelli, C., Gunnsteinsdottir, H., Me, A. (2003). "The status of statistics on women and men's entrepreneurship in the UNECE region", Paper presented at Workshop on Improving Statistics on SMEs and Entrepreneurship.

CHAPTER

12

Cross Cultural Challenges while doing Business in India

LEKHA NAGAR AND DR. AMRIK SINGH SUDAN

ABSTRACT

Owing to its rich and varied topography, vibrant culture and captivating With the globalisation of world business, India has become an appealing market for foreign investors. The problem of cross-cultural management arises as the cooperation between India and its culturally different Western partners continues to increase at an unprecedented rate. The 21st century is an era of the globalization of world economy. Cross-national business is facing great challenges in cultural differences.

In one survey entitled what is the major barrier in doing business in the world market, cultural differences ranked first in all eight items including "law, prize, information, competition, languages, delivery, foreign currency, time differences, and cultural differences." Hofstede (1993) believes that the spread of businesses into the global stage brings the issue of national and regional differences to the fore. 'There is something in all countries called 'management', but its meaning differs to a larger or smaller extent from one country to another" (Hofstede, 1993).

This paper presents an understanding on the general cultural differences between India, America and China by applying the cultural dimensions of Hofstede and Bond. It also discusses the impact of these cultural differences on their management practice from five aspects; cooperative strategies, conflict management, decision-making, work-group characteristics, and motivation systems.

INTRODUCTION

The 21s century is an era of the globalization of world economy. Cross-national business is facing great challenges in cultural differences.

In one survey entitled what is the major barrier in doing business in the world market, cultural differences ranked first in all eight items including "law, prize, information, competition, languages, delivery, foreign currency, time differences, and cultural differences." Hofstede (1993) believes that the spread of businesses into the global stage brings the issue of national and regional differences to the fore. "There is something in all countries called 'management', but its meaning differs to a larger or smaller extent from one country to another" (Hofstede, 1993).

It can also be observed that most of the failures faced by cross-national companies are caused by neglect of cultural differences. The globalization of the world economy, on one hand, has created tremendous opportunities for global collaboration among different countries; on the other hand, however, it has also created a unique set of problems and issues relating to the effective management of partnership with different cultures.

With the increasing importance of the India market in the world economy, many international companies rushed and planned to enter India to explore business opportunities. They enter the huge market by forming joint ventures or participating in mergers and acquisitions. This has spurred the need for cross-cultural research in India. It was reported that the great barriers caused by cultural differences like difficulty of communication, higher potential transaction costs, different objectives and means of cooperation and operating methods, have led to the failure of many Sino-foreign cooperation projects. The questions like "how to understand India" and "how to do business with Indian people" have occupied the minds of international business people who are planning to enter India.

GENERAL CULTURAL DIFFERENCES BETWEEN THE WEST AND INDIA

India, as the largest market and possibly the most appealing market in Asia, is entering into global collaboration with a wide range of foreign partners. It seems necessary to investigate the cultural differences between India and its international business partners in North America and Europe. To clarify the differences between India and the West, we will refer to Hofstede's four cultural dimensions and Bond's fifth dimension. Among researchers who have given a variety to definitions of culture. Hofstede is one of the first to adopt a pragmatic problem-solving approach in the field and relates culture to management. He defines culture as a kind of "collective programming of the mind, which distinguishes the members of one category of people from another" (Hofstede, 1980). He explained that culturally-based values systems comprised four dimensions: power distance, individualism/collectivism, masculinity/femininity, and uncertainty avoidance. Further research by Michael Bond (1989) identified a fifth "Eastern" dimension called long-term/short-term orientation.

By comparing some Western countries and India along these five dimensions according to their cultural dimension scores (Data source: hofstede, 1991), some tentative conclusions may be drawn. First, Western countries seem to be generally lower (United States of America 40, Canada 39, United Kingdom 35, Germany 35, and France 68) than India (80) in power distance.

Second, in terms of individualism, Western countries are generally much higher (United States of America 91, Canada 80, United Kingdom 89, Germany 67, and France 71) than India (20). Third, Western countries seem to have short-term orientation while India is considered to be long-term oriented.

Among these Western countries, America is frequently investigated in cross-cultural research, partly because of its economic power, partly because of its cultural representativeness. To a certain degree, the United States is considers as representing the so-called "Western Culture". Therefore, a comparison between USA and India seems to help clarify the cultural differences between the West and India and related cross-cultural challenges.

India and United States of America (USA) differ greatly with regard to their economic systems, political systems, social values, and laws, despite the substantial changes that have occurred in India during recent years. Table 1 shows the cultural dimension scores of United States of America (USA) and India (Data source: Hofstede, 1993). Some differences can be found. First, in terms of power distance, the scores of India are twice as those of United States of America (USA), which indicates that India is centralized (though it has shown some tendency toward decentralized power) while United States of America (USA) is relatively decentralised. Second, United States of America (USA) ranks first in individualism (strong individualism) while India is low in individualism (strong Collectivism). Third, United States of America (USA) has higher value than India in masculinity, which indicates that United States of America (USA) is medium masculinity while India is medium femininity. Fourth, India has higher values for uncertainty avoidance than United States of America (USA), which shows that Indian are relatively risk-avoiding while Americans are relatively risk-taking. Last, United States of America (USA) has a short-term orientation while India has a long-term orientation.

It has been widely accepted that cultural differences greatly affect human thinking and behaviour and thus, business organizations in which people interact on the basis of shared values. Management is embedded in a wider societal setting, and is heavily influenced by local historical and cultural norms (DiMaggio and Powell, 1983). The significant differences between United States of America (USA) and India seem to some affect of their management.

IMPACT OF CULTURAL DIFFERENCES ON COOPERATIVE STRATEGIES

Because entrepreneurs mature within a societal context, their attitudes toward cooperation are likely to be influenced by the underlying values of their society (Weaver, 2000). As discussed, United States of America (USA) is strong in individualism and medium masculine. They rely on their own view to determine what they should do. They tend to work alone and are reluctant to cooperate because their individualism and masculine culture view cooperation in general as a sign of weakness and place a high value on independence and control. India is a strong in collectivism and medium feminism. The Indians depend more on groups or institutions to determine what they should do and emphasis loyalty to the group. They are more likely to cooperate with others to avoid risks and reduce responsibilities. Their value systems appreciate duty to the group and harmony among its members while pursuing personal goals is viewed rather negatively in India.

In addition, in the process of cooperation, Americans place greater importance on contractual safeguards than the Indian. They believe that contracts can ensure that their partners' tendencies to focus on individual goals and aspirations do not interfere with their own individual goals and aspirations. But the Indians don't consider contracts as seriously as the American. They think there will always be changes and the contracts can be reasonably modified according to changes. Instead, they tend to pay more attention to relationships than contracts.

The two phenomena seem to be consistent with Weaver's findings. In his study of a group of seven-nation entrepreneurs about their attitudes toward cooperative strategies, he found that entrepreneurs from societies that are masculine and individualistic have a lower appreciations for cooperative strategies as compared to entrepreneurs from societies that are feminine and collectivist in nature. He also found that entrepreneurs from individualistic societies placed greater importance on contractual safeguards for maintaining effective cooperation than did those from collective cultures (Weaver, 2000).

IMPACT OF CULTURAL DIFFERENCES ON CONFLICT MANAGEMENT

The Indian and Americans tend to resolve conflicts in different ways. Since the Indian come from a strong collectivism and medium feminine society in which harmony and personal relationship are emphasized, they will try to use indirect ways to avoid direct and open conflict. When they face conflict, they prefer to use authority to suppress it, or settle things in private. They prefer to resolve conflict through negotiation and compromise. Individualistic and medium masculine American managers are used to confronting problems directly and bringing things out in the open. To resolve differences, American managers will prefer to use tactics that involve directly confronting others with rational arguments, factual evidence, and suggested solutions (Ting-Toomey, 1985). It is also consistent with the pragmatic short-term orientation and moderately low power distance in United States of America (USA). Indian managers use those tactics less than American managers, because using the tactics will provoke overt disagreement, which is considered highly undesirable.

In addition, American managers are reluctant to *invest* the time and effort required to enlist the help of other people (Yuki,. Falbe and Youn, 1993), when they have conflicts or problems with another party. In contrast, the strong collective orientation and uncertainty avoidance values in India encourage Indian managers to use indirect forms of influence that involve the assistance of a third party (Bond, 1991).

To deal with a difficult or controversial request, indirect forms of influence are preferred by Indian managers to avoid losing face and damaging others. When their Western partners propose to use direct and open ways to deal with the conflict, they may feel embarrassed. On the other hand, Western partners may get totally confused by the roundabout why the Indian use to solve seemingly simple problems. The different ways that Indian and American managers resolve conflicts seem to find support from Weaver's finding that feminine societies prefer to resolve conflict through negotiation and compromise (Weaver, 2000).

IMPACT OF CULTURAL DIFFERENCES ON DECISION-MAKING

Risk-taking/Risk-avoiding

Indian and American managers differ greatly in the attitudes toward risks when they make decisions for their different values in uncertainty avoidance. High uncertainty-avoidance Indian managers usually lack the adventurous spirit and the sence of risks. They dare not make immediate decisions if they feel the circumstance is uncertain, which may deprive them of the opportunity to compete in the market. In most cases, they would like to make comparatively safer and less risky decisions at the expense of the business opportunity. In contrast, low uncertainty-avoidance American managers are more likely to consider risks as natural and are volunteer to take the risks, especially in terms of developing new products, open a *new market* and applying new technology.

Levels of Participation in Decision-making

Managers from an individualistic or a collectivistic country prefer different levels of participation in decision-making. Another related cultural dimension is power distance. Individualistic American managers prefer making decisions individually or deferring to their supervisors instead of consulting with others. They come from a low power distance culture, and they value personal equality and believe in decentralization and empowered subordinates. But the situation in India is a little complicated. Some researches propose that decisions are typically participatory in collectivistic-high power distance countries (Smith *et al*, 1994). Other researchers think that employees in those countries should readily accept decisions handed down by their supervisors, and even resist participation in decision-making because of their unquestioning attitudes toward their supervisors (Graf *et al*, 1990).

In fact, the second phenomenon described by Graf *et al* (1990) predominates in Indian society while the first opinion is also true to a certain extent. That depends on many factors, the most important of which is "what kind of decisions are going to be made". Many Indian mangers especially those in state-owned enterprises, adopt the nonparticipatory approach to decision-making. Final decisions are usually made by higher level superiors without consulting their subordinates. Since some characteristics like high formality, low disclosure and openness as well as centralization are common in most companies. Indian employees seldom have the chance to really participate in the decision-making process. Fortunately, things are getting better after 20 years of reform in India. Participatory decision-making is beginning to be adopted in a number of modern Indian companies.

Impact of Cultural Differences on Work-group Characteristics

India is a relational-oriented country in which people place great importance on personal relationship. Indian managers may initially focus more effort on building social and interpersonal relations before entering into business or contractual relationship. They would like to spend time developing and maintaining guanxi during the process of interaction and consider it as a prerequisite to do business. In contrast, American managers may encourage their group members learn from each other, to focus on task rather than on social and interpersonal relations, and to build the confidence required for superior performance (Sosik and Jung, 2002). They place a much higher importance on the risk or business deal and hope to focus very quickly on specific business matters. They are achievement oriented, that is "work first". They don't think establishing personal relationship is necessarily involved in the work.

Living in a collectivism society, the Indian view differently as "in-group" and "out-group". They have much higher confidence in "in-group" members than "out-group" members. But the passing of time, and the development of gunanxi, the out-group members may turn into in-group members. If American managers want to do business with Indian managers, They should give their Indian partners enough time to know themselves and develop a personal relationship with them. Only when Indian mangers believe that they can trust their partners and consider them as part of the in group, will the business deal follow smoothly.

In contrast, individualists who exercise short-term thinking place more emphasis on pursuing hedonistic goals and may be unwilling to commit to long-term relationships developed through group interactions (Shamir, 1990). Many American managers are nøt patient enough to build personal relationships with Indian partners when they invest in India. They cannot understand how important the relations are to the Indian and their business.

Impact of Cultural Differences on Motivation Systems

The method of distributing pay to motivate employees may also depend on cultural values (Leung, 1997). As Aguinis (2002) stated, employees can be rewarded according to their performance (principle of equity), equally (principle of equality), or based on their needs (principle of need). In general, the equity principle is common in individualistic cultures while the equality principle is widely used in collectivistic cultures.

According to Triandis (1995), the basic distinction between collectivism and individualism is based on four universal dimensions. One is that the definition of the self is independent in individualism and interdependent in collectivism. Another is that personal and communal goals are not aligned in individualism but closely aligned in collectivism. These two may account for the great difference in motivation systems generally preferred by American companies and Indian companies, pay for performance in American companies, and pay equity in Indian companies.

In American value systems, great emphasis is placed on individual achievement, and they are expected to achieve success only by their individual efforts. They value competition, achievement and personal goals, and therefore, desire to have plans that recognize individual contributions. They generally see success as contingent upon their own efforts, so they prefer "pay for performance" systems which imply that an individual is solely responsible for what he has accomplished even though he may have had help from others. They consider these systems as effective means to motivate employees. The talents and job performance employees will be considered first by their superiors for salary increases and promotion.

Collectivistic cultures emphasise cooperation interdependence, and group goals, and thus, prefer plans that support group harmony (Aguinis, 2002). The Indian believe that the sense of belonging to the group and devotion to the group are important. They hold the view that one's success is mainly based on group work, so one cannot claim the reward just for oneself. In that case, the equality principle is reflected in the motivation system. When setting salaries. Indian managers will pay more attention to the working experience and academic qualifications of employees. In terms of promotion, one's political quality, history, and interpersonal relationship are given primary consideration.

CONCLUSION

By applying Hofstede's and Bond's cultural dimensions, a cultural comparison between America and India is made. The two countries differ greatly in nearly all the aspects. Based on the cultural difference, inferences are postulated on the impact of Sino-American cultural differences on some aspects of management is made is terms of characteristics, and motivation systems. There are several finding from this cultural comparison. First, Indian managers are more likely to favour cooperative strategies than American managers and American managers place greater importance on contractual safeguards than Indian managers. Second, when faced with conflicts, Indian managers tend to use indirect forms of influence that involve the assistance of a third party, while Americans prefer to use direct and open forms. Third, Indian managers tend to make less risky decisions than American managers. They tend to adopt the non-participatory approach to decision-making. Fourth, the Indian pay more attention to build social and interpersonal relations than Americans. Last, the equity principle is common in American companies while the equality principle is widely used in Indian companies.

Table 1

Cultural Dimension Scores of United States and India

Cultural Dimension Scores			
	Power Distance	*Individualism*	*Masculinity*
United States of America	40	91	62
India	80	20	50
Cultural Dimension Scores			
	Uncertainty Avoidance		*Long-term Orientation*
United States of America	46		29
India	60		118

Table 2

Cultural Differences Between America and India

Cultural Dimension	*USA*	*India*
Individualism/Collectivism Power Distance Uncertainty Avoidance Masculinity /Femininity Long/short-term Orientation	Strong Individualism Medium Risk-taking Medium Masculinity Short-term Orientation	Strong Collectivism Cantralised, Tendency toward democracy Risk-avoiding Medium Femininity Long-term Orientation

References

Aguinis Herman, 2002, "The Search for Universals in Cross-Cultural Organisational Behaviour". Chapter to appear in J. Greenberg (ed.), *Organisational Behaviour*. The state of the science (2nd ed.) Mahwah, N.J.: Lawrence Erlbaum Associates.

Bond, M. and G. Hofstede, 1989, "The cash value of Confucian values". *Human System Management*, 8, pp. 195-200.

Chang, S.J., 1995, "International expansion strategy of Japanese firms: Capability building through sequential entry". *Academy of Management Journal*, 38 (2), pp 383-407.

DiMaggio, P. and W.W. Powell, 1983, "The iron cage revisited: Institutional isomorphism and collective rationality in organizational field". *American Sociological Review*, 48: pp. 147-80.

Graf, L.A., M. Hemmasi, J.A. Lust and Y. Liang, 1990, "Perception of desirable organizational reforms in Indian state enterprises". *International Studies of Management and Organisation*, 20, pp. 47-56.

Hofstede, G., 1980, Culture's consequences: International differences in work-related values, Beverly Hills, CA: Sage.

———, 1991, Cultures and Organisations. New York: McGraw-Hill.

———, 1993, "Cultural constraints in management theories". *The Academy of Management Executives*, 7(1): pp. 81-94.

Leung, K, 1997, "Negotiation and reward allocations across cultures" in P.C. Early and M. Erez (eds.), New Perspectives on International Industrial/Organisational Psychology, pp. 640-75. San Francisco: the New Lexington Press.

Shamir, B., 1990, "Calculations, values and identities: The sources of collectivistic work motivation". *Human Relations*, 43. pp. 313-32.

Smith, P.B., M.F. Peterson, D. Akande, V. Callan, and N.G. Cho, 1994, "Organisational management in 14 countires: A comparison with Hofstede's dimensions" in A.M. Bouvy, F.J.R. van de Vijver, P. Boski and Schmitz (eds.), Journey into Cross-cultural Psychology, pp. 364-73. Amsterdam: Swets and Zeitlinger.

Sosik, John and Jung Dong, 2002, "Work-group characteristics and performance in collectivistic and individualistic cultures". *The Journal of Social Psychology*, Washington, pp. 5-23.

Ting-Toomey, S., 1985, Toward a theory of conflict and culture. In W.B. Gudykunst, L.P. Stewart, S. Ting-Toomey (eds.). Communication, Culture and Organizational Processes, pp. 71-86. Beverly Hills, CA: Sage.

Triandis, Harry C., 1995, Individualism and Collectivism, pp. 43-44. Boulder, Colorado: Westview Press.

Weaver, K. Mark, 2000, "Attitudes toward Cooperative Strategies: A Cross-Cultural Analysis of Entrepreneurs". *Journal of International Business Studies*, 31, Fourth Quarter, 2000: pp. 591-609.

Yukl, G., C.M. Falbe, J.Y. Youn, 1993, "Patterns of influence behaviour for managers". *Group and Organisation Management*, 18, 5-28.

CHAPTER

13

Women Entrepreneurship: Conceptual Study with Special Reference to Hazaribag City (Jharkhand)

DR. SHIVADEO SINGH AND DR. SAROJ RANJAN

ABSTRACT

Economic development of any country will be lop-sided without involving women who constitute half of the world population. Evidence has unequivocally established that entrepreneurial sprit is not a male prerogative. An entrepreneur is a person who has an enterprising quality with an eye on opportunities and an uncanny vision, commercial acumen and above all, a person who is willing to take risks because of the adventurous skill within, same holds good for women also. For women entrepreneurs, starting and operating a business involves considerable risks and difficulties regarding obtaining lines on credit, problems in marketing management, production, advertising and packaging and other socio-personal, managerial, technical, raw-material, inventory, financial and governmental assistance problems. This paper proposes to identify the major problems of women entrepreneurs in Hazaribagh city with the objectives of analyzing the various problems faced by them and by suggesting the remedial measures which will help in the acceleration of women entrepreneurship in the city.

Keywords: *Entrepreneurship, Women entrepreneurs Raw-material, Inventory.*

INTRODUCTION

The emergence of women entrepreneurs and their contribution to the national economy is quit visible in India. The no. of women entrepreneurs has grown over a period of time especially in the 1990 s women entrepreneurs need to be lauded for their increased utilization of modern technology, increased investment, finding niche in the export market, creating a sizable employment for other and setting the trend for other women entrepreneur in the organized sector.

In modern era, when women are showing their capabilities in various walks of life, the entrepreneurial classes of women are playing an increasingly prominent role in various sectors of Industrial growth and economic development. Entrepreneurship amongst women is a relatively recent phenomenon, which is gradually changing with the growing sensitivity of the roles, responsibilities and economic status of women in the society in general and family in particular. At the same time, it is also recognized that their challenges are immense and complex. For women entrepreneurs, starting and operating a business involves considerable risks and difficulties, because in the Indian social environment women has always lived as subordinate to men. There have been noticeable changes in the socio-psycho-cultural and economic norms of our society due to liberalized policy of the government of India, increase in the education levels of women and increased social awareness in respect of the role women plays in the society. It has now been recognized that to promote self-employment and to reduce the incidence of poverty, some drastic efforts have to be made to accelerate self-employment of women in various sectors.

OBJECTIVE FOR THE STUDY

Till now very few efforts have been made in exploring the women entrepreneurship in India. The data available on these aspects are scattered and not very authentic and doesn't help in accurate decision-making. However, it is also a fact that more and more women are getting into the domain of entrepreneurial activities. The efforts of Government of India through various welfare organizations, non-Governmental agencies, Ministry of Commerce and Industry and hosts of other agencies have started looking after and helping the efforts of women entrepreneurs. Encouraged with the efforts made by different agencies, this study on. Problems of women entrepreneurs in Sultanpur city (UP) with the following objectives have been undertaken:

1. To examine the awareness about women entrepreneurship.
2. To critically examine the problems faced by women entrepreneurs.
3. To suggest different remedial measures in helping and accelerating women entrepreneurs.
4. To examine managerial problem faced by women entrepreneurs.
5. To analyze the raw material and market problem.
6. To analyze the problem factors as well as problem of government assistance.

RESEARCH METHODOLOGY

For conducting this research a sample of 25 women entrepreneurs in Sultanpur city was selected out of total population 120 women entrepreneurs and for analysis of Table 8, researcher collected data of 20 male and 22 women entrepreneurs in Sultanpur city of Uttar Pradesh. Design of the study was exploratory type and data was exploratory type and data is collected from primary and secondary sources. Primary data was •collected through questionnaire/schedule/ personal interview method and secondary data collected through magazines, journals and newspaper (local newspapers).

ANALYSIS OF DATA

The data contained were analyzed as per the objectives of the study. The salient features of the studies are as follows:

1. Age of Women Entrepreneurs

Sl. No	*Age*	*No. of respondents*	*Percentage*
1.	20-30	14	56 %
2.	31-40	08	32%
3.	41-50	03	12%
	Total	25	100 %

Source: Primary Data..

Thus, on the basis of collected data it can be concluded that a good majority of women entrepreneurs falls in the age-group of 21-30 years.

1. *Marital Status*: 80% of the respondents were married.
2. *Family Background*: Majority of 72% belongs to the Upper Middle Class.
3. *Educational Qualifications*: Most of the respondents', i.e. 64% are either Graduates or PG.

2. Problem Faced by Women Entrepreneurs

SL No.	*Particular*	*No. of respondents*	*Percentage*
1.	Weak collateral position	10	40 %
2.	Affected family life	05	20 %
3.	Overcoming the belief that women are not serious as men	05	20 %
4.	Lack of confidence to start	03	12 %
5.	No support by family	02	08 %
	Total	25	100 %

Source: Primary Data..

Thus, on the basis of collected data it can be concluded that, most of the respondents stated that they have weak collateral positions followed by the problems that their family life is affected, overcoming the belief that women are not serious as men, lack of confidence to start business and no support by family.

3. Managerial Problem

Sl. No.	*Particular*	*No. of respondents*	*Percentage*
1.	Labour Absenteeism	10	40 %
2.	Transportation	07	28 %
3.	Labour turnover	04	16 %
4.	Lack of skilled worker	03	12 %
5.	Lack of managerial skill	01	04 %
	Total	25	100 %

Source: Primary Data..

Thus, on the basis of collected data it can be concluded that, Majority of the respondents faced the problems of labour absenteeism, followed by Transportation problem, problem of labour turnover, and of skilled labour and lack of knowledge of general management.

4. Technical Problem

Sl. No.	*Particulars*	*No. of respondents*	*Percentage* .
1.	Poor record-keeping	09	36 %
2.	Lack of management experience	07	28 %
3.	Inadequate availability of land	05	20 %
4.	Lack of guide agency	03	12 %
5.	Inadequate facility	01	04 %
	Total	25	100 %

Source: Primary Data..

Thus, on the basis of collected data it can be concluded that, the problem of poor record-keeping and control was ranked highest in technical problems followed by lack of management experience, inadequate availability of land, plots and premises and non availability of agency to guide.

5. Raw Material and Market Problem

Sl. No.	*Particulars*	*No. of respondents*	*Percentage*
1.	Poor knowledge of market and competition	10	40 %
2.	Poor inventory management	08	32 %
3.	Inadequate sales promotion avenues	05	20 %
4.	General difficulties	02	08 %
	Total	25	100 %

Source: Primary Data..

Thus, on the basis of collected data it can be concluded that, Majority of the entrepreneurs stated,

That they had poor knowledge of market, followed by problem of Inventory Management, and

Inadequate sales promotion avenues.

6. Financial Problem

Sl No.	*Particulars*	*No. of respondents*	*Percentage*
1.	Price fluctuation in raw material	12	48 %
2.	Lack of working capital	07	28 %
3.	Lack of loan facility	04	16 %
4.	Price fluctuation of finished product	02	08 %
	Total	25	100 %

Source: Primary Data..

Thus, on the basis of collected data it can be concluded that, most of the respondents faced the

Problems of price fluctuations and lack of availability of loan and subsidies facility.

7. Problems of Government Assistance

Sl. No.	*Particulars*	*No. of Respondents*	*Percentage*
1.	Advisory organization become exploitative	10	40 %
2.	Red tapism at various levels	06	24 %
3.	Lack of Govt. help in technical matter	05	20 %
4.	Poor Govt. support	04	16 %
	Total	25	100 %

Source: Primary Data.

Thus, on the basis of collected data it can be concluded that, majority of entrepreneurs stated that advisory organs became Dishonest and exploitative, there was poor government support.

8. Status of Problem Factor with Respect to Men

Problem factor	*Mean*	*S. Deviation*	*Mean*	*S. Deviation*	*Calculated value of t*	*Tabulated value of t*
Managerial problem	3.62	0.723	3.88	0.745	1.146	2.021
Technical problem	4.03	0.623	3.17	0.773	3.944	2.021
Raw material	2.54	0.692	2.32	0.712	1.014	2.021
Financial problem	4.13	0.691	3.71	0.719	1.9248	2.021
Government assistance	3.11	0.767	3.96	0.721	3.7018	2.021

There are 20 male and 22 females an entrepreneur has been taken as sample for the entrepreneurial development problem. This data obtained on the basis of Likert scale (H.S., S., D.S., H.D.). In such cases t-test being used by researcher for this purpose.

The hypothesis developed as:

H_o ; there is no significant difference between male and women entrepreneurs with respect to all problem factors independently.

H_1; Otherwise.

t statistic is defined as :

$$t=[\bar{x}_1-\bar{x}_2]$$

$$\sqrt{\sigma_p^2(\frac{1}{n_1}+\frac{1}{n_2})}$$

$$\text{when} \quad \sigma_2^p=\frac{(n_1-1)6_2^1+(n_2-1)6_2^2}{n_1+n_2-2}$$

x_1 = mean problem factor of male

n_1 = 20

and

x_2 = mean problem factor of women

n_2 = 22

σ_1 = S.D. of problem factor of male

σ_2 = S.D. of problem factor of women

Thus, on the basis of analysis of data obtained it can be

calculated that two factors, i.e. technical problem and government assistance are quite different (calculated value is greater than tabulated value) while male and women are employing their business in Sultanpur city of UP as indicated that men's are technically sound rather than women and secondly, the government policy help to women where they are employing her business. Moreover, all other factors are quite same for both men or women.

CONCLUSION

Entrepreneurship amongst women is a relatively recent phenomenon, which is gradually changing with the growing sensitivity of the roles, responsibilities and economic status of women in the society in general and family in particular. At the same time, it is also recognized that their challenges are immense and complex. Today there is a greater awakening among women, given an opportunity, they will deliver the results. In education, they have not only excellent but also become top achievers. Likewise in office and industry, may have shown brilliant results. Even in Rural India with education, women have shown better performance. Educating women is absolutely essential in strengthening her personality. The need of the hour is to provide an opportunity in a constructive atmosphere free from gender differences. The need for awareness motivation to be an active member of the society and courage to correct the faults of male counterparts are great challenges today. It is, therefore, encouragement of the growing intensity of motivation amongst educated young women for coming in the entrepreneurial stream and extend support with scientifically designed package of the technical and financial assistance, the non-government organizations have a bigger role in stimulating and nurturing the spirit of entrepreneurship amongst women towards this end, an integrated approach is necessary for making the movement of women entrepreneurship a success for this purpose, both the government and non-government agencies have to play a vital role.

SUGGESTIONS/RECOMMENDATIONS

(1). the women an entrepreneur have felt necessity of, "proper sheds and plots", financial subsidies and "availability of raw materials and adequate and proper transport facilities, etc. It is Assumed that these factors are needed to be considered mostly by the government and funding agencies for various purposes. But it also tells upon the efforts of the entrepreneurs in accelerating and making available various items of interest, which they could do with the help of agencies working keenly with them. However, it is recommended that necessary programmes for establishing and strengthening of these facilities may be given at proper time and place.

(2) There are many problems in relation to guidance as stated by the respondents. The Concentration of managerial problems and Socio-personal problems were found to be of serious nature, which may also be rectified by using the parameters associated to this study.

(3) Respondents have also communicated several financial problems in relation to price-fluctuations, non-availability of loan and funds and their release in time. Attention of the government as well as the voluntary organizations should be drawn in rectifying these problems. By making liberal assistance and gearing up the various facilities for enhancement of the status of enterprises.

(4) Finally on the basis of the study, several problems were faced with government assistance, which involves proper policy to help small entrepreneurs, red-tapism, non-availability of guidance and advisory service in time, etc.

(5) Obviously, these factors need to be considered in re-framing and re-organizing Entrepreneurial Development Programmes (EDPs) in the country, which will not only benefit the entrepreneurs but will be in the largest interest of the region and country, in achieving higher productivity and prosperity.

REFERENCES

Anil Agarwal (1984), "Environmental Change and Women in India," *Samayasakthi*, Vol. 1, 1984, p. 27.

Boserup, E. (1989), Women's Role in Economic Development: Earthscan, London.

Boserup, E. (1970), Women's Role in Economic Development.

Gupta, C.B., and Srinivasan, N.P., "Entrepreneurial Development", Sultan Chand and Sons, New Delhi, 2007.

Hisrich, R.D., The women entrepreneurs, characteristics, skills, problems and prescriptions for Success", in the *Art and Science of Entrepreneurship* (Mass Ballinger Publishing Co.), 2009.

Journal of Commerce and Management Studies, March 2011

Kent, C.A., D.L. Sexton, Karl, H. Vesper, "Encyclopedia of Entrepreneurship", (1982).

McClleland, D.C., "Need Achievement and Entrepreneurship: A Longitudinal Study", *Journal of Management*.

Personality of Social Psychology, 1, pp. 389-92.

Sinclair, M.T. (1991), Women, Work and Skill: Economic Theories and Feminist Perspectives.

CHAPTER

14

Sustainable Entrepreneurship in Small and Medium Enterprises

ROHINI GUPTA SURI AND DR. AMRIK SINGH SUDAN

ABSTRACT

Sustainable entrepreneurship is a spin-off concept from sustainable development that can be defined as the continuing commitment by business to behave ethically and contribute to economic development while improving the quality of life of the workforce, their families, local communities, the society and the world at large as well as future generations.

Sustainable entrepreneurship is an approach that is applied mostly by large, often industrial companies. In their wake, a whole range of sustainability certificates came about. Because of the proliferation of complex and costly procedures to obtain them, Small and Medium Enterprises have almost unanimously ignored and repudiated the idea of sustainable entrepreneurship. Since the gains can be substantial in terms of risk control, business relationships with large companies and positive reputation, the question is raised whether Small and Medium Enterprises can actually afford it to do business in a sustainable manner.

This paper seeks to discuss and highlight the following issues relating to sustainable entrepreneurship in Small and Medium Enterprises.

1. INTRODUCTION

Currently, there is a business hype in sustainable entrepreneurship. Every self-respecting company tries to brand itself as a sustainable entrepreneur. Business schools and employers' organisations devote whole conferences to the topic. Many terms are used like corporate social responsibility, ethical funds, eco-efficiency and so on. Although these words reflect different concepts, they all point at various aspects of sustainable development. There is also a booming business in (expensive) sustainable entrepreneurship certificates. Yearly sustainability reports are published almost exclusively by big companies. This raises the question whether Small and Medium Enterprises that constitute up to 90 percent of all businesses, actually can afford to be sustainable entrepreneurs?

In 1987, the World Commission on Environment and Development of the United Nations (the often cited Brundtland Commission) first described and defined sustainable development as a process in which the exploitation of natural resources, the allocation of investments and the process of technological development and organizational change are in harmony with each other for both current and future generations. Sustainability is a concept that is oriented towards the long-term and future generations as much as the present. Therefore, sustainability is at conflict with our fast consuming society and short-term and short-sighted policies. Sustainability fits nicely in the altering view on the prime responsibilities of companies and organisations and their stakeholders that go beyond the classical view in which a company's sole aim was profit maximization in the interest of the shareholdeRs.

Hence, sustainable entrepreneurship can be interpreted as a spin-off concept from sustainable development. Sustainable entrepreneurs are those companies that contribute to sustainable development by doing business in a sustainable way. As Kofi Annan pointed out (United Nations Global Compact Network):

> '... let's choose to unite the powers of markets with the authority of universal ideals. Let us choose to reconcile the creative force of private entrepreneurship with the needs of the disadvantaged and the requirements of the future generations...'

Sustainable entrepreneurship can be defined as the continuing commitment by businesses to behave ethically and contribute to economic development while improving the quality of life of the workforce, their families, the local and global community as well as future generations (World Business Council for Sustainable Development). From a sustainable entrepreneurship perspective, a company is a nexus of responsibilities towards the shareholders, but also towards nature, society and future generations. When the interests of these stakeholders are part of the decision making process in a company, we can genuinely speak about a whole new type of a company with a new type of operational management. A company does not operate on a deserted island, but is embedded in an economic, social, cultural and ecological environment. This offers possibilities and poses threats and obligations, and the theory and concepts of sustainable entrepreneurship try to find the right balance.

2. THE ESSENCE OF SUSTAINABLE ENTREPRENEURSHIP

Following a well-known marketing principle, sustainable development is said to deal with the 3 P's, which stand for People, Planet, Profit. All three aspects (including the last) have to be satisfied before an entrepreneurial activity can be labelled as sustainable.

The first aspect 'people' is about the behaviour of companies in social and ethical issues. How does a company treat their employees (or human resources) and does it promote social cohesion? The issues that need to be adequately dealt with are the protection of human rights, the non-indulgence towards fraud and corruption, the use of child labour, the gender relationships and discrimination on the work floor, labour participation in management and profits, behavioural codes and so on. While many labour regulations were imposed in the beginning of the 20th century by labour and socialist movements as well as in the golden sixties, voluntary, self-imposed systems in this area are most recent, for example, SA 8000 and AA 1000.

Secondly, sustainable entrepreneurship takes care of the natural environment. 'We did not inherit the earth from our ancestors; the earth is on loan from our children', says the Indian adage that clearly summarizes individual and corporate responsibility towards the natural environment. The second aspect 'planet' raises the question of the effect and remedy of the impact of a company on natural resources and the landscape. Environmental care, chain management, eco-efficiency, clean products, sustainable technology development, sustainable industry fields and eco-design are concrete examples of these issues. The consequence for business behaviour is that either environmental integrity becomes a business goal next to profit-seeking or that environmental protection becomes a real constraint on profit maximization (the mathematical calculus in business optimization model leads in both instances to same outcome). The integration of environmental concerns into business practices is driven by both regulation (environmental legislation) and self-regulation (ISO 14000 and so on).

Finally, the third aspect 'profit' does not – as one might expect - relate solely to the purely financial results of an enterprise. Profit is also about the use and allocation of value added for employment, investments in machines and infrastructure and sponsoring and about the distribution (e.g. labour participation). The definition of sustainable entrepreneurship is not a static one since the world and the ideas that emerge are by nature dynamic. But, while the former two aspects of sustainable entrepreneurship (which relate to the material and immaterial contributions of dynamic corporate behaviour to nature and to global and local communities which shape our natural and social environment (Bos, 2002)) are subject to changing views on people and planet, the last aspect is the very essence of a business enterprise.

Janssen (2001) provided a list of ten ground rules for becoming a sustainable entrepreneur:

1. The corporation should start reducing the environmental damage, respecting human rights and treating its employees with great care;
2. Sustainable entrepreneurship has to be a self-initiated process and should not simply be a response to external pressure;
3. If a corporation wants to practice sustainable entrepreneurship, it should identify clear aims and targets;
4. The aims should be closely related to the corporation's practice and should match the corporate values and its primary activities;
5. The aims have to be closely related to the consumers' needs;
6. The corporation has to be capable of explaining the relationship between sustainability and its activities and production process;
7. The corporation should adhere to these aims on a long-term basis;
8. Consumers and pressure groups should have a transparent overview of investments made by the corporation related to sustainable entrepreneurship;
9. Sustainable entrepreneurship practiced by the corporation should not shifted to the consumers via a price increase; and
10. A corporation should not attempt to overemphasize its efforts.

Bos (2002) added an additional rule to the list:

1. A corporation should make sure that its practices are shared by the corporation as a whole, and that they are not solely efforts of the management.

3. SUSTAINABLE ENTREPRENEURSHIP AND SMALL AND MEDIUM ENTERPRISES

Sustainable entrepreneurship is a business approach that seems almost exclusively reserved for large industrial companies. They certainly took the lead in this development after Shell decided in 1995 to dump the Brent Spar Oil Platform in the Atlantic Ocean. The pressure from Greenpeace and other environmental organisations as well as the heated protests by consumers and ecological activists alike forced the Shell Oil Company to reconsider and change its plan. Although initially unintended, it is fair to say that the decision about the final destination of the Brent Spar was made in consultation with Shell's stakeholders, in particular public opinion and environmental interest groups. The affair had some further consequences: it changed Shell's global style of doing business. Sustainable entrepreneurship became their business approach and, from previously irresponsible polluter, Shell is now clearly one of the (self-declared) market leaders in sustainability. The old policy of Shell, summarized as 'Decide Act Defend' was transformed into a business model 'Dialogue Decide Act' that determines the process of important decisions. Many (big) companies followed suit: Nike, Levis, Philips, Unilever and so on. More large companies are engaging in sustainable entrepreneurship policies, more specifically the ones that focus on product and chain-directed environmental care.

Sustainable entrepreneurship presupposes that companies are fully aware of the impact of their behaviour on the material and immaterial situation of their direct and indirect environment (see Bos, *supra*). It involves not only the scanning of opportunities and threats in the market, but also of the environment. The fact that sustainable entrepreneurship is up till now largely a large companies' playground is partially due to the more limited resources of Small and Medium Enterprises. However, as we will show later on, it is impossible for them to ignore this new economic reality.

Small and medium-sized enterprises make a substantial contribution to economic growth and employment in most countries around the world (Organization for Economic Co-operation and Development, 1997). The individual impact of Small and Medium Enterprises is relatively small, but their collective impact is substantial. Small and Medium Enterprises typically represent about 95% of all private sector firms in most modern nations, and so form a major portion of all economic activity (Schaper, 2002). Furthermore, they account e.g. for 35% of exports from Asia and approximately 26% of exports from developed countries including the United States (Organization for Economic Cooperation and Development, 1997). In selected countries such as Italy, South Korea and China, Small and Medium Enterprises contribute as much as 60% of total national exports (Knight, 2000, pp. 12-13). In Belgium, they represent 95% of all businesses and employ 40% of the labour force in private companies.

There is no unequivocal definition of a SME. A whole set of definitions exists that are used in various institutions. The European Commission (Commission of the European Communities, 1996) recommends the following definition that guides all its measures aimed at micro-, small and medium-sized enterprises. Small and Medium Enterprises are defined as enterprises which:

- have fewer than 250 employees;
- have either an annual turnover not exceeding 40 million Euro, or an annual balancesheet total not exceeding 27 million Euro; and
- conform to the criterion of independence.

When it is necessary to distinguish between small and medium-sized enterprises, the former are defined as enterprises which:

- have fewer than 50 employees;
- have either an annual turnover not exceeding 7 million Euro, or an annual balancesheet total not exceeding 5 million Euro; and
- conform to the criterion of independence as defined in the footnote *supra.*

4. THE GAINS OF SUSTAINABLE ENTREPRENEURSHIP FOR SMALL AND MEDIUM ENTERPRISES

4.1. Gains for All Enterprises

Bos (2002) states there are two main reasons for corporate organizations to take into account the socio- and eco-ethical aspects of their behaviour. The first consequence for not doing so is bad publicity. When a corporate enterprise is perceived by the general public as unethical, this will damage its corporate reputation, which, in turn, may result in a loss of income, profits and share value from a conscious consumers' boycott or unconscious bypass of its products. It is important to remember, however, that many benefits from sustainable entrepreneurship are being defined as costs or losses avoided. Secondly, Bos emphasizes that idealism drives more and more organizations that consider themselves to be more than solely profit-making ventures. If not led by idealism, however, corporate enterprises can still enhance their public reputation by showing respect for people and planet instead of only for profit.

Sustainable entrepreneurship gives companies an opportunity to distinguish themselves from others. So far, the results have been encouraging as shown by the returns on funds of companies that are engaged in sustainable development. Examples are the RG Sustainable Shares Fund, ABN-AMRO Sustainable World Funds and the SNS Eco-Shares Funds. Also the Dow Jones Sustainability Group Index, launched in 1999, shows that sustainable companies financially outperform otheRs. The Dow Jones Sustainability Group Index includes the best performing companies with regard to financial results, social and environmental accountability.

4.2. Gains for Small and Medium Enterprises

A general gain of adapting sustainable entrepreneurship in Small and Medium Enterprises is the internal dynamics that sustainable approaches introduce in the production process and human resource management. It is likely to lead to a more bold investment policy in both technology and personnel that will yield results in the long-run. A direct and obvious gain accrues to those Small and Medium Enterprises that supply their products to large companies which themselves are sustainable entrepreneurs and require from their supplier to be so as well in order to be in business. Such large companies may have a direct interest in product and chain directed environmental care as well as labour conditions. Small and Medium Enterprises that do not anticipate to these developments and requirements, seriously risk loosing business to small and medium-sized competitors that do or did invest in sustainable production methods. Another argument for adopting sustainable entrepreneurship lies in the concentration trend of big, global companies. Small and Medium Enterprises obviously cannot compete with these international players. Small and Medium Enterprises should therefore focus on their surroundings, an essential part of the definition of sustainability. Being involved in the local community may prove a sustainability technique for Small and Medium Enterprises that global companies will find hard to copy which, for that reason, turn to locally well embedded Small and Medium Enterprises. The other benefits of sustainable entrepreneurship can be summarized as follows:

- A positive image and reputation;
- Lesser dependency on depleted resources;
- Higher motivation of employees and attractiveness for new employees;
- Efficient production due to superior technologies and better skilled staff;

- Superior insight in market preferences and opportunities;
- Risk control (environmental accidents, scandals, bad publicity, etc.);
- Lower burden from changes in (environmental and social) legislation;
- Corporate social responsibility; and as mentioned before
- Internal business dynamics;
- Business partnerships with other sustainable entrepreneurs;
- Business partnerships with global players.

Sustainable entrepreneurship requires an ongoing dialogue between shareholders and stakeholders. Since a healthy financial basis remains essential, not only will shareholders have to live up to their social and environmental responsibility, but—especially in the case of Small and Medium Enterprises—will stakeholders have to understand that sound financial results are essential for the survival of the company. If not, there is no company, let alone a sustainable one. After listing the potential gains of sustainable entrepreneurship for SME, the key question remains: can they afford it?

5. THE INSTRUMENTS OF SUSTAINABLE ENTREPRENEURSHIP

As mentioned earlier, sustainable entrepreneurship became en vogue after the successful 1995 protests against the dumping of the Brent Star Oil Platform in the open sea. The owner of the platform, the Dutch multinational Shell, withdrew its decision and started to consult its stakeholders. Sustainable entrepreneurship became the new approach and buzz word. Many big companies like Nike, Levis, Philips, and Unilever followed suit and they all now publish their yearly sustainability reports.

Standards for production have been available for quite a while. They were originally created to define product quality. More recently, standards for human resource management at a company level came about that were meant to induce and maintain good housekeeping. Whereas it is rather easy to decide whether or not a product complies with a standard, processes are more difficult to assess. Consequently, these types of standards are less easy to interpret and compare. The first international management standard, the quality standard ISO 9000 came into use in the beginning of the 1990s and was followed by the environmental standards, the ISO 14000 series (now ISO 19000). The standards, however, were considered in some cases to be too complex or too general. As a result, specific standards (QS 9001) or simplified ones (ISO 14001-light) were developed. Different industries also take a different approach to management norms and have adopted their own versions. There are also new and modified versions of international standards emerging such as the new version of ISO 9000 and an unofficial version of ISO 14000 for the working environment. Certification is also beginning to occur in relation to social accountability, information security, ethical trade, equality in the workplace and fire prevention (Martensson, 2000). In the next paragraph, we will briefly discuss some techniques and procedures.

5.1. Audits

5.1.1. Environmental Audits

An environmental audit is one of the procedures in sustainability support devices. An environmental audit often has a descriptive character. The organisation as a whole and its environmental management in particular are being evaluated (competences, responsibilities, communication, education, etc.). In 1996, ISO developed three standards for environmental auditing: ISO 14010, ISO 14011 and ISO 14012 that were recently superseded by ISO 19K. In general, ISO is an audit process that assesses environmental control systems. These audits are executed by an audit team by order of a client. This client can be the company itself or a body that has the legal or contractual authority to request an audit. The client determines the goals. The audit team should be independent and not related to the client or the activities they audit; it should be objective and free from bias and conflict of interest throughout the process. The use of external or internal audit team members is at the discretion of the client. Other conditions are: sufficient information about the subject that needs to be audited, sufficient cooperation from the enterprise under scrutiny and a systematic way of working. There is a standard procedure that describes the way of preparing and executing the audit (kick-off meeting, gathering of information, end meeting, etc.), the way of composing the documents (reports of meetings, procedures and checklists to assess the different elements of the environmental control system, etc.) and it also describes the lay-out of the report. In conclusion, it is fair to say that an environmental audit costs a lot of time and effort. It can also become an expensive process when the company audited is unfamiliar with ISO and/or not environmentally conscious.

5.1.2. Social Audits

Contrary to environmental audits, there is to our knowledge no internationally recognized norm for social audits. Nevertheless, AA 1000 and SQAS (the latter for the transportation industry) are well known. A social audit is defined as the process by which an organization reflects on, measures, evaluates and reports on its social impact and ethical behaviour and adjusts them according to its goals and values and those of its stakeholders. A social audit has also been described as a learning process with four major building stones (Borgo, Mazijn and Spillemaeckers, p. 46):

- Dialogue with the stakeholders;
- Use of quantitative and qualitative performance indicators and benchmarks. This requires a social bookkeeping system. Examples of indicators: absenteeism, dismissals and resignations, labour accidents, total earnings and so on;
- External verification (necessary for the credibility of the next step); and
- Reporting of and communication about goals, efforts and results.

5.2. Codes

Corporate codes of conduct are defined as a statement of principles by which a business agrees to abide voluntarily over the course of its operation. Such a code of conduct creates and continuously evaluates benchmarks for the company's management. It more or less forces the company to behave according to its own principles for its own good and for the better of their direct and indirect environment. Famous international companies have drawn up a corporate code of conduct, primarily in reaction to their critics and activists from consumers' organizations and other pressure groups. Well-known examples are the Levi-Strauss global sourcing and operating guidelines, the Nike code of conduct and the GAP code of vendor conduct. Most of these codes are rather vague and idiosyncratic. Recently, there is a tendency to create more uniform codes.

5.3. Management

A management system can be defined as the organizational structure, responsibilities, procedures, processes and operational duties necessary to carry out certain goals. General management is concerned with the overall operational, financial and strategic management. However, different management systems can be put in place for various goals like environmental care, quality assurance, safety, etc. The best known standard for quality management is probably ISO 9000/9001. Examples of environmental management systems are ISO 14001 and EMAS. The best known social management system is SA 8000.

5.4. Labels

A label is put on a product when that product distinguishes itself in a specific product category. The criteria for labelling should be well defined and transparent. Following the ISO approach, environmental labels are classified into three types:

- Type I environmental labelling programs are voluntary programs where an independent labelling authority sets the criteria based on the lifecycle approach of the concerned product group;
- Type II labelling is about environmental labels that form an integral part of the marketing effort for the product. ISO 14021 sets the standards under which conditions a product can be labelled as environmentally friendly; and
- No criteria are set for a type III labelling, but an independent labeller is involved. It is barely used.

6. THE AFFORDABILITY OF SUSTAINABLE ENTREPRENEURSHIP FOR SMALL AND MEDIUM ENTERPRISES

The proliferation of new types of management systems and their certification techniques seems to create its own problem of sustainability especially for Small and Medium Enterprises. While management systems supposedly have the effect of introducing a systematic approach to the issues of environmental protection, health and safety of employees and neighbours, quality management, etc., there is the pertinent risk that the sheer variety of systems creates overlap and waste. These instruments then require more material and financial resources that a small, individual company can afford. Another risk is that companies, especially Small and Medium Enterprises, become disenchanted by the whole idea of sustainable management systems. While an SME can afford to invest in one ISO-certificate, to obtain all certificates that prove (or signal to the outside world that) the company is a sustainable entrepreneur may prove prohibitively expensive for an SME. While the advantages are many, can Small and Medium Enterprises afford to become sustainable entrepreneurs?

Well, the Small and Medium Enterprises seem to think that they can not. The vast majority is sitting on the sidelines when it comes to the ISO game, let alone sustainable entrepreneurship. In 2000, only 0.1% of European non-micro companies had an accredited environmental management system (EMS) such as ISO 14001 or EMAS. Although eco-efficiency schemes have reached tens of thousands of companies, they only represent a tiny fraction of the millions of European Small and Medium Enterprises. Even fewer companies are aware of eco-design and sustainable manufacturing concepts.

Attitudes to the environment and understanding of sustainable development remain very poor, but cost and time pressure seems the crucial factors. Small and Medium Enterprises as a group have not been persuaded that spending (quite some) money to obtain an ISO-certificate is a sound business decision in terms non-specific promises of cost savings. For most Small and Medium Enterprises, shortage of resources—time, personnel as well as money—is the rule. Therefore, the tendency is to ignore complicated, new systems is easy to understand.

Small and Medium Enterprises face the following problems that Small and Medium Enterprises with regard to sustainable development (Hilton, 2000):

- Lack of resources, time and money;
- Lack of capabilities, skills and knowledge;
- Lack of awareness of issues, risks, regulation;
- Lack of training needs analysis (TNA);
- Lack of awareness of tools and techniques;
- Lack of awareness of provisions and their benefits;
- Lack of strategic and holistic thinking;
- Lack of internal communication and integration;
- Lack of work floor staff involvement;
- Lack of flexibility and fear of change;
- Lack of external communication (networking); and
- Mistrust of other companies in groups.

While there has been a large increase in the number of initiatives aimed at Small and Medium Enterprises, the measures of support have not always been correctly tailored to the needs of Small and Medium Enterprises. Support systems are often found to be too abstract and impractical, too general (insufficiently apt for specific industries), too passive, superficial or lacking in quality, too expensive, time consuming or inflexible or poorly targeted or promoted. There has also been an overemphasis on environmental management systems and certification at the expense of eco-efficiency, eco-design, integrated approaches and sustainable production.

Nonetheless, these initiatives can become successful by means of at least putting sustainable development on the agenda of Small and Medium Enterprises and their federations. Also regional partnerships can help to ensure proper co-ordination, provide pragmatic support, reduce confusion, develop individual relationships and provide all the right incentives for Small and Medium Enterprises. Finally, Small and Medium Enterprises should be encouraged to form or join self-help networks, for example, environment business, eco-efficiency and joint-EMS implementation clubs. But whatever the approach taken, it should be simple and transparent.

7. CONCLUSION

Sustainable entrepreneurship is a spin-off concept from sustainable development that covers many new evolutions in business like corporate social responsibility, ethical entrepreneurship, ecological care, stakeholder participation and so on. It can be defined as the continuing commitment by business to behave ethically and contribute to economic development while improving the quality of life of the workforce, their families, local communities, the society and the world at large as well as future generations.

Sustainable entrepreneurship is an approach that is applied mostly by large, often industrial companies. In this article, the question was raised whether Small and Medium Enterprises could afford it to do business in a sustainable manner. The gains are clear: business relationships with large companies that require sustainable partners, a positive reputation that attracts and motivates employees and risk control.

This evolution led to the creation of a whole battery of certificates to promote sustainable entrepreneurship. However, the proliferation of complex and costly procedures to obtain them has caused the almost complete repudiation of sustainable entrepreneurship by the SME community. Given the opportunities, this is short-sighted strategy.

REFERENCES

Bos, A. (2002), 'Sustainable Entrepreneurship in a Changing Europe: Pedagogy of Ethics for 'Corporate Organizations in Transformation', in: G.F. Simons, D. Min *et al.* (eds.). *EuroDiversity: A Business Guide to Managing Differences*, Oxford: UK and Woburn, USA: Butterworth-Heinemann, 16.

Cecora, J. (2000), 'Entrepreneurs and Small and Medium Enterprises in Regional Economies: Policy Issues for Sustainable Development in a Globalizing Economy', *International Review of Sociology*, 10(1): 83-100.

Commission of the European Communities (1996), 'Commission recommendation of April 3, 1996 concerning the definition of small and medium-sized enterprises', Brussels, C(96) 261.

Hilton, M. (2000), 'SME Support for Sustainable Development: Principles and Practice', in: *European Foundation for the Improvement of Living and Working Conditions, Sustainable Development, Small and Medium Enterprises and New Enterprises* (Conference Report), Luxembourg: Office for Official Publications of the European Communities, 2001, 25-27.

Janssen Groesbeek, M. (2001), Sustainable Entrepreneurship—Theory, Practice, Instruments, Amsterdam: Business Contact.

Krishna, K.L. (1999): "Industrial Growth and Diversification" in Uma Kapila (ed.) Indian *Economy Since Independence*, Academic Foundation, New Delhi.

Knight, G. (2000), 'Entrepreneurship and Marketing Strategy: The SME under Globalisation', *Journal of International Marketing,* 8(2): 12-32.

Martensson, M. (2001), 'Management systems, certificates, labelling: how many can a small company manage?', in: *European Foundation for the Improvement of Living and Working Conditions, Sustainable Development, Small and Medium Enterprises and New Enterprises* (Conference Report).

Luxembourg: Office for Official Publications of the European Communities, 10-11.

Mezher, T. and H. Ramadan (1999), 'The costs and benefits of getting the ISO 9000 certification in the manufacturing sector in Saudi Arabia', *Quality Assurance,* 6: 107-22.

Organization for Economic Co-operation and Development (1997), Globalisation and Small and Medium Enterprises, Paris: OECD.

Peeters, L., P. Matthyssens and L. Vereeck (2002), Stakeholder Synergie, Leuven: Garant.

Prizinsky, D. (1997), 'Small Manufacturers Slow to Climb aboard ISO Train', *Crain's Cleveland Business,* 18(11): 2-5.

Schaper, M. (2002), 'The challenge of environmental responsibility and sustainable development: Implications for SME and entrepreneurship academics', in: U. Füglistaller, *et al.* (eds.), *Radical changes in the world – Will Small and Medium Enterprises Sour or Ccrash?*

Schaper, M. (2002), 'The Essence of Ecopreneurship', In: M. Schaper (ed.) *Environmental Entrepreneurship: Greener Management International (GMI),* 38, Sheffield: Greenleaf.

Srivastava, Vivek (2000): "Impact of India's Economic Reforms on Industrial Productivity, Efficiency and Competitiveness", Part I, National Council of Applied Economic Research, New Delhi.

Srinivasan, T.N. (1994): "Foreign Trade Policies and India's Development", in T.N. Srinivasan (ed.), *Agriculture and Trade in India and China,* ICEG, San Fransisco.

Suri, V. and D. Chapman (1998): "Economic Growth, trade and energy: Implications for the Environmental Kuznets Curve", *Ecological Economics,* 25.

CHAPTER

15

Designing the Green Supply Chain

Rohini Gupta Suri and Dr. Amrik Singh Sudan

Abstract

The supply chain has been traditionally defined as a one-way, integrated manufacturing process wherein raw materials are converted into final products, then delivered to customers. Under this definition, the supply chain includes only those activities associated with manufacturing, from raw material acquisition to final product delivery.

However, due to recent changing environmental requirements affecting manufacturing operations, increasing attention is given to developing environmental management (EM) strategies for the supply chain.

The supply chain concept grew out of the recognition that the process of transforming raw materials into final products and delivering those products to customers is becoming increasingly complex. As such, it became increasingly apparent that analysis (and subsequent improvement) of the individual supply chain stages did not lead to improvement of the chain as a whole. Thus, the concept of the supply chain emerged to describe all production stages from raw material acquisition to final product delivery. Changes in the state of the environment, leading to subsequent public pressure and environmental legislation have necessitated a fundamental shift in manufacturing business practices. No longer is it acceptable or cost-effective to consider only the local and immediate effects of products and processes; it is now imperative to analyze the entire life cycle effects of all products and processes. Therefore, the traditional structure of the supply chain must be extended to include mechanisms for product recovery. This extension presents an additional level of complexity to supply chain design and analysis; more specifically, the addition of the product recovery mechanism gives rise to numerous issues affecting strategic and operational supply chain decisions. Consequently, the extension of the traditional supply chain requires the establishment and implementation of new performance measurement systems.

These new measurement systems will serve as the centerpieces of environmentally-conscious implementation plans, based on continuous improvement, that will enable organizations to become and remain competitive while achieving sustainable processes.

The objectives of this paper are to:

(1) *investigates the environmental factors leading to the development of an extended environmental supply chain,*
(2) *Describes the elemental differences between the extended supply chain and the traditional supply chain,*
(3) *Describes the additional challenges presented by the extension,*
(4) *Presents performance measures appropriate for the extended supply chain, and*
(5) *Develops a general procedure towards achieving and maintaining the green supply chain.*

1. INTRODUCTION

Environmental change is upon us. We are seeing turbulent weather patterns—everything from record heat waves, rains, snowfall and hurricanes—that are becoming extreme in every sense throughout the world. "People should be concerned about what we are doing to the climate," said Jay Lawrimore, Chief of the Climate Monitoring Branch of the National Oceanic and Atmospheric Administration. "The burning of oil and other fossil fuels releases carbon dioxide, which rises, blankets the earth and traps heat." Brenda Ekwurzel, a Climate Scientist with the Union of Concerned Scientists, a public interest group, states, "When you look at temperatures across the globe, every single year since 1993 has been in the top 20 warmest years on record."

Not only do we have climate problems but we are also dealing with a resource depletion issue. India and China are growing economically at double digit rates, and the population of the world continues to grow creating shortages of many resources that we use to take for granted. "The World Bank reports that 80 countries now have water shortages that threaten health and economies while 40 percent of the world—more than 2 billion people—have no access to clean water or sanitation." We will ultimately run out of natural resources if we keep up this torrid pace of use. "Humankind has inherited a 3.8 billion per year store of natural capital. At present rates of use and degradation, there will be little left by the end of the next century."

Many of us may not be around by the turn of the next century but our children and our children's children will be suffering from our over-indulgence.

The United States will soon start to see more legislation enforcing stricter environmental regulations. Europe has already passed several laws with the most notable being the Restriction of Hazardous Substances directive (RoHS). This legislation restricts the amount of certain substances in electrical and electronic equipment. There are also other initiatives being pushed in Europe to control and curtail resources that may damage or deplete the environment. China has also passed its own RoHS law which is more stringent than the European version. The Chinese law restricts all of the same substances that the European law does but it allows no exemptions and requires "laboratory testing and labeling requirements."

In the next several years you will be hearing more about Supply Chain Sustainability or the Green Sustainable Supply Chain. A Green Sustainable Supply Chain can be defined as "the process of using environmentally friendly inputs and transforming these inputs through change agents - whose byproducts can improve or be recycled within the existing environment. This process develops outputs that can be reclaimed and re-used at the end of their life-cycle thus, creating a sustainable supply chain." The whole idea of a sustainable supply chain is to reduce costs while helping the environment. Many people would argue that being environmentally friendly increases your costs. In the past, most companies were focused on reducing unit costs. Many companies later evolved into looking at total landed costs with the on-set of global trade. Companies also started looking at the usage costs with a piece of equipment (i.e. what are my cost per copy when using a copier). In today's "sustainable" world the thinking should be what is the lifecycle costs of this part, piece of equipment or supply chain process.

The chart shows on next page what a Green Sustainable Supply Chain looks like within a company. Sustainability could be a tremendous weapon for companies to reduce costs. There are many facets of the supply chain that could be improved by looking at it from a sustainability standpoint. The first issue that sustainable companies are focusing on is the design and production of the product. "Interface Corporation is a leading maker of materials for commercial interiors. In its new Shanghi carpet factory a liquid had to be circulated through a standard pumping loop similar to those used in nearly all industries.

A top European company designed the system to use pumps requiring a total of 95 horsepower the Interface engineer working on the

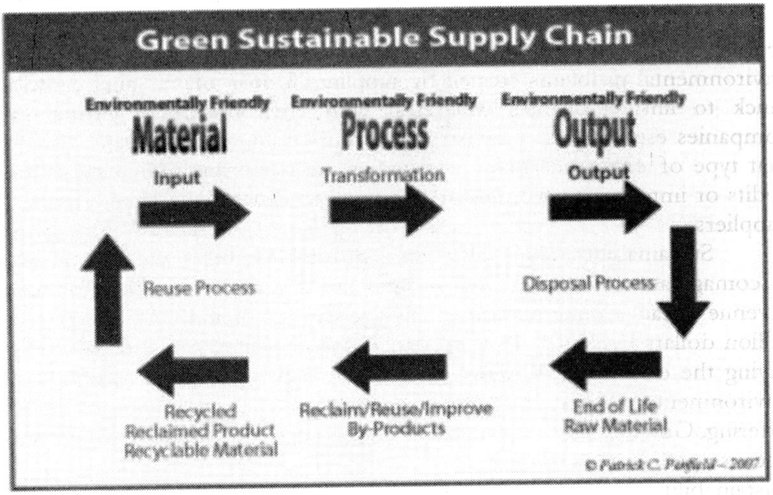

project re-designed the system by using "fatter pipes" and straightening/shortening the length of the pipes used in the system. The overall horsepower requirement was now 7 horsepower or a 92 percent reduction. His re-designed system cost less to build, involved no new technology and worked better in all respects. This engineer used whole system thinking which can help managers find small changes that lead to big savings that are cheap and free."

Many forward thinking companies are using the environmental issues to their advantage. They are innovating and coming up with cutting edge solutions that help them become more profitable while helping the environment. In the book Green to Gold the authors talk about 3M and their program Pollution Prevention Pays (3P). "Anything not in a product is considered a cost, it's a sign of poor quality. As 3M execs see it, everything coming out of a plant is either product, byproduct (which can be reused or sold), or waste. Why then should there be any waste?" If every company thought along those lines would we have the landfills and environmental problems that we have today? Many companies are also focusing on their indirect purchases (packaging and transportation) to reduce environmental issues. Reducing the amount of cardboard or filler by designing "smart packages" can save companies money. "Filling your trucks as full as possible. Dell has upped its average truck load from 18,000 to 22,000 pounds and worked with UPS to optimize delivery strategies. 3M developed an innovative system to install adjustable decks in trucks. Placing pallets on two levels allowed one 3M facility to reduce the number of daily truckloads by 40 percent and allowed them to save $110,000 per year."

One of the bigger issues facing companies these days is the actions of suppliers. Companies today are being held accountable for environmental problems created by suppliers. Unfortunately, the press is quick to link companies who deal with environmentally unfriendly companies especially if a catastrophe occurs. Can your company afford that type of exposure? Many companies are performing environmental audits or implementing "rules of conduct" to check the actions of their suppliers.

Sustainability can also be profitable. GE now has an "Ecomagination program where they are focused on growing their revenue stream from environmentally friendly products to the tune of 20 billion dollars by 2010." They recognize the opportunity associated with saving the environment. Many utility companies are offering customers environmentally produced power and charging a premium for that offering. Grocery stores are able to charge a higher price on organic food because people are willing to pay a premium for food grown organically. Sustainability can be a competitive advantage for many companies. If you can develop a sustainable supply chain think of the money that can be saved by not having to dispose of harmful by-products, reducing obsolescence, decreasing the amount of money spent on scrap and the resources spent on adhering to regulatory issues. Several companies have developed new revenue sources on the by-products they used to throw out! Many companies are using sustainability as a competitive advantage to grow market share within their industry.

So how would your company develop a sustainable supply chain? The United States Environmental Protection Agency (EPA) has written a guide called "The Lean and Green Supply Chain: A Practical Guide for Materials Managers and Supply Chain Managers to Reduce Costs and Improve Environmental Performance." This is an outstanding guide that provides a systematic approach to implementing a Green Supply Chain. "It's a four step decision-making process. The first step is to identify environmental costs within your process or facility. The next step is to determine opportunities which would yield significant cost savings and reduce environmental impact. The third step is to calculate the benefits of your proposed alternatives. The last step is to decide, implement and monitor your improvement solutions." The manual also gives several great examples of what companies have done to "green" their Supply Chain.

In the future companies will be moving to a sustainable supply chain. The harsh reality is that we need to change what we are doing from a supply chain standpoint in order to ensure that future generations will have resources to use in their lifetime. The benefit of implementing a green sustainable supply chain is that we can improve the profitability of our company and help the environment. Green can not only be profitable, but the right thing to do.

Years ago, the concept of environmental quality was almost non-existent in the United States.Then, the concept came to mean cleaner air and cleaner water. Now, environmental quality has come to mean ..safe drinking water, healthy ecosystems, safe food, toxic-free communities, safe waste management, and the restoration of contaminated sites (Council on Environmental Quality, 1996). Concurrently, there has been increasing public attention placed on the overall condition of the natural environment. This attention may be largely attributed to information provided by the media, through growing numbers of environmental and consumer interest groups (Fiksel, 1996). The most commonly perceived enemy to environmental protection is manufacturing and production operations. That is, manufacturing and production processes are viewed as the culprits in harming the environment, in the forms of waste generation, ecosystem disruption, and depletion of natural resources (Fiksel, 1996). Indeed, waste generation and natural resource use, primarily attributed to manufacturing, contribute to environmental degradation by outstripping the earth's ability to compensate and recover, and thus, are not sustainable by the earth's ecosystem.

The current state and trend of environmental degradation (from regulatory, consumer, and moral standpoints) indicate a need for a change in manufacturing philosophy. That is, there must be a fundamental shift in the way production systems operate. There must be a move towards sustainability, achieved through vast reductions in resource use and waste generation, and a move away from one-time use and product disposal. The first step in such a move is to extend the structure of the current one-way supply chain to a closed loop, including supply chain operations designed for end-of-life product and packaging recovery, collection, and re-use (in the forms of recycling and/or remanufacturing). The objectives of this research are to: (1) describe the current state of the natural environment, (2) investigate the environmental factors leading to the development of an extended environmental supply chain, (3) describe the additional challenges presented by the extension, (4) present performance measures appropriate for the extended supply chain, and (5) develop a general procedure towards achieving and maintaining the green supply chain.

2 THE STATE OF THE ENVIRONMENT

2.1. Solid and Hazardous Waste

The amount of solid waste generated in the United States has been growing steadily over the past 30 years and is expected to continue to grow (Council on Environmental Quality, 1996).According to the

United States Environmental Protection Agency (EPA), approximately 12 billion tons of industrial waste (and approximately 208 million tons of municipal waste) is generated every year in the United States. Over 4 billion tons of the total waste generated is hazardous waste, and is increasing at a rate of 10% annually (Environmental Protection Agency Office of Solid Waste, no date, and Fiksel, 1996). This translates into approximately 10 pounds of total waste per person per day (approximately 4.3 pounds of municipal waste per day). Although disposal fees vary by region, the national average waste disposal fee has increased dramatically during the span of 1985 to 1995, rising from $8.20 per ton in 1985 to $32.19 per ton in 1995 (Council on Environmental Quality, 1996). These costs are largely a result of the fact that, according to EPA estimates for municipal solid waste, only 56 million tons (27%) was recovered by recycling or composting and 33.5 million tons (16%) was incinerated, while 118.5 million tons (57%) was landfilled (Environmental Protection Agency Office of Solid Waste, no date).

2.2. Natural Resource Use

The United States extracts an increasing amount of material from United States lands and territories annually, currently in excess of ten tons of material per person (United States Congress Office of Technology Assessment, 1992). In fact, material consumption has increased by a factor of four since the turn of the century (while population has increased by a factor of three during the same period) (United States Congress Office of Technology Assessment, 1992). During this time, the largest increases in natural resource extraction were derived from mining operations (metals and non-metallic ores) and from organics (plastics, and petrochemicals) (United States Congress Office of Technology Assessment, 1992). Additionally, the types of resources extracted have shifted from agricultural and forestry resources in the early 1900.s to mining and organics today (United States Congress Office of Technology Assessment, 1992). It is important to note here that modern product designs are generally more efficient (requiring much less material to produce) and result in products that are lighter in weight; however, these modern products are also highly complex, making them generally more difficult to repair, recycle and/or remanufacture.

2.3. Water and Air Pollution

Water

Although the rivers, lakes, and coastal waters of the United States are cleaner today than they were in the early 1970.s, water pollution is still a very real concern. For example (Council on Environmental Quality, 1996):

- Nearly 40% of all U.S. waters are still too polluted to support all of their designated functions.
- Contaminated fish advisories or bans were issued in 1995 for over 1,700 bodies of water (representing a 14% increase over the previous year) to protect the public from eating contaminated fish.
- More than 4,000 beaches were closed in 1995 due to harmful levels of bacteria and other pollutants.
- Approximately 20% of the population receive water from a facility that is in violation of at least one national safety requirement.

Air

Similar to water quality, air quality in the United States has undergone considerable improvement in recent years. However, also similar to air quality, some troubling facts still remain:

- In excess of 2/3 of the global urban population (primarily in developing countries) breathes air that has unhealthy particulate levels at least part of the year (Percival, *et. al.*, 1992).
- It has been estimated that air particulate levels in the United States are responsible for approximately three percent of all deaths in the U.S. (corresponding to 60,000 deaths per year) (Percival, *et. al.*, 1992).
- In November of 1993, the EPA designated 42 U.S. areas as non-attainment areas for carbon monoxide (41 of which were classified as moderate; Los Angeles was classified as serious) (Council on Environmental Quality, 1993).
- Approximately 59 million people in the U.S. live in counties in which pollution levels failed to meet at least one air quality standard in 1993 (Council on Environmental Quality, 1993).

3. ENVIRONMENTAL POLICY

3.1. Public Pressure

In the United States of America, an estimated 75% of consumers claim that their purchasing decisions are influenced by a company.s environmental reputation, and 80% would be willing to pay more for environmentally friendly goods (Lamming and Hampson, 1996). On a worldwide level, a recent 22-country survey of environmental attitudes found that (Elkington, 1994):

- In half of the countries surveyed, the environment was considered one of the three most serious problems.
- In most countries, the majority of the citizens surveyed said that the state of the environment affects their health, and an even greater majority say that the environment affects the health of their children.
- In 16 of the 22 countries, citizens said that they avoid products that are harmful to the environment.

Thus, in the USA, and worldwide, there is an overall awareness of the worsening state of the environment, as well as a desire to reverse that trend, even if it costs more to do so.

3.2. Environmental Legislation

In the Constitution of India it is clearly stated that it is the duty of the state to 'protect and improve the environment and to safeguard the forests and wildlife of the country'. It imposes a duty on every citizen 'to protect and improve the natural environment including forests, lakes, rivers, and wildlife'. Reference to the environment has also been made in the Directive Principles of State Policy as well as the Fundamental Rights. The Department of Environment was established in India in 1980 to ensure a healthy environment for the country. This later became the Ministry of Environment and Forests in 1985.

The constitutional provisions are backed by a number of laws – acts, rules, and notifications. The EPA (Environment Protection Act), 1986 came into force soon after the Bhopal Gas Tragedy and is considered an umbrella legislation as it fills many gaps in the existing laws. Thereafter a large number of laws came into existence as the problems began arising, for example, Handling and Management of Hazardous Waste Rules in 1989.

Following is a list of the environmental legislations that have come into effect:

1986—The Environment (Protection) Act authorizes the central government to protect and improve environmental quality, control and reduce pollution from all sources, and prohibit or restrict the setting and/or operation of any industrial facility on environmental grounds.

1986—The Environment (Protection) Rules lay down procedures for setting standards of emission or discharge of environmental pollutants.

1989—The Objective of Hazardous Waste (Management and Handling) Rules is to control the generation, collection, treatment, import, storage, and handling of hazardous waste.

1989—The Manufacture, Storage, and Import of Hazardous Rules define the terms used in this context, and sets up an authority to inspect, once a year, the industrial activity connected with hazardous chemicals and isolated storage facilities.

1989—The Manufacture, Use, Import, Export, and Storage of Hazardous Micro-organisms/Genetically Engineered Organisms or Cells Rules were introduced with a view to protect the environment, nature, and health, in connection with the application of gene technology and micro-organisms.

1991—The Public Liability Insurance Act and Rules and Amendment, 1992 was drawn up to provide for public liability insurance for the purpose of providing immediate relief to the persons affected by accident while handling any hazardous substance.

1995—The National Environmental Tribunal Act has been created to award compensation for damages to persons, property, and the environment arising from any activity involving hazardous substances.

1997—The National Environment Appellate Authority Act has been created to hear appeals with respect to restrictions of areas in which classes of industries, etc. are carried out or prescribed subject to certain safeguards under the EPA.

1998—The Biomedical Waste (Management and Handling) Rules is a legal binding on the health care institutions to streamline the process of proper handling of hospital waste such as segregation, disposal, collection, and treatment.

1999—The Environment (Siting for Industrial Projects) Rules, 1999 lay down detailed provisions relating to areas to be avoided for siting of industries, precautionary measures to be taken for site selecting as also the aspects of environmental protection which should have been incorporated during the implementation of the industrial development projects.

2000—The Municipal Solid Wastes (Management and Handling) Rules, 2000 apply to every municipal authority responsible for the collection, segregation, storage, transportation, processing, and disposal of municipal solid wastes.

2000—The Ozone Depleting Substances (Regulation and Control) Rules have been laid down for the regulation of production and consumption of ozone depleting substances.

2001—The Batteries (Management and Handling) Rules, 2001 shall apply to every manufacturer, importer, re-conditioner, assembler, dealer, auctioneer, consumer, and bulk consumer involved in the manufacture, processing, sale, purchase, and use of batteries or components so as to regulate and ensure the environmentally safe disposal of used batteries.

2002—The Noise Pollution (Regulation and Control) (Amendment) Rules lay down such terms and conditions as are necessary to reduce noise pollution, permit use of loud speakers or public address systems during night hours (between 10:00 p.m. to 12:00 midnight) on or during any cultural or religious festive occasion

2002—The Biological Diversity Act is an act to provide for the conservation of biological diversity, sustainable use of its components, and fair and equitable sharing of the benefits arising out of the use of biological resources and knowledge associated with it.

Table 1

Environmental Regulations in India

Year	*Environmental Regulations*
1974	Water (Prevention & Control of Pollution Act) Amendments, 1988
1975	The Water (Prevention & Control of Pollution) Rules
1977	The Water (Prevention & Control of Pollution) Cess Act
1978	The Water (Prevention & Control of Pollution) Cess Rules
1981	The Air (Prevention & Control of Pollution) Act, Amendments, 1987
1982/ 1983	The Air (Prevention & Control of Pollution) Rules
1986	The Environment (Protection) Act, Amendments (1989, 1990, 1993, 1996, 1997, 1998, 1999, 2000, 2001)
1986	The Environmental (Protection) Rules
1992	E (P) Act Notification – "Environment Statement"
1994	E (P) Act Notification – "Environmental Clearance"
1997	Amendments in the Environment Clearance, Notification – "Public Hearing" made mandatory
1989	The Hazardous Wastes (Management and Handling) Rules, Amendments, 2000, Draft Amendments 2002
1989	Manufacture, Storage and Import of Hazardous Chemical Rules, Amendments, 1994, 2000
1991	The Public Liability Insurance Act/Rules, 1992
1995	The National Environment Tribunal Act
1997	Prohibition on the Handling of Azo dyes
1997	The National Environment Appellate Authority Act
1998	The Bio-Medical Waste (M&H), Rules
1999	Notification for making 100% Utilization of Fly-ash made mandatory
2000	Municipal Solid Waste (M&H) Rules
2000	Ozone Depleting Substance (R&C) Rules
1999	Regulation on recycling of Waste Oil and Non-ferrous scrape
2000	Noise Pollution (Regulations and Control)
2001	Batteries (M&H) Rules

3.3. Jurisdiction of Environmental Legislation in India

Central Government	*State Government*
Ministry of Environment and Forest	State Deptt. of Environment
• Regional Offices	
Central Pollution Control Board	State Pollution Control Board/State Pollution Control Committee
• Zonal Offices (6)	—Regional Offices

4. THE SUPPLY CHAIN RE-DEFINED

The new environmental era represents a new challenge to manufacturing and production enterprises worldwide. The challenge is to develop ways in which industrial development and environmental protection can symbiotically coexist. The first step in meeting this challenge is to re-define the basic structure of the entire supply chain, by accommodating environmental concerns associated with waste and resource use minimization.

4.1. The Traditional Supply Chain

The traditional supply chain is defined as an integrated manufacturing process wherein raw materials are manufactured into final products, then delivered to customers (via distribution, retail, or both). Figure 1 below illustrates the structure of the traditional supply chain.

FIGURE 1

The Traditional Supply Chain

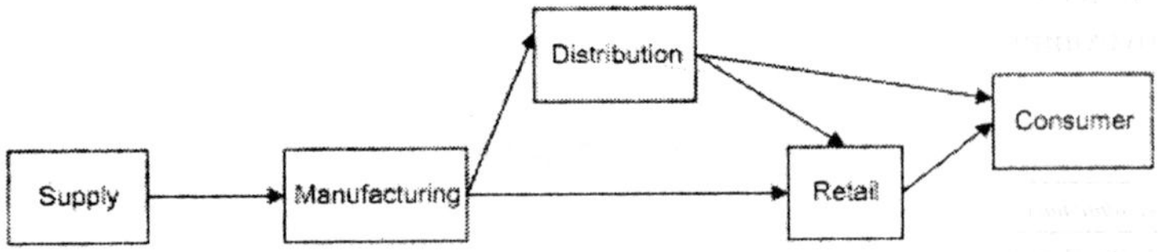

Design, modeling, and analysis of the traditional supply chain has primarily focused on optimizing the procurement of raw materials from suppliers and the distribution of products to customers. The issues considered within this scope of analysis include (Beamon, 1998):

- *Production/Distribution Scheduling:* Scheduling the manufacturing and/or distribution schedule.
- *Inventory Levels:* Determining the amount and location of every raw material, sub-assembly, and final assembly storage.

- *Number of Stages (Echelons):* Determining the number of stages (or echelons) that will comprise the supply chain. This involves either increasing or decreasing the chain.s level of vertical integration by combining (or eliminating) stages or separating (or adding) stages, respectively.
- *Distribution Center (DC)—Customer Assignment:* Determining which DC(s) will serve which customer(s).
- *Plant-Product Assignment:* Determining which plant(s) will manufacture which product(s).
- *Buyer-Supplier Relationships:* Determining and developing critical aspects of the buyer-supplier relationship.
- *Product Differentiation Step Specification:* Determining the step within the process of product manufacturing at which the product should be differentiated (or specialized).
- *Number of Product Types Held in Inventory:* Determining the number of different product types that will be held in finished goods inventory.

4.2. The Extended Supply Chain

The ultimate objective of extending the traditional supply chain is to allow consideration of the total immediate and eventual environmental effects of all products and processes (known as product and process stewardship, respectively). The stewardship concept is based on the recognition that the environmental effects of an organization include the environmental impacts of goods and processes from the extraction of raw materials, to the use of goods produced, to the final disposal of those goods (Lamming and Hampson, 1996).

The evolution of manufacturing enterprises from traditional, problem-solving environmental management techniques to fully integrated environmental management (EM) is described in Table 2 below..

TABLE 2

Environmental Regulations in India

Evolutionary Stage	*Characteristics*
1. Problem Solving	Traditional approaches View regulatory compliance as a burdensome cost of doing business
2. Managing for Compliance	Primitive attempts at EM coordination and integration Compliance-oriented
3. Managing for Assurance	Visionary/long-range planners Utilize risk management to balance potential future environmental liabilities versus costs
4. Managing for Eco-efficiency	Pollution prevention instead of pollution control Waste minimization and source reduction
5. Fully Integrated	Environmental quality viewed as an aspect of Total Quality Management (TQM) Global concern about processes and entire product life cycle

(Adapted from (Fiksel, 1996)).

Thus, in the earliest evolutionary stages of environmental management, organizations separate environmental performance from operational performance. However, as organizations evolve, they begin to integrate environmental objectives within the framework of their existing operational objectives. In this way, the following potential benefits may be realized:

- Reduced product lifecycle costs increased profitability. More specifically, effective environmental management results in the avoidance of the following costs (Cattanach, *et. al.*, 1995).
- Cost avoidance of purchasing hazardous materials as inputs, which reflect the internalized costs associated with environmental harm.
- Cost avoidance of storing, managing, and disposing process waste, particularly as waste disposal becomes increasingly expensive.
- Cost avoidance of stigmatization or market resistance to environmentally harmful products.
- Cost avoidance of public and regulatory hostility towards environmentally harmful organizations.
- Reduced environmental and health risks reduced liability risks (Cattanach, *et. al.*, 1995 and Zhang, *et. al.*, 1997).
- Safer ,cleaner factories (Zang, *et. al.*, 1997).

The fully-integrated, extended supply chain contains all of the elements of the traditional supply chain (Figure 1), but extends the one-way chain to construct a semi-closed loop that includes product and packaging recycling, re-use, and/or remanufacturing operations. The extended supply chain is illustrated in Figure 2. Figure 2 represents the traditional supply chain links as solid lines, and the links corresponding to the extended supply chain as dashed lines.

The W's enclosed by diamonds represent waste (or disposed) materials.

4.2.1 Recycling and Re-use

Recycling is the process of collecting used products, components, and/or materials from the field, disassembling them (when necessary), separating them into categories of like materials (e.g., specific plastic types, glass, etc.), and processed into recycled products, components, and/or materials. In this case, the identity and functionality of the original materials are lost. (Thierry, *et. al.*, 1995). The success of recycling depends on: (1) whether or not there is a market for the recycled materials, and (2) the quality of the recycled materials (since most

FIGURE 2

The Extended Supply Chain

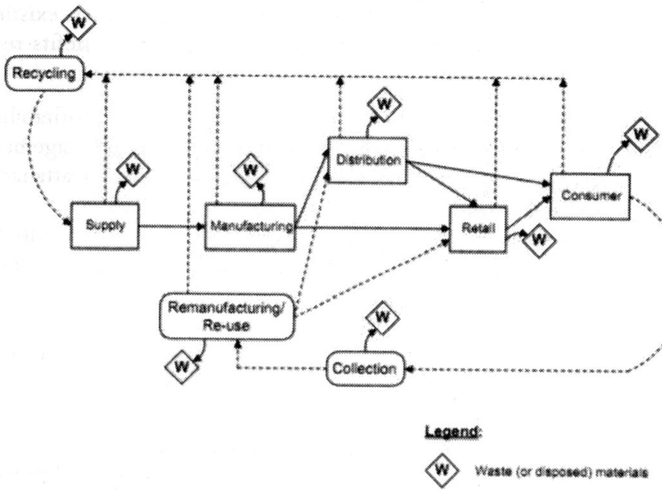

recycling processes actually reduce the value of the material from its original value, as the material itself has degraded). Re-use is the process of collecting used materials, products, or components from the field, and distributing or selling them as used. Thus, although the ultimate value of the product is also reduced from its original value, no additional processing is required.

4.2.2. Remanufacturing

The process of remanufacturing consists of collecting a used product or component from the field, assessing its condition, and replacing worn, broken, or obsolete parts with new or refurbished parts. In this case, the identity and functionality of the original product is retained. The resulting (remanufactured) product is then inspected and tested, with the goal of meeting or exceeding the quality standards of brand new products. Thus, in some cases, the remanufactured product can exceed the original product in quality and/or function. This is due to the fact that during the remanufacturing process, the design of the replaced parts and/or components may have been improved since the original product was manufactured. The unique advantage of remanufacturing is that, unlike recycling and re-use, the process of remanufacturing does not degrade the overall value of the materials used.

4.3. The Extended Supply Chain: Operational and Strategic Issues

Extending the supply chain to include recovery operations, such as remanufacturing, recycling, and re-use adds an additional level of complexity to supply chain design, and a new set of potential operational and strategic considerations. These new considerations arise from two basic problems:

(1) uncertainty associated with the replacement/recovery process (in time requirements, quality, and quantity of returned products, packaging, and/or containers), and

(2) the reverse distribution process itself (collection and transportation of used products, packaging, and/or containers).

Examples of operational and strategic issues associated with recoverable product systems are:

- Inventory control policies (including lot-sizing, scheduling, and safety stocks) given highly uncertain timing, quality, and quantities of replenishments (Guide, *et. al.*, 1997a, Guide, *et. al.*, 1997b, Haynsworth and Lyons, 1987, and Perry, 1991).
- Impact of uncontrollable recovery processes on inventory composition, production planning, and scheduling (i.e., the demand and recovery processes are not perfectly correlated, potentially resulting in uncontrolled growth of unwanted parts and non-availability of critical parts) (Van der Laan, *et. al.*, 1996a and Van der Laan, *et. al.*, 1996b).
- Disassembly planning (including scheduling, sequencing, and disassemblability analysis) for material recovery (Gupta and Taleb, 1994 and Johnson and Wang, 1995).
- The number and location of collection/recovery facilities.
- Collection procedures and customer incentive systems for retrieval operations.
- Effects of traditional supply chain strategies (e.g., decentralized *versus* centralized business functions, facility location, purchasing strategies) on environmental performance (e.g., energy use, solid waste, pollution, product recovery).
- Simultaneous operational/environmental supply chain optimization; merging environmental and operational goals into traditional analysis.

- Level and location of buffer inventories must be considered on both sides of the extended supply chain (forward and reverse) (Fleishmann, *et. al.*, 1997).
- New criteria for vendor selection and certification.

5. PERFORMANCE EVALUATION

An important component in supply chain design and analysis is the establishment of appropriate performance measures. A performance measure, or a set of performance measures, is used to determine the efficiency and/or effectiveness of an existing system, or to compare competing alternative systems. Performance measures are also used to design proposed systems, by determining the values of the decision variables that yield the most desirable level(s) of performance.

5.1. Traditional Supply Chain Performance Measures

Available literature regarding traditional supply chain systems identifies a number of performance measures as important in the evaluation of supply chain effectiveness and efficiency. These measures are typically concerned with: (1) customer satisfaction, service, or responsiveness or (2) cost. The interested reader is referred to Beamon (1996) and Beamon (1998) for a discussion of these measures.

5.2. Performance Measures for the Extended Supply Chain

Although a number of performance measures appropriate for traditional supply chains have been developed, these existing measures are inadequate for use in the extended chain. The existing measures are inadequate in capturing the dual extended supply chain objectives of economic efficiency and environmental protection. This identifies a need to develop new, more inclusive, measures to describe supply chain performance. ISO 14000 identifies the need for these measures implicitly in its certification requirements. In fact, these certification requirements (as previously identified), refer directly to requiring environmental impact analysis and assessment, continuous measurement, targets, and monitoring procedures.

Table 3 on next page describes and classifies the existing performance measures for the extended supply chain.

The types of performance measure(s) used by an organization will largely depend on their evolutionary stage in Environmental Management. Table 4 lists each of the organizational evolutionary stages (as previously shown in Table 2) *versus* the performance measure type(s) with which that stage is most likely associated.

Table 3

Extended Supply Chain Performance Measures

Performance Measure Classification	*Performance Measure* *(Measured over Product and Process Life Cycle, except where indicated)*
Resource Use	Total energy consumed Total material consumed (e.g., water, timber, steel, etc.)
Product Recovery **Remanufacturing** **Re-use** **Recycling**	Time required for product recovery % recyclable/reusable materials (volume or weight) available at end of product life % product volume or weight recovered and re-used Purity of recyclable materials recovered % recycled materials (weight or volume) used as input to manufacturing % product disposed or incinerated Fraction of packaging or containers recycled Material Recovery Rate (MRR) Core Return Rate (CRR) Ratio of virgin to recycled resources Ratio of materials recycled to materials potentially recyclable Materials Productivity: economic output per unit of material input
Product Characteristics	Useful product operating life Total mass of products produced
Waste Emissions and Exposure Hazard	Total toxic or hazardous materials used Total toxic or hazardous waste generated Solid waste emissions % product (weight or volume) disposed in landfills Concentrations of hazardous materials in products and by-products Estimated annual risk of adverse effects in humans and biota Waste ratio: the ratio of wastes to all outputs.
Economic	Average life-cycle cost incurred by the manufacturer Purchase and operating cost incurred by the consumer Average total life-cycle cost savings associated with design improvements
Economic/Emissions	Ecoefficiency: adding the most value with the least use of resources and the least pollution. Generally, the ability to simultaneously meet cost, quality, and performance goals, reduce environmental impacts, and conserve valuable resources

Sources: (Fiksel, 1996, Guide, *et. al.*, 1997b, Krupp, 1992, and Schmidheiny, 1992)

6. TOWARDS THE GREEN SUPPLY CHAIN

In general, the impact of manufacturing operations on the environment may be categorized as follows: (1) waste (all forms), (2) energy use, and (3) resource use (material consumption). In order to achieve the green supply chain, manufacturing organizations must follow the basic principles established by ISO 14000. In particular, organizations

TABLE 4.

Evolutionary Stage vs. Performance Measure Classification

Evolutionary Stage	*Performance Measure Classification*
1. Problem Solving	Waste Emissions and Exposure Hazard; Economic
2. Managing for Compliance	Waste Emissions and Exposure Hazard; Economic; Product Characteristics
3. Managing for Assurance	Economic; Product Characteristics; Economic/Emissions
4. Managing for Eco-efficiency	Product Characteristics; Economic/Emissions; Resource Use
5. Fully Integrated	Product Characteristics; Economic/Emissions; Resource Use; Product Recovery

must develop procedures that focus on operations analysis, continuous improvement, measurement, and objectives. An implementation procedure for extending the supply chain includes the following tasks.

Identify Processes. For each product within the supply chain, identify all inputs, outputs, by-products, and resources.

Develop a Performance Measurement System. Given the complexity of most supply chains, a single performance measure will likely be inadequate in assessing the true performance of the supply chain. Thus, a system of performance measures will be necessary. Such a performance measurement system must include measures for the three environmental categories given above, as well as existing operational measures. The interested reader is referred to Beamon (1998) for a discussion of performance measure criteria and selection.

Measure the Supply Chain System. Calculate the actual composite performance at each step in the supply chain process for each product. The composite performance, as calculated at each supply chain process step, will be a function of the performance measures developed above. The composite performance, therefore, may be a single numerical value, or (more likely) a vector of numerical values.

Prioritize. After all processes for all products have been measured, prioritize the process steps in order of increasing composite performance, as calculated above.

Develop Alternatives and Select Approach. Develop alternatives for performance improvement (targeting first those process steps exhibiting the worst composite performance,based on prioritization above), and select a preferred approach.

Establish Auditing and Improvement Procedures. Establish schedules and procedures for auditing and continuous improvement, including emergency and non-compliance procedures.

7. CONCLUSION

The supply chain concept grew out of the recognition that the process of transforming raw materials into final products and delivering those products to customers is becoming increasingly complex. As such, it became increasingly apparent that analysis (and subsequent improvement) of the individual supply chain stages did not lead to improvement of the chain as a whole. Thus, the concept of the supply chain emerged to describe all production stages from raw material acquisition to final product delivery. Changes in the state of the environment, leading to subsequent public pressure and environmental legislation have necessitated a fundamental shift in manufacturing business practices. No longer is it acceptable or cost-effective to consider only the local and immediate effects of products and processes; it is now imperative to analyze the entire life cycle effects of all products and processes. Therefore, the traditional structure of the supply chain must be extended to include mechanisms for product recovery. This extension presents an additional level of complexity to supply chain design and analysis; more specifically, the addition of the product recovery mechanism gives rise to numerous issues affecting strategic and operational supply chain decisions. Consequently, the extension of the traditional supply chain requires the establishment and implementation of new performance measurement systems.

These new measurement systems will serve as the centerpieces of environmentally-conscious implementation plans, based on continuous improvement, that will enable organizations to become and remain competitive while achieving sustainable processes.

Reference

S. Shanthakumar: Introduction to Environmental Law : 2008, 3rd edition.

CHAPTER 16

Profitability of Coconut Production—A Case Study of Sholavandhan

DR. SHOBANA NELASCO AND RUTH REBECCA

ABSTRACT

The coconut palm, supposed to be one of the five legendry 'Deva Vrikshas' has been eulogized as 'Kalpa Vriksha' the all giving tree in the classics of India. Its fruit is called 'Lakshmi Phal' and is used on social and religious functions in India irrespective of whether the palm is locally grown or not. The palm which is generally believed to belong to the Indian Archipelago, is now found to be growing prolifically in the tropical region. This is a highly profitable crop. But due to the ignorance the farmers.

DESCRIPTION OF THE SHOLAVANDHAN AREA

Sholavandhan is fertile village lying at Vadipatti Taluk in Madurai District. Now Sholavandhan is a selection grade and Town panchayat. This was first inaugurated on 16-8-1968 before that Sholavandhan had been the village panchayat. Now Sholavandhan has eighteen wards. The nearby villages are Thiruvedagam in the East, Kuruvithurai in the west, Vadipatti in the North and Narayanapuram in the South.

TABLE 1

Census Information of Sholavandhan

Particulars	*1971*	*1981*	*1991*	*2001*
Town Area (in sq. km) Population Number of households	15.24 16,451 4,050	15.24 18,230 4,820	15.24 22,328 5,320	15.24 21,661 5,900

Source: Data format for Project Towns Report, Town Panchayat Office, Sholavandhan.

The total population of Sholavandhan was 21,661 in 2001. Out of these 11,798 males and 9863 females. As far as the literacy concerned, more than 25 percentages are educated, at least up to 10th standard. From the total population 15 percentage of the people occupied Government jobs and most of them are depend upon Agriculture

TABLE 2

Slum Information of Sholavandhan

Particulars	*Numbers*
Total number of slums	4
Total slum population	1800
Number of slum households	450

Source: Data format for Project Towns Report, Town Panchayat Office, Sholavandhan

Agriculture

In Sholavandhan bulk of income comes from agriculture and nearly 85 percentages of the people depend on agriculture eighty directly or indirectly for their livelihood.

TABLE 3

Area and Production of Principle Crops in Sholavandhan upto 10.08.03

Crops	*Area*	*Production (in tonnes)*
Paddy	2739	3.690
Cholam	125	1.000
Cumbu	15	1.750
Cotton	20*	1.000*
Sugarcane	375	20.000
Coconut	801	8789#
Gingely	150	0.700
Black gram	20	0.450
Green gram	250	0.400

Source: Crops Cultivation Register, Agricultural Office, Sholavandhan.

Note: * in quintal

per hectare

Table 4

Seasons for Crop Cultivation

Crops	*Season*	*Month*
Paddy	Kariff	June -August
Paddy	Samba	September-December
Paddy	Thaladi	December-February
Pulses	Late Samba/Thaladi	July-August
Kind of Pulses	Late Samba/Thaladi	July-August
Oil Seeds	Late Samba/Thaladi	July-August

Source: Crops Cultivation Register Agricultural Office, Sholavandhan.

Table 5

Rainfall in 2002-03

Year	*Month*	*Rainfall in (mili meter)*
2002	January	• -
	February	53.4
	March	52.2
	April	21.1
	May	68.2
	June	28.8
	July	-
	August	100
	September	95.1
	October	176.3
	November	120
	December	18.8
		733.9
003	January	
	February	
	March	
	April	
	May	
	June	
	July	
	August	
	September	-
	October	24.6
	November	92.3
	December	25.0
		20.9
		12.0
		122.6
		44.9
		83.0
		1.1
		-
		426.4

Source: Crops Cultivation Register Agricultural Office, Sholavandhan.

The data relating month wise rainfall, during the period 2002-03 in Sholavandhan has been given with above particulars. In the year 2002 except January and July the rainfall were favorable for the increase of agricultural production. But in the year 2003, the total rainfall was only 426.44-milli meter. The quantity of rainfall has been adequate for the coconut production. The unfavorable monsoon condition in the month of January, February and December affected little in the production of coconut.

Industry

Sholavandhan has some small-scale industries like coir factory, coconut oil mill, cottage industries, basket weaving, poultry farming, soap making and matchbox making, etc.

Health and Service

There is one public and so many private hospitals are functioning to prevent and cure various diseases like polio, cholera, small pox, etc.

Transport

Availability of transportation facility is a combination factor to stimulate the production habits among the people.

TABLE 6

Details of Roads

Road type	*Length in km*
Concrete	5.701
Black top	5.540
WBM	0.200
Others	5.400
Total	16.841

Source: Data format for Project Towns Report, Town Panchayat Office, Sholavandhan.

Infrastructure

In Sholavandhan post office, railway station, police station, 2 Government higher secondary Schools, several union and private nursery schools, etc. A Vivekanandhar college is also functioning there.

There is also one PRC depot it operates 36 buses to so many villages and towns. In addition private bus services like SPS, PRP, Balashakthi, Nallamani, etc. and Pandian mini bus service also functioning. Railways to both Madurai and Dindigul also link Sholavandhan. Therefore, it makes possible to transport agricultural products easily.

Apart from these all other facilities like electricity, drinking water, sanitation, Xerox, telephone facilities are also available here.

TABLE 7

Average Cost and Return Structure of Coconut Cultivation per Acre of Small, Medium and Large Farms

(In Rs./Acre)

Cost Components	*Small Farm*	*Medium Farm*	*Large Farm*
Cost A	2346.73	2,779.93	2,138.94
Labor	752.94	926.00	887.85
Chemical fertilizer	633.30	327.27	340.91
Pesticides	128.89	129.37	156.64
Ploughing	2,772.20	2,856.54	2.672.27
Farm manure	1,253.85	1,315.72	2,118.42
Irrigation	108.30	121.26	118.26
Seeds	702.15	571.43	418.20
Transportation			
Cost A — Total	8698.36	9027.52	8,851.49
Cost C (cost A Plus)	-	1,578.95	2,093.02
Rent on land	300.00	714.29	412.50
Taxes			
Total cost (cost C)	8998.36	11320.76	11,357.01
Yield per acre in nuts	5320	6720	7958
Yield per acre in Rs.	15960.00	20,160,00	23,874.00
Net returns in Rs.	6.961.61	8.839.24	12,516.99

Source: Computed from survey data.

The above Table reveals that yield per acre of large farmm was more than that of small and medium farms. The yield per acre in rupees was also more in large farms it was Rs. 23784 compared to small farm and medium farm of Rs. 15960 and Rs. 20160 respectively. Total cost of production was also high for the same community that was Rs.11357.01, where as the small and medium farm of Rs. 8998.36 and Rs. 11320.76.

All of the three farms spent much for the farm manure. The amount spent on this, accounted for Rs. 2772.20, Rs. 2856.54 and Rs. 2672.27 of the total cost for small farm, medium farm and large farm.

Secondly, labour cost occupies the maximum amount from the total cost that was Rs. 2346.73 for small farm, Rs. 2779.93 for medium farm and Rs. 2138.94 for the large farm. After this, rent paid for land constituted Rs. 1578.95 and Rs. 2093.02 for medium and large farm. But this was nil in case of small farm because each and every farmer in this farm group had their own land for cultivation.

Other than the above costs chemical fertilizer, pesticides, ploughing, irrigation, seeds, transportation and tax also included in the total cost of production that was Rs. 752.94, Rs. 633.30, Rs. 128.89, Rs. 1253.85, Rs. 108.30, Rs. 702.15, Rs. 300 for small farm.

The medium farm faced different rate for the same costs that weans Rs. 926, Rs. 327.27, Rs. 129.37, Rs. 1315.72, Rs. 121.26, Rs. 571.43, Rs. 714.29, likewise the large farm spend Rs. 887.85, Rs.340.91, Rs. 156.64, Rs. 2118.42, Rs. 118.26, Rs. 418.20 and Rs. 412.50 respectively.

It was concluded that the large farm spend much for this crop cultivation and at the same time they received much out of this cultivation. The net return was Rs. 12516.99 than the small and medium farm that was Rs. 6961.64 and Rs. 8839.24 per acre.

TABLE 8

Average Cost and Return of Coconut Cultivation Per Acre of All the Farms

Cost Components	*Averages of all the farms*	
	In rupees	*In percentage*
Cost A	2421.81	22.94
Labor	855.60	8.10
Chemical fertilizer	433.82	4.11
Pesticides	138.30	1.31
Ploughing	2,767.00	26.21
Farm manure	1,562.66	14.80
Irrigation	115.94	1.10
Seeds	563.93	5.34
Transportation		
Cost A—Total	8,859.12	83.91
Cost C (cost A Plus)	1223.99	11.59
Rent on land	475.60	4.50
Taxes		
Total cost (cost C)	10,558.71	100.00
Yield per acre in nuts	6,666	
Yield per acre in Rs.	19,998	
Net returns in Rs.	9439.29	

Source: Computed from Survey Sata.

It was observed from the table that all the farmers obtained a yield of 6666 nuts per acre and worth being Rs. 19, 998 per acre. The net income over the total cost turns out to be Rs. 9439.29 per acre.

Out of the total cost of cultivation of Rs. 10558.71, operating costs (cost A) amounted for 83.91 percent (Rs. 8859.12) and fixed costs 16.09 percent (Rs. 1699.59)

Out of this cost component, expenditure on farm manure was the major input and it was 26.21 percent of the total cost. The second largest share goes to human labour that was 22.94 percent of total cost. It was followed by irrigation and. They accounted for 14.80 percent and 11.59 percent of total cost respectively. Expenditure on pesticides, ploughing, seeds, transportation and taxes constituted nearly 4.11 percent, 1.10 percent, 5.34 percent and 4.50 percent of the total cost.

SUGGESTIONS

- The farmers should identify some other varieties, which is suitable for the climatic conditions.
- Inter cropping should be encouraged in the coconut garden which will helps the farmers to earn additional income.
- It is suggested to promote irrigation facilities in coconut gardens with special emphasis on adoption.
- Farmers reported that various pest and diseases affected their coconut gardens. Hence strong Research and Development efforts should be made for devising effective control measures for controlling pest and diseases.
- It is suggested that in the study area a coconut growers committee consisting of representatives from agricultural department may be formed to discuss the production and marketing problems regarding coconut.
- The crops planted in inherited lands are not looked after well by their children. To get more yields, more attention should be given for this crop.
- The major form support to be extended to farmers is not only financial, but also organizational at local levels. This purpose is to mobilize the farmers and instill interest and confidence in their minds in achieving higher levels of production and farm level income.
- The government should publicize the availability of coconut crop insurance facilities.
- The government should arrange minimum support price towards this crop in this area. Then only the price fluctuation will not affect the farmers.
- The various programmers and facilities given by the Coconut Development Board should be known to the farmers by the agricultural department.

CONCLUSION

In the research study, the problems faced by the coconut growers were identified, for this the researcher has made some suggestions. Such measures would help the coconut growers to make their cultivation of coconut a grand success.

REFERENCES

Books

Daniel Sundararaj, D. and Thulasidas, G., *Botany of field crops*, Mac Millan India Ltd., 1993.

Jaiswal, P.L., *Handbook of Agriculture*, Council of Agricultural Research, New Delhi, 1980.

Pathi, P.C., *Principles and practice of farm costing with farm studies*, Government Central Press, Bombay 1933, p. 81.

Randhawa, M.S., *A History of agriculture in India*, 1986.

Sambamurthy, A.V.S.S. and Subramanyan, N.S., *A text book at Economic Botany*, Wiles Eastern Ltd., 1989.

———, *Crop production guide*, Tamil Nadu Agricultural University, Coimbatore, 1999.

Journals

G.D. Agarwal, "Principle of Evaluation and Apportionment of items of cost", *Indian Society of Agricultural Economics*, Bombay, 1960, pp. 46-50.

A. Likhan, H.S. and Rajagopalan, V., "An economic analysis of coconut production in Tiptur Talut of Tamkuna district", Karnataka (Mysore), *The Madras Agricultural Journal*. Vol. 1.66(9), 1979, pp. 582-86.

Barman, S.C., "Production and prices of coconut and its oil in Bangladesh", *Economic Affairs*, Vol. 47, December 2002, p. 221.

Dr. Chidambaram, K., Dr. Punithavathi and G. Thenmozhi, "The problems of coconut fiber industry", *Yojana*. May 15, 1991, p. 24.

Gopalan, K. and Venkataraman, M.S.., "Cost of Cultivation of coconut in Travancore", *The Indian Coconut Journal*, 1986.

Jacob John and Achuthan Nair, "Differences with holding in size in the composition of coconut based homesteads of southern Kerala", *Indian Coconut Journal*, September 2002, pp. 7-9.

Kahlon, K.S., and Karan Shing, Farm efficiency measures, Economics of farm management in India, New Delhi, 1981, pp. 75-103.

Muthuchamy, I., Senthilvel, S., "Sub-surface drip irrigation for coconut", *Kissan World*, February 2001, p. 15.

Nagaraj, N., "An Economic Evaluation of Drip irrigation for coconut plantation", *Indian Journal of Agricultural Economics*, p. 284.

Nagarajan, S.S., "Potted Coconut Seedling in Rajapalayam", *Kissan World*, January-June 1985, p. 11.

Narayana, D. and Nair, K.N., "Trends in Area, production and productivity of Coconuts in Kerala", *Indian Journal of Agricultural Economics*, Vol. 44, No. 2, April-June 1989, p. 159.

Parameswara Gupta, E.A., "Trends in Area, production and productivity of Coconuts in India", *Southern Economist*, December 1996, pp. 26-28.

Parameswara Gupta, E.A., "Processing and consumption of coconut in India," *Southern Economist*, February 1997, p. 20.

Ramakumar, R., "Cost and Margins in Coconut Marketing", *Indian Journal of Economics*, Vol. 56, No. 4, October-December 2001, p. 668.

Ramanathan, T., "A promising variety of coconut" *Kissan World*, April-May 1984, P-29-30.

Rath, R., Gosh, P.K., and Sahoo, S.C., "Intercropping in coconut," *Kissan World*, January 19, p. 46.

Rathiha, R., and Edwin Gnanadas, M., "Strategies for Coconut Price Stabilization", *Southern Economics*, June 1, 2002, p. 17.

Shanmugan, K.S., "Coconut fertilization", *Kissan World*, Vol. 7, No. 4, April 1980, pp. 28-30.

Singh, K. and Kahlon, A.S., "Resource Productivity Comparison of varying levels of technology in Punjab", *Indian Journal of Agricultural Economics*, 28(20), 1973, pp. 12-13.

Sivanandam, Kanadasamy and Ramasamy, "Returns to investment in forest plantation", *Financing Agriculture*, 1981, pp. 21-26.

Sundaravaradan, S., "Micro-nutrients—A must for coconut tree", *Kissan World*, May 1985, p. 48.

Thampan, P.K., "Gains from organic farming", *The Hindi Survey of Indian Agriculture*, 1997.

Thomas Mathew, "A study on the consumption of Tender Coconuts in Tamil Nadu", *Indian Coconut Journal*, September 1991, p. 5.

Report

Directorate of Economics and Statistics, Studies on the Economic of Farm management in Tanjure, Tamilnadu, Report for the year 1969-70, Ministry of Agriculture, Government of India, New Delhi, 1974, p. 23.

Thesis

Chandran, C., "A Study on Coconut Marketing in Tamilnadu:, M.Phil. Dissertation Submitted to Madurai Kamaraj University, 1992.

CHAPTER

17

Business Education at Cross Roads

DR. ANIL GUPTA, DR. AMRIK SINGH SUDAN AND KULDEEP SINGH MANHAS

INTRODUCTION

To unravel the snarl into which our education system finds itself, it is necessary to answer the basic question: What is education? The irrefutable answer is that education is preparation for Life. Life is precious and there lies the necessity for long years of careful and formal preparation for it. In the present world it is the crucial component of human development as it contributes to strengthen the moral fibre of the society, helps in building up a strong and vibrant nation, creates social cohesion and thus, helps in bringing about National unity, developing tolerance and establishing a more human social and economic order. In simple, education is the process of learning and knowing of different things unknown to human beings.

For twenty years after independence the nationalist Government in the colonial education policy of the British rulers made no change. In 1964, the need was felt to change the education policy to provide the education, which is based on the development of personal qualities of human beings and developing abilities and skills among the population of the country. The change in education policy was no doubt in accordance with the requirements of the country and it now has become easy to see what goal of education is! Education by definition is forward looking.

It is to accelerate the production of knowledge and its dissemination through sophisticated communication channels such as Internet and T.V. overcoming the difficulties. In countries like Finland, Norway and Germany there is a tradition of helping youngster to develop career lines without going to university. Hence such kind of version is really of great importance for young students.

LINKAGE BETWEEN EDUCATION AND ECONOMY

The education will be meaningful and realistic if the proper linkage is developed between education and business houses. The concept of education as an investment has been increasingly accepted in 1966. The Report of Education Commission (1964-66) referred to education as the only instrument of peaceful social change. Higher education and intelligentsia nurtured by it have a special role in determining the quality of the overall environment. Higher Education also supplies a wide range of increasingly sophisticated and ever changing variety of manpower needed in industry, agriculture, administration and services. This is true so for as the concept of job oriented education is concerned. It will not be too late to start the process of removing constrains and updating the system of education both from supply side and demand supply. There is an oriental saying that a journey of thousand miles begins with but a single step. Socrates said the unexamined life is not worth living. In the changing technological, social and economic environment, education has assumed a life long learning, close association with industry and other sectors of the economy.

EMERGING TRENDS

India is an emerging economy and a society in transition. Economic liberalization has strengthened the number of middle class. The crux of Globalization in education sector is that the teaching learning is no longer for "Building of Nations" but businesses for "Profit Making:. Students, Professors, Institutional Organizers are the parties affected by the globalization of educational services. We have strengths in some areas as there are 300 Universities and deemed Universities, more than 16,000 Colleges and 6.5 million students. At the commencement of the 11th Five Year Plan, there were 19 Central Universities, Seven Indian Institutes of Technology, 20 National Institutes of Technology, 4 Indian Institutes of Information Technology, 2 Indian Institutes of Scientific Education and Research, 6 Indian Institutes of Management and one school of Planning and Architecture besides other centrally funded educational institutions. We have weakness in some areas that there is all-around deterioration in quality education. World-Class faculty are not attracted and retained in our system. Our educational institutions are functioning as mere examination conducting bodies. It is also seen that most of the colleges and universities are facing the common problems of resource crunch, paucity of requisite faculty and inadequate infrastructure. When the number of Ph.D's available in the country is far less than the total Ph.D's required to fill the vacant posts, it would be futile to expect that qualified teachers would emerge from no where. Moreover, higher education is accorded to low priority by the government both in respect of infrastructural development including faculty and flow of funds.

In the face of these weaknesses Indian Education System is force to accept a change with the dawn of economic globalization in Engineering, Management, Computer Technology and other feasible areas. The Vice-Chancellors, Political Leaders, Administrative Heads and Policy-makers at the Central and State Levels have taken bold and creative steps to turn threats into opportunity. Industry-Institutions interaction, Project Linkage, Exchange of Faculty and Students, Development of Joint Courses for Job-Market, International Exchange of Scholars between home and host Countries, etc., are some of the emerging trends in education.

Our honorable President Sh. A.P.J. Abdul Kalam very rightly said that Education in its real sense is the pursuit of truth. It is an endless journey through knowledge and enlightenment. Such a journey opens up new vistas of development of humanism.

"Inclusion" is the growth mantra for policy-makers today—not just in India but all over the world. And for inclusive growth, what could be more crucial in put than business education. Good and Wholesome Business Education empower one, opens up opportunities for individual and group entitlements, productive employment and participation in various areas of social life. In all business education catalyzes economic growth. If India, with its predominantly young population wants to reap demographic dividends, then India is to ensure quality of education, to promote the spirit of quest and the workout meaningful curricula aligned with job market. In the following sections, we discuss some of the challenges being experience by business as Business Education is imparted to bridge the gap between they are able to meet the growing needs of a rapidly expanding economy.

BUSINESS EDUCATION

Keeping in view the education in India it is stated that enterprises today require professionally educated and trained persons in order to meet crucial challenges of tomorrow in view of the irreversible trend of political, economic, commercial and technological interdependence. Initially education was confined to Government and Government Sponsored Institutions and Universities. Now there are various categories of colleges and universities in India like State, Central, Private, Deemed and upcoming Foreign Colleges and Universities. Through proper Education the achievement of economic and social development can be further expedited as Human Resources Development has multiplier effect on the utilization of all other resources. That is why the concept of marketing education has assumed great importance in the human world today as we have been investing huge amount of money and deploying a good number of personnel and sophisticated technologies to educate the masses. Therefore, it is necessary to know the results. If we invest more for removing illiteracy (inputs) but the literacy (output) is not increasing proportionately then the efforts of the country to make it free from illiteracy are not yielding the desirable results. As such the education policy is cot-ineffective. The very essence of marketing is to make our policies and programs cost-effective so that we are able to remove the illiteracy without putting more financial burden.

To de-emphasis college education, which interestingly would increase its value, structure would need to be created in the job market, which promise decent pay attached to career ladder. That requires co-operative activities between Government, Educators and Business says Lazerson. It will help to prepare some students for colleges and allow other the option to enter the job market. The fusion of liberalized and vocational education would create people more able to cope with a transitional and evolving world.

For the last one and half decade the Indian Economy has under gone a tremendous transformation due to the forces of globalization, liberalization and information technology. These forces have revolutionized the rules of trade, commerce and industry. In business environment a complete shift has taken place from National to Global, from Protection to Competition, from Over-regulated to Free Market, from Manufacturing to Service Sector, from Quantity to Quality, from Traditional Commerce to E-Commerce, from GATT to WTO, etc. All these forces taken together have a profound impact upon the commercial and business practices. This has made the demands for new skills. Thus, the institutions imparting commerce and business education are responding to these changes and supply the manpower as per new requirements after re-tooling their institutions and redesigning their Curricula.

BUSINESS EDUCATION IN NEW MILLENNIUM

In the new millennium the business education embraces information technology and E-Commerce, Commerce Education, Human Resource Development, Corporate Governance, International Trade, Finance and Accounting, etc. for industries of all types, service sector and even for non-profit-making organizations. The area of education makes clear that Management Education and Commerce Education are the two branches of Business Education. Information technology has revolutionized the trade, Commerce Education and all spheres of life. However, the digital gap exists between developed and developing countries as 4-5% business is transacted at Internet. Business education in India at School, College and University level has not been able to cater to the demand for the market. We are in 2011. Still India has approximately half of the world's illiterates. Even those who stay full 12 years in school do not get the education in the proper sense. The education must move beyond the class-room and outside the text books to make the learning job-oriented. Coming to the higher education only 10-11% of India's youth aged between 18-23 years have access to it as compared to 34% in Singapore and 65% in United States of America (USA) and around 70% in developed countries. Mushrooming growth of private institutes have been offering professional courses in Management and Commerce education across the country but these are not able to lead to employability of Commerce and Management graduates due to weak base of +2 level, lack of interaction with business houses, non-preference of commerce graduate and post-graduates in banks and insurance sector. Due consideration should be given to these issues by all concerned.

Time demands that commerce and management graduates and post-graduates should be exposed to global environment issues like World Trade Organization (WTO), Foreign Direct Investment (FDI), ISO Quantity Standard, Indian and International Accounting standards. So long as we are not able to re-structure the curriculum and to have interaction with industry, business education will not make a much head way. A paradigm shift in the mind set and thinking is essential for effective utilization of human resources in view of the dynamic competition posed as a result of globalization, liberalization and information technology.

The business education in India can be furthered with the help of components of marketing mix such as product mix, promotion mix, place mix and price mix. These components of marketing mix are related to the product. The process, people and physical distribution are the three other components of marketing mix in relation to the service sector. These three components have been contributing significantly the awareness among the students in today's competitive world. Business education that use to be recognized as public good is now increasingly being viewed as private good.

Commercialization of business education at higher level is on the increase where students are seen as customer.

Education market is more deregulated, diverse and private. Despite all these changes education in seen not just as a subset of society but as a super set that can transform society into a wealthy nation. Just as industrial revolution changed the face of the 19th century. Business education marketing can change and rejuvenate society in the 21st century. It will elevate the socio-economic status of the people by developing abilities and skills.

EXAMINATION

There are many forms of evils and mal practices and many ways have been devised to indulges in it. Now-a-days and alarming development has been noticed in some institutions and that is to bring weapons in the examination hall to intimidate the invigilators. Some time a disorder is created in the examination hall alleging the question paper to have been set outside the prescribed syllabus. Students, blame that teachers have become commercial minded. They do not teach properly and regularly in the class room. Pedagogies on the other hands say that the prevailing education system of distributing degrees is mainly responsible for mass coping and other mall practices.

Allegations and counter-allegations have become the routine matter. The question is how to improve the examination system as the pattern of question papers and the type of questions have considerable impact on the conduct of examination. To control the coping menace researches have shown that if students are divided in groups and two or three different question papers are set from prescribed syllabus, and are given to each student one by one to answer the question. This will eliminate the chances of mass coping and smuggling of answer sheets from outside. Another advantage of this systems is that if a question paper is leaked out it will have hardly any effect as the candidate does not know which question paper he is likely to get.

On line examination system for professional courses is also defective as the candidate alongwith four-five friends and bundle of books, answers the question to pass the examination. Great care needs to be exercised to select and draft questions to make on line examination a great success otherwise on line education is inconceivable without a certain minimum degree of integrity.

If things have to be improved and particularly in a sensitive and crucial area like the integrity of examinations, the matter would have to be considered thoroughly and a set of proposal evolved are to be acceptable to every one. The basic approach should be that if a teacher helps in or connives at any form of copy, he is not worthy of the trust repost in him. After a warning and so on, if he commits the same offence again then he would be asked to resign himself or he will be shown the door forever—otherwise such a bad teacher would become a farce and a pretense, indeed a breeding ground for corruption in the society. Academician should earnestly accept the truth that any chaos in the society will be a reflection of them. They do not necessarily impart opinion but are supposed to kindle the mind.

CHALLENGES AND SUGGESTIONS

The Indian economy has opened up the Global Competition and business education as such has become a major instrument of this change. Still there are challenges before this sector. Some of the challenges Business Education facing today is as under:

The institutes of higher learning have to learn how to seek gifts and how to build relationship with alumni and industry. Successful alumni are willing to support their alma-mater by providing gifts, endowments and sponsorships to minimize the financial burden on the government.

University should review and redesign the curriculum by involving representative of professional like Company's Secretaries, Charted Accountants, Cost Accountants, Executives, etc.

Modern business world over is influenced greatly by computer aided technology and India as a part of world trade organization cannot afford to remain uninfluenced by.the cybernetics.

There is a need to bridge the gap that exists between the Indian Accounting Standards and International Accounting Standards and that of the world in order to ensure better harmonization of accounting practices required in view of the present global economic scenario.

Strong and viable rapport between academic and industry must be developed to expose the learner to the job situation.

Traditional teaching methodology be done away with and replaced by learner centered teaching methodology like case studies, role playing project report, assignments, quiz programs, business games, simulations, etc. This will developed interest and total involvement of students in the teaching learning process.

Low standard of education and the percentage of failures and dropouts is very high. This involves a huge loss to the society. Therefore, there is need to take appropriate steps to improve the standard of education and to reduce the rate of dropouts.

Shortage of staff (teachers) in views of introduction of new courses. This leads to hiring of "Contract Teacher" who are not fully involved in institution building activities.

Impact of global changes and national economic scenario to meet this challenge, the industry requires modernization, up-gradation of the competence of the human resource, utilization of modern management techniques, etc. Our education system needs to be made to respond effectively to these emerging challenges.

The fee structure, which holds the key to the regulation of admission, has been totally neglected more due to the fact that a fee rise can affect adversely a political party then to its regulatory efficiency.

Policies encouraging privatization in education need to be made with caution.

In the above discussion it has been mentioned that academician should earnestly accept the truth that any chaos in the society will be reflection on them.

Bringing the learners into the class room and sustaining their interest till the last day of intrusion is the toughest job requiring ingenuity of high order. The smallest hurt to the learner's ego or barest reflection on his ability to learn turns him away from class room. Therefore, in the 21st century teaching methods should be made high tech to create the interest of the learners.

Higher education in India is still ill-equipped, fragmented and outdated. Anachronistic affiliating system must vanish from Indian soil.

Without trained teachers even the fanciest buildings, best textbooks and facilities will not do.

Comparing the number of Universities in India with other parts of the world, which are considered as the knowledge countries, India has a very small size of Universities and colleges to meet the emerging needs of advanced research. This situation calls for serious remedial action and the creation of new era in the field of knowledge economy. Besides, the acute shortage of funds is another area posing a serious threat to knowledge economy.

No doubt globalization concept has made an impact of minds of young generations and they are heading towards global business knowledge, it would be a sufficient if teacher so teach that the young minds will learn how to discover the methods are solving problems that are known or unknown. Still the young must be taught how to think when the methods to be used is not known and the problem to be solved has not been formulated as mere education about global business and degrees in MBA, BBA or any other equivalent course will not be enough to make India to developed nation. On the other hand teachers have to develop the intellectual capacity to see beyond what seems to be the obvious. Good, bad or indifferent teaching has become a common, denominator. In the present times when the teachers' pay scales are much better, they are still identifying as discontented persons. Most of them look upon teaching as job for which the pay cheque is the only reward. We must appreciate that higher education in India is still a luxury. Unfortunately India's limited resources and manpower are being frittered away leading to gallowing devaluation of standard in education and in academic integrity.

As stated above the education policies have also been redesigned in view of technological, social and economic changes but no attempt has been made to find ways for inducing the faculty to blend teaching as life long learning. It is suggested that the curriculum has to be such that students need to be educated as specialist and generalist, not exclusively as one or the other. Non-cognitive skills such as communication skills, negotiating skills, and interpersonal skills have been included in the curriculum. These have been included but not at the cost of technical or cognitive knowledge. Some teachings of ethical and moral values be included in syllabus to inculcate the value of life as these values are eternal values of truthfulness, honesty, sincerity and integrity.

CONCLUSION

Business education at secondary and higher levels in various forms formal and informal is a measuring tool for economic changes that would lead to enrich the education system. Development experts and planners have been reshaping the educational system to meet the development needs.

The innovation in curriculum as well as techniques of teaching bringing positive changes in the quality of business education. Students coming out with the degrees are well acquainted with the up to date knowledge of computers, accounting standards, knowledge of corporate world, etc. as the very essence of business education is to make our policies and programmes cost effective and result-oriented to overcome socio-economic, political and moral problems.

Further the quality of Business Education can be improved largely by the optimum use of all resources available. Corporate housed need to take part in education reforms by investing in professional development programmes for teachers as it will give the opportunity to implement creative ideas which will intrun improve the education.

References

Jha, S.M., Marketing Services, 2004, Himalaya Publishing House, New Delhi.

The Indian Journal of Commerce, Vol. 57, July/September 2004.

Yojana Journal, Vol. 38, January 26, 1994 and Vol. 42, August 1998.

The Economic Times, February 4, 2005, p. 11, Candice Zachariahs.

The Indian Journal of Commerce, Vol. XLIX, December 1996.

Kurukshetra—A Journal of Rural Development, Vol. 53, No. 8, July 2005, pp. 18-19.

Yojana—A Development Monthly, Vol. 49, September 2005, September 2009, January 2010 and June 2010.

The Tribune, August 24, 2010, p. 12.

CHAPTER

18

Role of Management in Indian Economy

MINALI MEHRA

ABSTRACT

From years ago, Rural has been identified as a 'deprivation', rural being a landmass without access to continuous electricity, water, the stock market. Indian rural market it has such a vast size and demand base offers great opportunities to marketers. Two-thirds of countries' consumers live in rural areas and almost half of the national income is generated here. It is only natural that rural markets form an important part of the total market of India. Rural marketing is an evolving concept, and as a part of any economy has untapped potential; marketers have realized the opportunity recently. Improvement in infrastructure and reach, promise a bright future for those intending to go rural. Rural consumers are keen on branded goods nowadays, so the market size for products and services seems to have burgeoned. The rural population has shown a trend of deficient to move into a state of gradual urbanization in terms of exposure, habits, lifestyles and lastly, consumption patterns of goods and services. There are dangers on concentrating more on the rural customers. Reducing the product features in order to lower prices is a dangerous game to play. How corporate are making a foray in rural areas to promote business. Future lives in rural India. This seems to be true for India.

Keywords: *Deprivation, non-urban, rural marketing, rural management,*

INTRODUCTION

Concept of rural Marketing has come into picture recently. The concept of rural marketing is usually misunderstood by people. Rural marketing was considered normally as a marketing of agricultural products which is actually not. It's all about implementation of business activities and flow of goods from urban sectors to the rural sectors of the country for the marketing of various products. In real sense rural marketing covers two aspects like selling of agricultural produce into urban market and urban products into rural market. Our nation is classified in around 450 districts, and approximately 6,30,000 villages, which can be sorted in different parameters such as literacy levels, accessibility, income levels, penetration, distances from nearest towns, etc. 70% of India's population lives in 6,27,000 villages in rural areas.

According to the NCAER study, there are almost twice as many 'lower middle income' households in rural areas as in the urban areas. Marketers today define rural as people living a different lifestyle as opposed to that of those who have settled in the bigger cities and towns. Anything that is not urban, "Rural India comprises all places that are not urban." "Urban India constitutes places with a population of more than 5,000, a population density above 400 per square kilometer, all statutory towns, that is, all places with a municipal corporation, municipal board, cantonment board, notified area council, etc. and with 75% of the male working population engaged in non-agricultural employment. All non-urban is rural." Marketing of products or services or goods in rural market is not so easy. We have to formulate various strategies for the same. We have to face various constraints while capturing rural market.

Rural Market, such a vast market occupies an important role in Indian Economy. **This has been come into picture from last few years. Rural sector in spite of lion's share has been completely ignored until now.** Rural markets in India constitute a wide and untapped market for many products and services which are being marketed for the urban masses. But no one has given a thought over it from the last. Earlier government was responsible for serving the rural customers. Now-a-day's corporate are also considering rural market as a lucrative opportunity wants to enter there. Customers will buy only if it within their reach of the customers. Later on affordability comes. Moreover marketing basics have totally changed now. **Earlier from production era now it is customer-oriented. Company has to manufacture as per the needs and wants of the buyer. After manufacturing we have to create market of our products. Not simple marketing in fact innovative ways of marketing are find out to grab sales. Rural India buys small packs, as they are perceived as value for money.** There is brand stickiness, where a consumer buys a brand out of habit and not really by choice. Brands rarely fight for market share; they just have to be visible in the right place. Even expensive brands, such as Close-up, Marie biscuits and Clinic shampoo are doing well because of deep distribution, many brands are doing well without much advertising support—Ghadi, a big detergent brand in North India, is an example.

India: 7.808.308.909.309.407.308.606.305.80

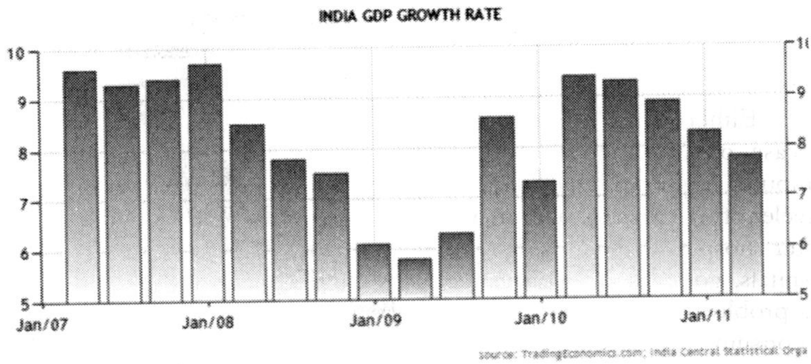

India GDP Growth Slows to 7.8%. Census 2011 covered 35 States/Union Territories, 640 districts, 5,924 sub-districts, 7,935 Towns and 6,40,867 Villages. In Census 2001, the corresponding figures were 593 Districts, 5463 sub-Districts, 5161 Towns and 6,38,588 Villages. There is an increase of 47 Districts, 461 Sub-Districts, 2774 Towns (242 Statutory and 2532 Census Towns) and 2279 Villages in Census 2011 as compared to Census 2001.

Rural market has following attributes and the following facts substantiate this:

Estimated annual size of the rural market—

Population	742 million people
FMCG	Rs. 65,000 Crore
Durables	Rs. 5,000 Crore
Agri-Inputs (including tractors)	Rs. 45,000 Crore
Two/four wheelers	Rs. 8,000 Crore

Like Thoreau and Tolstoy, Gandhi, revered as the father of modern India, believed that ***the country's future lay in her villages.*** These days, every marketer would agree. Marketing is the pivot of economic development in rural areas. It is an essential component in income and employment generation in farm and non-farm sectors. Besides, ensuring efficient coordination between production, transportation, communication and storage facilities, we have to realize the significance of credit financing and the value of market information for decision making.

For making sales in rural areas we have to face lot of hustles. These are:

* Poor working conditions.
* No proper facilities of electricity.
* No proper roads connectivity.
* No effective means of communication.
* Poor means of transport.
* Scope for infrastructure, many more things, etc.

Either it may be goods and services they have to face difficulty. In case of goods like electronic goods (mobile phone, refrigerator, computer, iron, washing machine, DVD player may be anything, four-wheeler, two wheeler, and many more also. In the name of services better hospital, hotels lodging facility, education, maternity centers, hospitals, good school colleges, employment opportunities, etc. In fact this problem is a critical issues for both the parties either manufacturer or consumer also.

PRODUCTS FOR RURAL MARKETS

From Four P's to Four A's

The traditional **four P's of marketing—product, price, place and promotion,**—have been replaced by a different framework for analysis. "A number of companies have worked on various elements of the marketing mix to improve the **four A's—affordability, awareness, availability and acceptability—for rural markets.**

To solve market issues in rural management we have to work out on the **4 A's model of Availability, Affordability, Acceptability and Awareness** which provide us idea of developing appropriate strategies to tackle the marketing issues for rural marketing. Among this model last A is comparatively important. The basics of the marketing are creating awareness of the product. Rest 3 A's follow the first one. Until now only government was serving the rural customers. With this emergence of new opportunities in the rural market companies are looking prospect to enter into the rural market. They may do their business and earn profits. This will be a win-win situation for both the parties. Moreover, entry of companies into the rural market will help them in developing rural business and give an employment to the rural people. With this the rural population will be in a position to enjoy the fruit of this new step.

Opportunities

- Infrastructure is improving rapidly. In 50 years only, 40% villages have been connected by road, in next 10 years another 30% would be connected.
- More than 90% villages are electrified, though only 44% rural homes have electric connections.
- Rural telephone density has gone up by 300% in the last 10 years; every 1000+ pop is connected by STD.
- Social indicators have improved a lot between 2000 and 2011.
- Number of "pucca" houses doubled from 22% to 41% and "kuccha" houses halved (41% to 23%).
- Percentage of BPL families declined from 46% to 27%.
- Rural literacy level improved from 36% to 59%.

Low penetration rates in rural areas, so there are many marketing opportunities:

Durables	Urban	Rural	Total (% of Rural HH)
CTV	30.4	4.8	12.1
Refrigerator	33.5	3.5	12.0

FMCGs	Urban	Rural	Total (% of Rural HH)
Shampoo	66.3	35.2	44.2
Toothpaste	82.2	44.9	55.6

Marketers can make effective use of the large available infrastructure:

Post Offices	1,38,000
Haats (periodic markets)	42,000
Melas (exhibitions)	25,000
Mandis (agri markets)	7,000
Public Distribution Shops	3,80,000
Bank Branches	32,000

Proliferation of large format Rural Retail Stores, which have been successful also:

* DSCL Haryali Stores
* M & M Shubh Labh Stores

* TATA/Rallis Kisan Kendras
* Escorts Rural Stores
* Warnabazaar, Maharashtra (Annual Sale Rs. 40 crore)
    - o Scope for health insurance, crop insurance micro-Insurance and other insurance products.
    - o Scope for quality standards of the seeds.
    - o Scope for latest equipment for farming.
    - o Scope for sale of credit card, visa card and master card, etc.
    - o Scope for fitness zones, beauty saloons.
    - o Scope for proper medical facilities, proper equipments.
    - o Scope for modern schools, colleges.

BENEFITS FROM COMPANY'S PERSPECTIVE

Rural Jobs Can Provide Momentum to the Wheel of the Economy. Companies see Bright Prospects in Rural India. They look for better prospectus from jobs and business perspective. Large customer database is there in rural India. Moreover, companies will get tax rebate. Development of that particular backward area takes place. The nearby population gets employment. They can earn their livelihood by working in these places. The rural economy has provided a cushion. Overall sentiment in the country was different from other parts of the world where each household had at least one person with a pink slip." Reliance Industries plans to pump in $5.5 billion over the next few years to create a farm-to-storefront infrastructure for a pan-India retail network. Reliance Info com, a mobile services provider plans to double the rural coverage to 400,000 villages, making up 50% of the rural population.

- The rural initiative of the Mumbai-based $1.3 billion **House of Godrej—Godrej Aadhaar** plans to set-up 1000 stores across India in the next five years.
- Delhi-based telecom major **Bharti Airtel** chairman Sunil Mittal has tied up with **Wal-Mart**, which will need its supply chain.
- From the **Goenkas to the Gulabchands**, from the **Tatas to the Thapars**, every major Indian business group has plans to move into the hinterland.

CREATING BRANDS FOR RURAL INDIA

Rural markets are really powerful. Certain adaptations are required to cater to the rural masses; they have unique expectation and warrant changes in all four parameters of product, price, promotion and distribution. Only the need is to emphasize on adapting the product and price in terms of packaging, in sachets, priced to suit the economic status of the rural India in sizes like rupees 5 packs and Re. 1 packs that are perceived to be of value for money. This is a typical penetration strategy that promises to convert the first time customers to repeated customers. The promotion strategies and distribution strategies are of paramount importance. Ad makers have learnt to leverage the benefits of improved infrastructure and media reach. The television airs advertisements to lure rural masses, and they are sure it reaches the target audience, because majority of rural India possesses and is glued to TV sets! Dynamics of rural markets differ from other market types, and similarly, rural marketing strategies are also significantly different from the marketing strategies aimed at an urban or industrial consumer. This, along with several other related issues, have been subject matter of intense discussions and debate in countries like India and China and focus of even international symposia organized in these countries.

Rural markets and rural marketing involve a number of strategies, which include:

* Client and location specific promotion
* Joint or cooperative promotion
* Bundling of inputs
* Partnership for sustainability

Client and Location specific promotion involves a strategy designed to be suitable to the location and the client. Joint or co-operative promotion strategy involves participation between the marketing agencies and the client. 'Bundling of inputs' denote a marketing strategy, in which several related items are sold to the target client, including arrangements of credit, after-sale service, and so on. Media, both traditional as well as the modern media, is used as a marketing strategy to attract rural customers.

Partnership for sustainability involves laying and building a foundation for continuous and long lasting relationship. Innovative media can be used to reach the rural customers. Radio and television are the conventional media that are reaching the rural audience effectively. But horse cart, bullock cart and wall writing are the other media, which can carry the message effectively to the rural customers.

ROLE OF MANAGEMENT IN RURAL INDIA

NABARD has a crucial role to play in the agricultural financing sector. Another important aspect in the rural marketing is market information. Information regarding production, market arrivals, day-to-day prices, changes in the stock with prices is essential for the farmers in order to equip them to withstand the prevailing situation.

Government schemes like **(NREGS) National Rural Employment Guarantee Scheme,** which guarantees 100 days of employment to one member of every rural household. This has reduced rural underemployment and raised wages. Several factors have led to an increase in rural purchasing power.

The increase in procurement prices the government set **the minimum support price—MSP—**for many farm products that has contributed to a rise in rural demand. A series of good harvests on the back of several good monsoons boosted rural employment in agricultural and allied activities. The increase in rural purchasing power is reflected in rural growth across a number of categories. For example, in the financial year 2009 April-March sale of FMCG (fast moving consumer goods) rural volume growth is estimated to be 5% to 12% higher than urban growth across a number of categories.

Buying power in the hands of the rural rich is higher than the buying power of the urban rich. The purchasing power of rural India is more than half for fast moving consumer goods. **Maximum number of durables and automobile sales is from rural economy.** The rural market is witnessing a 15% growth rate. Some 60% of the durables market lies in rural India. Telecom in rural India is growing at 31%. Across product categories, however, there seems to be a lot of action. Media—particularly TV—has been a great leveler. Even in small villages, people who have seen the urban lifestyle on television seem to want similar goods and services. Companies have realized this and are going all out to tap this latent demand.

INTRODUCTION OF TINY ACCOUNT

The **State Bank of India (SBI)** has started a zero-balance bank account program for villagers. Called the SBI Tiny account, there are no physical branches or officials, just a paid volunteer who is equipped with a small box and a cell phone. The box enables biometric fingerprints, at the time of opening the account to confirm the account holder's identity. The advantage for the villagers is that they can withdraw money from their accounts at any time of the day or night. SBI hopes to cover 100,000 villages by 2012. The bank has tied up with India Post for some services.

Maruti Suzuki, India's leading automobile manufacturer, today sells 5% of its vehicles in the rural markets. The company expects this number to rise to 15% in the next two years. "This is not just our wish, but reflects market demand,"

In telecom, service providers are making a beeline for the villages. That's where the growth in what is now the world's fastest growing mobile market lies. According to industry estimates, 70% of all new subscribers will come from rural areas.

Success story of Nokia had earlier launched **a basic handset with a torch (large parts of rural India don't have electricity) and an alarm clock.** In December 2008, it went one step further with the launch **of Nokia Life Tools. "Nokia Life Tools is a range of agriculture, education and entertainment services designed especially for the consumers in small towns and rural areas of emerging markets**. Mobile device manufacturers are also tailoring their products to this market.

The **Tata** have launched a mass-market jewellry brand—**Gold Plus.** The Tata train unemployed youth and send them to the villages as brand ambassadors. The problem with gold in India is that it is often adulterated. In rural areas, gold jewelry is not for ornamentation; it is a safety net for emergency situations.

There is substantial scope to create products that are oriented towards rural sectors. **Chik shampoo** created the jasmine variant in tune with the culture of women using jasmine flowers to style their hair in a few parts of the country. The sachet is as much a packaging (product) strategy as a price strategy. "**The shampoo in sachets created a new product segment.**

Distribution and promotional channels also need to be different for rural markets. Companies are getting their act together here, too.

Hindustan Unilever has project **Shakti,** under which women's self-help groups act as the last link of the retail chain. On one side are the fast-moving consumer goods (FMCG) and the consumer durables companies. On the other are consumers in rural India, potentially the largest segment of the market. Finally, the two are coming together.

The fact that this has not happened in the past is not for want of trying. In Mumbai and New Delhi corner offices, executives have long recognized that to build real sales volumes they will have to reach outside the big cities. In several categories, rural India already accounts for the lion's share.

According to MART, a New Delhi-based research organization that offers rural solutions to the corporate world, rural India buys 46% of all soft drinks sold, 49% of motorcycles and 59% of cigarettes. This trend is not limited just too utilitarian products: 11% of rural women use lipstick.

Other numbers are equally revealing. ***According to the National Council of Applied Economic Research (NCAER), an independent, non-profit research institution, rural households form 71.7% of the total households in the country. Spending in this segment is growing rapidly and consumption patterns are closing in on those of urban India.***

Soft drinks giant **Coca-Cola** is growing at 37% in rural markets, compared with 24% in urban areas. In media television mostly based on rural. Rural mass has been the recipe of Indian television.

Private sector companies like ITC have set-up the **IT-enabled e-Choupal network.** This has eliminated the role of middlemen. **ITC e-Choupal** is a hub location services a cluster of e Choupals. This is also an effective approach for the welfare of Indian Farmers. The farmers can directly sell their produce directly rather than selling to anyone else. They get better prices for their products. Moreover in case of borrowing money they will get money at reasonable rates. Marketers can directly in touch with the producers they will get bulk at cheaper rates. They can easily get farm produce at better prices. It works as a centralized hub for buying and selling crops. Moreover they introduce innovative ways for agriculture. By purchasing directly from the farmer, ITC significantly improved the efficiency of the channel and created value for both the farmer and itself. While the e-Choupal network was initiated to facilitate more efficient and effective procurement, the connectivity—both physical and informational—between the farmer and the market that it facilitated has allowed ITC to use it for distribution of goods and services from the market to the farmer. It has thus, evolved into a business platform.

The ***e-Choupal infrastructure consists of:***

* A kiosk with Internet access in the house of a trained farmer, called a *Sanchalak*. This kiosk is within walking distance of target farmers.
* A warehousing hub managed by the former middleman, called a *Samyojak*. This is within a tractor-driveable distance of target farmers. (The former middlemen were given a role to avoid resistance to the project. They joined because they could see that their traditional business was in jeopardy.)
* A collaborative network of companies orchestrated by ITC with a pan-India presence.

This is, of course, a simplified structure. And there has been a stream of new initiatives. For instance, in August 2004, ITC introduced the **Choupal Sagar,** a rural retail outlet at the hub. The first was set-up at Sehore in Madhya Pradesh. "This 7,000 sq. ft. mall sells consumer goods as well as agri-products. In a recent move, ITC has set-up its first urban outlet, the other end of the **e-Choupal chain,** to retail fresh fruit and vegetables. The e-Choupal project is already benefiting more than 3.5 million farmers,

Several **European multinational firms**—and a few U.S. firms—have been making inroads into rural India for years. Companies such as

Unilever, Phillips and Nestle have long been known to India's rustic *dukaandaars*, or merchants. Among U.S. firms, companies such as Colgate and Gillette have made considerable headway.

The **Vani project**, however, is operational in more than 20,000 villages in states like Madhya Pradesh, Karnataka, Chhattisgarh and Andhra Pradesh. **Hindustan Lever has also tied up with partners such as Tata Consultancy Services, India's largest software firm, which is actively involved with the Shakti portal, and ICICI, a financial services institution that is involved with providing micro-credit loans.**

To sort the problems of rural India like poverty, early marriage, low level of education, ill-health, etc.

- **Godrej** introduced three brands of Cinthol, Fair Glow and Godrej (soap) in 50-gram packs.
- **Adidas and Reebok** increased their sales by 50% in rural markets by reducing prices. Size and design changes.
- **Videocon** introduced a washing machine without a drier.
- **Philips** launched a low-cost smokeless chulha (stove).
- **DCM Shriram** developed a low-cost water purifier especially for rural areas.
- **LG Electronics** developed a customized TV (cheap and capable of picking up low-intensity signals) for the rural markets.
- **Coca-Cola** provided low-cost iceboxes as regular power outages meant families could not depend on refrigerators.

REASONS FOR IMPROVEMENT OF BUSINESS IN RURAL AREA

- Socio-economic changes (lifestyle, habits and tastes, economic status),
- Literacy level (25% before independence—more than 65% in 2001),
- Infrastructure facilities (roads, electricity, media),
- Increase in income,
- Increase in expectations, and
- The penetration is still low. Coke, for instance, reaches barely 25% of the rural market. This means the upside potential is huge for companies that develop effective rural marketing strategies. According to NCAER, the low penetration rates can be attributed to three major factors: low income levels, inadequate infrastructure facilities and different lifestyles. But income levels are going up,

infrastructure is improving and lifestyles are changing. Almost a third of the rural population now uses shampoo. Now old rural has been revitalized today.

CONCLUSION

Distributing small and medium sized packets through poor roads, over long distances, into deep pockets of rural India and getting the stockiest to trust the mobility is a Herculean task. Giving the confidence those advertisements will support. Sales force is being trained to win the confidence of opinion leaders. Opinion leaders play an important role in popularizing the brand. They sometimes play the role of entry barriers for new products.

The method of promotion needs to be tailored to suit the expectations of the market. Techniques that have proved to be successful are Van campaigns, entertainment films, generating word of mouth publicity through opinion leaders, colourful wall paintings. The wide reach of television has exposed the other wise conservative audience to westernization. Panchayat televisions in Tamilnadu carries message that are well received and contribute to community development. There is huge mass is waiting for us in rural areas for serving them. The need of the hour is how to reach, how to win their confidence in order to serve them.

REFERENCES

http://business.mapsofindia.com/india-gdp/2010-2011.html

http://knowledge.wharton.upenn.edu/india/article.cfm?articleid=4386

http://www.world-agriculture.com/agricultural_marketing/rural_marketing.php

http://www.indiaprwire.com/

Rural Marketing, Ravindranath, V. Badi and Naranyansa, V. Badi, Himalaya Publishing, 2004.

The Hindu-Business Line.

www.deccanherald.com

www.indiantelevision.com

http://knowledge.wharton.upenn.edu/india/article.cfm?articleid=4172

McKinsey Report estimates that by 2020.

CHAPTER

19

Genesis of Coir Industry in India: Export Practices

K. AMUTHA

INTRODUCTION

India is the largest coir producer in the world accounting for more than 90 per cent of the total world production of coir fibre. Coir is popularly known as the 'golden fibre.' It is a natural fibre extracted from fibrous husk of the coconut shell. The coir sector in India is diverse and involves households, cooperatives, NGOs, manufactures and exporter. The coir industry employs more than 6,50,000 persons, of whom majority are from rural areas belonging to the economically weaker sections of the society. Nearly 80 per cent of the coir workers in the fibre extraction and spinning sectors are women. Though India accounts for more than two-third of the world population of coir and coir products, its exports are around 700 million. Range of products husk is used as fuel, as surface and seeding pit, and as a fertilizer in coconut gardens due to the high potash content. The most important commercial utilisation of husk is in the manufacturing of coir. There are eighteen varieties of coconut palms, but husk from only three varieties is suitable for coir production. Coir is used to make a wide range of products such as ropes, brushes, brooms, mats, mattresses, baskets, etc.

Today the Coir Geo-Textiles or coir Bhoovastra has shot into the lime light which are mainly due to its eco-friendly characteristics. They are being usually effectively used for improving soil behaviours, preventing soil erosion, and in helping consolidation of soil. Coir is a 100% organic and bio-degradable fibre which has great water absorbency and has a definite edge over synthetic goeo textiles, in the environmental aspect and issues. The greatest advantage of Coir is that it provides an ecological niche for rapid establishment of vegetation.

MEANING

Coir is a product of the *coconut* tree, *Cocos nucifera*, and it is sometimes known as coco fiber. The substance is extracted from the hairy husk of coconuts, and used to create a variety of products such as mats, carpets, upholstery stuffing, and brushes. Coir, pronounced KOY-er, is a very coarse, stiff fiber, and it is also extremely resistant to rot and salt water, making it an ideal material for situations in which other fibers would decay.

FEATURES OF COIR PRODUCTS

- Provides excellent insulation against temperature and sound.
- Not easily combustible.
- Flame-retardant.
- Unaffected by moisture and dampness.
- Tough and durable.
- Resilient; springs back to shape even after constant use.
- Totally static free.
- Easy to clean.

USES OF COIR

- Besides its main use as floor covering and in rope making, coir fibre finds extensive use as packaging material to protect goods against shock in transport.
- Coir fibre finds its use in the production of activated carbon, artificial horse hair, paper pulp, roofing tiles, writing boards, thermal insulations, high stretch paper, manufacture of olive oil filters, etc.
- In Germany, coir is rubberized for making cushion seating for automobiles and railways. The rubberization is brought about by Splashing layers of coir fibre and rubber with the help of a specialized machine.

- Coir yarn has been found to be ideal lead for hop wines which is used in brewery in U.S.A Coir bags are used in tea estates for collecting tea leaves and for transportation: and also for lifting coal from mines.
- Coir yarn is used for making fenders which are attached to ships and boats for preventing collision and shock.
- Coir mats are used for commercial packaging purposes and circular brush mats are used for packing.
- Coir mattings after bituminisation offer possibilities of being used as floor covering in godowns to withstand moisture adsorption by stored goods.
- Rubber backed coir mats are sol-proof, sound absorbent and do not scratch polished floor. Heavy matting made out of thick coir rope is being used for transporting gas cylinder. It is used as a strainer in tube wells in place of wire mesh.

FUNCTIONS OF COIR BOARD

The Coir Board is a statutory body established under the Coir Industry Act, 45 of 1953 for promoting the overall development of the coir industry and improving the living conditions of the workers engaged in the traditional industry. Some of the functions of the Coir Board for the development of coir industry include:

- Undertaking scientific, technological and economic research and development activities.
- Collection of statistics relating to exports and internal consumption of coir and coir products.
- Development of new products and designs.
- Publicity for promotion of exports and internal sales.
- Marketing of coir products in India and abroad.
- Preventing unfair competition among producers and exporters.
- Assisting in the establishment of units of the manufacturers of products.
- Promoting cooperative organizations among producers of husk, coir, fibre, yarn and manufactures of coir products, ensuring remunerative returns to producers and manufactures, etc.
- Ensuring remunerative returns to producers and manufactures.

- Promoting exports of coir yarn and coir products.
- Advising on all matters relating to the development of the coir Industry

PRODUCTION

The global annual production of coir fiber is about 350,000 metric tonnes. The world's top two producers are India and Sri Lanka together account for about 90 per cent of the world production. In India the coir industry started and flourished in Kerala, which has a long coast line, lakes, lagoons and backwaters providing natural conditions required for carry out retting process. The industry has also picked up in Tamil Nadu, Karnataka, Tripura, Pondicherry and the union territories of Lakshadweep and Andaman & Nicobar Islands, through the efforts of Coir Board. Coir is the rubbery husk of the coconut shell. Being tough and naturally defiant to seawater, coir guards the fruit sufficiently to endure months of floating on ocean currents to be washed up on a sandy shore, where it may bud and mature into a tree, condition it has adequate fresh water, since all the extra nutrients it requires have been passed along the seed. This uniqueness makes the thread fairly functional in

TABLE 1

Production of Coir and Coir Products during 2000-01 to 2009-10

(Metric tonnes)

Year	*Coir Fibre*	*Coir Yarn*	*Coir Products*	*Coir Rope*	*Curled Coir*	*Rubberised Coir*
2000-01	3,64,000	2,33,400	71,500	51,300	31,000	51,000
2001-02	3,69,400	2,36,900	72,575	52,700	31,800	51,800
2002-03	3,53,700	2,26,800	75,750	50,000	28,000	50,250
2003-04	3,64,000	2,32,500	77,900	50,000	29,500	51,000
2004-05	3,85,000	2,45,500	98,000	50,000	36,500	60,000
2005-06	4,10,000	2,70,000	1,20,000	50,000	38,000	60,200
2006-07	4,30,000	2,88,000	1,70,000	52,000	48,000	68,000
2007-08	4,37,800	2,90,000	1,72,000	52,470	49,540	70,840
2008-09	4,91,000	2,92,900	1,73,550	52,470	49,540	70,840
2009-10	3,91,450	2,21,900	1,30,300	39,400	37,200	53,200

Source: Annual Report of the Ministry of MSME, 2009-10.

floor and outdoor mats, aquarium filters, cordage and rope and garden mulch. Coconut husk is the basic raw material for coir products and only 50 per cent of the husk is used as fuel in rural areas. Hence, there is a lot of scope for the growth of coir industry. This renewable resource is under-utilised as the local coir mills process only fraction of the available husk, which accrues year round as a waste during coconut processing.

The highest production of coir fibre has 4, 91,000 MTs in the year 2008-09. It has decreased 3,91,450 MTs in the year 2009-10. The coir yarn has witnessed an increase up to 2,92,900 MTs during the year 2008-09. The coir products was to be 1,73,550 MTs in the year 1, 73,550. The production of coir rope and curled coir was 52,470 MTs and 49,540 MTs during the year 2007-08 and 2008-09. The production of rubberized coir has witnessed an increase to 70,840 MTs. However, the production of coir and coir products has decreased during the year 2009-10. Because the reasons were behind that, non-availability, shortage of raw materials and the highest price of raw materials.

CONSUMPTION OF COIR PRODUCTS

The domestic consumption of coir fibre has 4, 72,000 MTs in the year 2008-09. There was a rapid domestic consumption of 5, 07,000MTs

TABLE 2

Details of Consumption of Coir and Coir Products from 2000-01 to 2009-10

(Metric tonnes)

Year	*Coir Fibre*	*Coir Yarn*	*Coir Products*	*Coir Rope*	*Curled Coir*	*Rubberised Coir*
2000-01	37000	110000	24000	51000	1900	48500
2001-02	39800	120000	28000	52700	2300	51600
2002-03	89200	125300	26500	49700	27500	49700
2003-04	38400	196959	19465	49692	29424	50538
2004-05	365750	222500	22500	49500	35700	54000
2005-06	405000	235000	38000	49500	37000	60000
2006-07	418500	248000	45000	49000	40000	67000
2007-08	421500	250000	46000	51000	49000	68500
2008-09	472000	252500	46410	51460	41450	69320
2009-10	507000	257500	47010	51910	41900	70020

Source: Annual Report of the Ministry of MSME, 2009-10.

in the year 2008-09. Domestic consumption of coir yarn has slowly increased every year from 2000-01—2009-10. The consumption of coir products witnessed an increase to 47010 MTs during the year2009-10. The consumption of curled coir has 2300 MTs in the year 2001-02, suddenly it has a witnessed an increase to 27,500 MTs during the year 2002-03. The following years it has slowly increased. The consumption of rubberised coir has slowly increased every year. However, suddenly it has increased 60,000 MTs to 67000 MTs in the year 2006-07. Ever though, the domestic consumption of all the coir products has slowly increased every year.

EXPORT

The Coir Industry in India has a strong position in the export orientation since the early days, when trade was dominated by European companies. A wide range of coir and coir products are exported from the country. These include coir yarn, coir pith, coir fibre, rubberised

TABLE 3

Export of Coir and Coir Products from India during 2000-01 to 2009-10

(Metric tonnes)

Year	*Matric Tonnes*	*Rs. in Million*
2000-01	67,493	3136.6
2001-02	71,335	3136.6
2002-03	84,183	3527.0
2003-04	102,253	4074.9
2004-05	122,927	4734
2005-06	136,027	5084.5
2006-07	168,755	6051.7
2007-08	187,567	5928.8
2008-09	199,925	6399.7
2009-10	170,000	5431.5

Source: Annual Report of the Ministry of MSME, 2009-10.

coir, curled coir, coir rope, coir geo-textiles, coir rugs, handloom mats and coir, handloom matting, power loom mats, etc. Coir products are exported to more than 100 countries all over the world. At the beginning of the independence of the local trade in the hands came a turning point in the history of the coconut industry in India. With the level of fiber and yarn exports, India is an exporter of goods with added value. This in turn leads to a significant change in total volume and value of exports. During the year 1950 the average export volume dropped to 73,200 tonnes per year and then in the year 1960 the export volume decreased to 62,300 tonnes per year. It was back to 45,700 tonnes per year in the 1970's and 26,700 tonnes in 1980. From an average annual export realization of Rs.80 million in 1950 of Rs. 120 million in 1960, Rs. 210-Rs. 300 million in 1970 and 1980. Until the mid-90's it was Rs. 1,710 million in sales to Rs. 3527 million in 2003.

Coir exports from the country have already exceeded the target in 2005-06 and have also registered an increase of 11% in quantity and 7% in the year 2004-05. Exports had touched an all time high level of 1,36,027 tonnes in volume and Rs. 5084.47 million. In the year 2004-05, exports were of the order of 1, 22,927 tonnes at Rs 4734 million. In the year, 2008-09, the export of coir and coir products was 1,99,295 MTs valued at Rs. 63997.7 million. In the year, 2009-10, the export of coir and coir products was 2,94,508 MTs valued at Rs. 80405.22 million. It is the all time high export performance from the country. At present, coir and its products are exported to more than 100 countries and the US is the single largest market with a share of more than 40% in the total export. The global annual production of coir fiber is about 3,50,000 metric tonnes. The world's top two producers are India and Sri Lanka together account for about 90 per cent of the world production.

DEVELOPMENT OF COIR INDUSTRY

The thrust areas for the development of coir sector in the country are as follows:

1. Modernation of production infrastructu re by means of appropriate technology without displacement of labour.
2. Promotion of coir exports and new products through undertaking of market promotion measures abroad.
3. Expansion of domestic market through publicity and propaganda.
4. Development of skilled manpower through training.

5. Promotion of research and development activities like process improvement, product development and diversification, elimination of drudgery and pollution abatement.
6. Extension of research and development findings through field demonstrations.
7. Development of coir industry in all coir producing states in association with the state governments.

CHAPTER

20

Technology and Change: Need for Developing New Pattern of Human Relations

DR. ALKA GUPTA AND DR. AMRIK SINGH SUDAN

ABSTRACT

Technology has been the pace setter for structural changes in socio-economic system which led to forging a new pattern of relationship between European countries and the continents of Asia, Africa and Latin America since the advent of Industrial Revolution. It gave Europe the status of a great economic force and an impregnable citadel of political power and authority which made European countries economic and political masters of more than two-third human population in the world. The emergence of Multi National Corporations on the world scene signifies the beginning of a new international order with multiple ramifications. Their enormous command over economic resources and their enviable possession of managerial skill and Technology have unlimited potential for accelerating the pace of economic development in developing countries on the one hand and to build bridges of understanding between the developing and developed countries on the other. Their role in technology transfer in catalytic and their influence on employment generation is encouraging.

PRESENT SCENARIO

Under the present economic global order, every country-whether developed or under developed, member or non-member of a particular Regional Economic Bloc is expected to interact with each other under the provisions which almost all the countries are its members. Under this new World order, the member counties are expected to dispense with the regime of controls, quotas and restriction in international trade relations. Member-counties have agreed to work under an accepted producer and code of conduct which may be the first step towards globalization of industry and trade. This change requires an effective system of communication through which most procedures experienced by member- countries could be resolved to the advantage of all concerned.

In addition, management shall have to make efforts by motivating members of the organization to dispense with all irritants coming in the way of cultural fusion. Some positive and negative reinforcements ought to be built in the system apart and from encouraging informal organization system in which superior and subordinates and co-workers with different cultural background could come closer to each other and understand each other's cultural niceties to pave the way for harmonious relationship at work place. Further, the dichotomy in cultures could be converted into uniformity in values, beliefs and attitudes towards collegues and organizations by establishing uniformity in their thinking and organizations by establishing uniformity in their thinking and action. Organizations may take such steps as wearing of only one uniform by every worker—big or small—during working hours, establishment of a common eating place where both the superior and subordinate or co-workers belonging to different cultural groups not only eat together but also eat of at each other's places. World is again witnessing structural changes since 1990's.

In other words, globalization is instrumental; in establishing cultural fusion by bringing together people with different cultural backgrounds. It seeks to bring about changes in individual and group behaviour, their perceptions, beliefs, value system which may be helpful in establishing the world free from political tension and turmoil.

New Paradigms for Technology Development in the Knowledge Millennium

Twenty-first century is facing new challenges for the very existence of human race. The problem of global warming though created by human beings all over the world is standing with its gigantic open mouth to swallow every human being on this planet earth. Likewise, the impending water shortage owing to imbalance in ecology and environment is posing a serious threat to all living creatures throughout the world. In addition, the problem of human health is becoming more and more serious inspite of meaningful researches to tighten the noose on dreadful diseases like cancer, AIDS, Swine Flu apart from equally disastrous diseases like blood pressure and diabetes which discard the barriers of rich and poor. These human problems can be tamed by the stick of technology in the form research and development and other forms of innovations and inventions. In addition to solving the above stated problems, the other challenges before the technology mission may be control over population growth, chaining of recession and the overall reduction of the pangs of unemployment which is adopting universal posture in the wake of recession surfacing in practically every nook and corner of the world.

These are the new challenges which have to be overcome by the advocates of technology mission so that the entire human race may be protected against the endangered human extinction.

INTRODUCTION

Technology has been the pace setter for structural changes in socio-economic system which led to forging a new pattern of relationship between European countries and the continents of Asia, Africa and Latin America since the advent of Industrial Revolution. It gave Europe the status of a great economic force and an impregnable citadel of political power and authority which made European countries economic and political masters of more than two-third human population in the world.

Likewise, United States, Germany, Japan and the former Soviet Union turned out to be major power centers on global scene owing primarily to technological innovations and advancement which led to the formation of new wet of political relationships in the form of NATO and Warsaw Pact countries. United States and the former Soviet Union acquired the status of super powers and the lesser powerful became followers of their blocs.

Technological traveled to Asia via United States in the middle of 19th century and other continents thereafter gave confidence and power to the people of third world which together with other socio-political and economic forces dismantled the political supremacy of European powers.

The emergence of Multi National Corporations on the world scene signifies the beginning of a new international order with multiple ramifications. Their enormous command over economic resources and their enviable possession of managerial skill and technology have unlimited potential for accelerating the pace of economic development in developing countries on the one hand and to build bridges of understanding between the developing and developed countries on the other. Their role in technology transfer is catalytic and their influence on employment generation is encouraging.

Any institution endowed with such oligopolistic power and influence is apt to surpass the frontiers of its legitimate programme of establishing a new international relationship on reciprocal basis.

The experience of 1960's and 1970's ushered in the beginning of a new relationship between the MNC's and the host of developing countries in 1980's. This insulated further in 1990's by the elements of reciprocity and understanding of each others "needs and capacities". Developing countries have the resources and markets, they have also developed over the period, the bargaining capacity. Consequently, the developing countries have constant demand for the inflow of capital, will and technology, whereas the developed countries have the availability of all these three components of development.

Present Scenario

World is again witnessing structural changes in 1990's. Disintegration of the former Soviet Union and the other East European countries have proved irrelevance of Marxist-Leninist model. The Soviet or Russian leadership has reconciled to secondary role in the world and the former Soviet Union has disintegrated into small economically or perhaps politically unviable independent countries. All these newly independent countries are passing through political instability and grave economic crisis which may be overcome through communication technology.

Transformation of communist political system has put an end to the old war hysteria in two blocs; it has however established a new system of political and economic relationships between the NATO and erstwhile Warsaw pact countries of greater inter-dependence and competitiveness. Apart from technological developments in manufacturing, communication, information, transport and other infrastructural areas and the discoveries in biotechnology, new material and composites, other developments forging new relationships are: (i) pluralism in international investment; (ii) internationalization of financial markets; (iii) gaining importance of invisible items on the conspectus of international trade; and (iv) vast untapped Soviet and East European markets.

In addition, a dichotomous situation has arisen with the popularization of the philosophy of globalization and liberalization and the formation of Regional Economic blocs—the most powerful being the European community and ASEAN. Similar blocs, though not so powerful have emerged in the continents of Africa and Latin America, apart from SAARC in India Ocean countries.

Under the present economic global order, every country-whether developed or under developed, member or non-member of a particular Regional Economic Bloc is expected to interact with each other under the provisions which almost all the countries are its members. Under this new World order, the member countries are expected to dispense with the regime of controls, Quota and restrictions in international trade relations. Member countries have agreed to work under an accepted procedure and code of conduct which may be the first step towards globalization of industry and trade. This change requires an effective system of communication through which most procedures experienced by member-countries could be resolved to the advantage of all concerned.

Technology and the New World Order

Emergence of new relationships in the areas of defence, production of goods and services, international trade and R&D may be managed with the help of technological innovations including communication behavious in different areas such as micro-electronics, bio-engineering and discoveries of new materials and composites.

Globalization of defence is possible through information and communication technology apart from internet system. As most countries in their respective regions in the continents of Europe, Asia, Africa and Latin America are involved in like situation, it is taking heavy toll of their scarce resources which could otherwise be diverted for socio-economic development.

Though globalization in one go is not possible, the first step in this direction may be the system of Regional Defence Regime (RDR) with limited material and financial support from some developed country on a limited material and financial support from some developed country on a limited scale. This support may be needed in the beginning till the countries in the region are able to forge the relationship of inter-dependence and Cooperative Defense System (CDS) through an effective communication system.

Experiment with new political relationship in European community owing to bitter experience of second world war, though economic corridors of one currency, one system of political administration and common market could fructify the dream of effective European Union with such technology of production and distribution that is capable of fulfilling the needs and expectations of member-countries on the one hand and their counterpart in other regions with whom they have to interact with on the other. Second economic relationships will tread the way for an effective political system of an improved form of political confederation. Such a new form of politico-economic relationship shall have to be premised on the edifice of production and distribution technologies and communication and information technologies.

Other regional blocs like ASEAN and SAARC have been formed with the limited objective of Regional Common Market (RCM). RCM will not obliterate unhealthy competitions among the countries of the region but will also ensure a better deal to most of the poor and backward countries against the developed and powerful countries. The entire region may have to be developed on "Systems Model" in production in which the means of production and distribution shall have to be located in the light of locational and other operational benefits engaged by a country in the region. Success of RCM will depend on the proper use of technology in the area of production, distribution, information and communication.

The concept of globalization "Just Economic World Order" may be implemented by the newly emerging force. The Transnational Corporations (TNCs) by: (i) undertaking innovative research and development in production technology; (ii) ensuring optimal efficiency in production and distribution at global level by pooling resources of hard and software on north and south and their exploitations and distribution of end products with the help of appropriate technology; and (iii) establishing linkages between the prosperity of the developed and poverty of the developing countries through information and communication. This new relationship between the home and host countries of the TNCs may prove to be the right step towards establishing the "Just Economic World Order".

New Pattern of Human Relations

In this backdrop of structural changes, the behaviour of human beings at work has to be chiseled to new expectations of individual and group behaviour within the relationship framework of organizational structure. In other words, a new pattern of organizational culture in terms of values, beliefs and attitudes towards work, colleagues, superiors, subordinates and the organization has to be developed.

Spread of industrial activity without national limits will bring in individuals and groups with numerous cultural diversities reflecting different behaviour styles. Some are highly independent and assertive and others are dependent and respectful; some belong to the category of "I am OK you are not OK", and an equally large number to "I am not OK you are OK". These diversities in individual behaviour unless they are put together in a running thread of commonality and organizational goals may not be achievable. Likewise, the changes in group behaviour and organizational structure in terms of authority relationships, span of control and flow of authority and responsibility are needed to establish uniformity in group's objective in changed scenario.

These changes could be possible through proper communication regarding the needs to ensure effective interaction with each other to understand each other's sentiments and expectations within the framework of organizations goal and objectives. To obtain better interaction some short-term training programmes may e organized so that each of them is equipped to appreciate the niceties and sensitivities and the attitudinal characteristics of different cultural groups comprising the unit. The irritants coming in the way of better understanding of each other view ought to be removed to ensure harmony and better relationship. With constant communication both formally and informally the group with multiple cultures develops ultimately a uniform code of conduct, value system, beliefs and faiths which will not only help in realizing organizational goals of production and marketing but more importantly the global peace and harmony to ensure better world.

In addition, management shall have to make efforts by motivating members of the organization to dispense with all irritants coming in the way of cultural fusion. Some positive and negative reinforcements ought to be built in the system apart from encouraging informal organization system in which superior and subordinates and co-workers with different cultural background could come closer to each other and understand each other's cultural niceties to pave the way for harmonious relationship at work place. Further, the dichotomy in cultures could be converted into uniformity in their thinking and action. Organisations may take such steps as wearing of only one uniform by every worker-big or small-during working hours, establishment of a common eating places where both the superior and subordinate or co-workers belongings to different cultural groups not only eat together but also eat food at each other's places.

World is again witnessing structural changes since 1990's.

In other words, globalization in instrumental in establishing cultural fusion by bringing together people with different cultural backgrounds. It seeks to bring about change in individual and group behaviour, their perceptions, beliefs, value system which may be helpful in establishing the world free from political tension and turmoil.

New Paradigms for Technology Development in the Knowledge Millennium

Twenty-first century is facing new challenges/problems for the very existence of human race. The problem of global warming though created by human beings all over the world is standing with its gigantic open mouth to swallow every human being on this planet earth.

Likewise, the impending water shortage owing to imbalance in ecology and environment is posing a serious threat to all living creatures throughout the world. In addition, the problem of human health is becoming more and more serious inspite of meaningful researches to tighten the noose on dreadful diseases like cancer, AIDS, Swine Flu apart form equally disastrous diseases like blood pressure and diabetes which discard the barriers of rich and poor. These human problems can be tamed by the stick of technology in the form research and development and other forms of innovations and inventions.

In addition to solving the above stated problems, the other challenges before the technology mission may be control over population growth, chaining of recession and the overall re-education of the pangs of unemployment which is adopting universal posture in the wake of recession surfacing in practically every nook and corner of the world.

These are the new challenges which have to be overcome by the advocates of technology mission so that the entire human race may be protected against the endangered human extinction.

References

Beery John, W., Human Ecology and Cognitive Style, Beverly Hills Cal, Sage/Halsted.

Bhattacharya, S.K. and N. Venkataramin, Managing Business Enterprises, Strategies, Structures and Systems, Vikas Publishing House, New Delhi.

Gupta, N.S., Multi National Corporations, Himalaya Publishing House, New Delhi.

Katz, Deniel and Kahn, Robert, L., The Psychology of Organisation, 2nd Edition, Wiley, New York.

Michael, E. Porter, The Competitive Advantage of Nations, Macmillan, New Delhi.

Weber, M., Theory of Social and Economic Organisation, N.Y., Free Press.

CHAPTER

21

Customer-Relationship Management: A Case Study of Commercial Banks in Jammu and Kashmir

DR. AMRIK SINGH SUDAN

INTRODUCTION

The biggest management challenge in the new millennium of liberalization and globalization for a business is to serve and maintain good relations with the king—the customer. In the past, producers took their customer for granted, because at that time, customers were not demanding nor had many alternative sources of supply or suppliers. Since he was a passive customer, the producer dictated terms and had little customer commitment. But today, there is a radical transformation. The changing business environment is characterized by economic liberalization, increasing competition, high consumer choice, enlightened and demanding customer, more emphases on quality and value of purchase, etc.

All these changes have made today's producer shift from traditional marketing to modern marketing. Modern marketing calls for more than developing a product, pricing it, promoting it and making it accessible to target customers. It demands building trust, a binding force and value added relationship with the customers to win their hearts. The new age marketing aims at wining customers for ever, where companies greet the customers create products to suit their needs, work hard to develop like time customers through the principles of customers delight, approval and enthusiasm.

Customer relationship management is emerging as a core marketing activity for business operating in this fiercely competitive environment. World Wide Statistics reveal that by 2010, 70% of the companies will revamp their customers process, 53% will use e-commerce to transform relationships and as many as 49% will organize themselves by customer type.

The above statistics reveal that the future is customer relationship management because when a company manages its customer relationships well, the benefits got by it for exceed the cost. It is interesting to note that attracting a new customer is seven times more costlier than retaining an old one and if a customer is once satisfied he will advocate regarding the product and the company and in a span of ten years he will bring to the company 100 new customers. But on the other hand if he is dissatisfied with the company product or service, he may not only switch to the competitors but also tells others regarding his problems which may restrict the future prospects to move up the ladder of loyalty. So a customer is like a double-edged knife who should be handled carefully.

In Customer-Relationship Management (CRM) the company tries to identify that small percentage (20%) of key account holders who's contribution to the company revenues is high (80%). So from this point of view, Customer-Relationship Management (CRM) is also known as key Account Management.

Customer Satisfaction is one of the formidable challenges in marketing of services in Banking Industry in our country. The thrust on customer service has increased after liberalization of Indian economy. Thus, customer satisfaction is an urgent need of the hour. This can be achieved by efficient customer service and immediate tactful handling of customer's grievances. Better Customer-Management Relationship leads to customer satisfaction and good public image is to win over the customers. Hence, public relation should be one of the strategies to achieve customer satisfaction.

The new organization's emphasis is on partnerships between firms; multiple types of ownership and partnership and partnering within the organization (division, wholly owned subsidiaries, licenses, franchises, joint ventures, etc.); teamwork among members of the organization, often with team members from two or more cooperating firms; sharing of responsibility for developing, converging and overlapping technologies; and often less emphasis on formal contracting and managerial reporting, evaluation, and control systems. The best visual image of these organizations may be a "Wheel" instead of a pyramid, where the spokes are "knowledge links" between (focal) organization at the hub and strategic partners around the rim (Badaracco, 1991). These forms were pioneered in such industries as heavy construction, fashion, and computers, where markets often span geographic boundaries, technology is complex, products change quickly, and doing everything oneself is impossible. These confederations of specialists are called by many names including "networks" (Miles and Snow, 1986; Thorelli 1986), "value-adding partnerships" (Lawrence, 1988), "Alliances" (Ohmac 1989), and "shamrocks" (Handy, 1990). All are charactersied by flexibility, specialization, and an emphasis on **Relationship Management** instead of market transactions. They depend upon administrative processes but they are not hierarchies (Thorelli, 1996); they engage in interactions within ongoing relationships and they depend on negotiation, rather than market based process, as a principal basis for conducting business and determining prices.

MANAGING CUSTOMER LOYALTY AND DEVELOPMENT

Managing customer-development process is one of the critical dimensions of relationship marketing. Basically it involves a twin focus customer catching, and customer keeping. Customer catching is the process of attracting new customers (inviting new blood), while the customer keeping aims at the process of retaining the existing customers (encouraging old blood).

Customer catching process will have to effectively deal with the suspects, prospects, and disqualified prospects. To start with, suspects include everyone who might possible buy the organisation's product or service. But the firm is interested in sensing, serving, and satisfying the prospects who possess both the ability and willingness to buy the product or service. Accordingly, the marketer must be available to precisely assess and identify the prospects among the suspects in the market and at the some time exclude the disqualified prospects who are unprofitable for the organization to serve.[1]

The relationship centered organization should undertake proper and adequate efforts to convert many of its qualified prospects into first-time customers, satisfied first-time customers into repeat customers, repeat customers into client.

Commitment and trust lead directly to cooperative behaviour. This is conducive to relationship marketing success. This is precisely due to the following main reasons:

1. Relationship commitment and trust encourages marketers to works at preserving relationship investments by cooperating with exchange partners;
2. Resist attractive short-term alternatives in favour of the expected long-term benefits of staying with existing partners, and
3. View potentially high-risk actions as being prudent because of the belief that their partners will not act opportunistically.[2]

Banking system has a significant role to play in the rapid growth of economy. The banks that doe well in 1990s and 2000 A.D. will be those which rediscover the customer through cutting the edge known as relationship Banking.[3] Effectiveness of Banking system depends upon the customers satisfaction and infact, customer is the kingpin of the banking industry. Customer satisfaction is dependant on the climate in which contacts between customers and the bank take place. Healthy climate promotes pleasant transactions for both the parties. No industry in the service sector can afford to ignore its customers and as Banks are required to give top priority in providing satisfactory and efficient service to their customers.

OBJECTIVES

The main objectives of the paper are:

(1) To study the actual level of customers satisfaction in four Commercial Banks, i.e. State Bank of India, Punjab National Bank, Central Banks of India and Jammu & Kashmir Bank Ltd. By analyzing their attitude toward:
   (a) Bank
   (b) Employees
   (c) Management

(2) To make a comparative analysis of level of customers' satisfaction in four banks.

METHODOLOGY

For preparing the present paper 111 customers were selected on stratified random sampling method which formed a sample of more than 20% of the employees, taken up for study. Customers have been split up in three categories viz. Businessmen, Service-Class and Practitioners. 33 customers from State Bank of India, 26 from Punjab National Bank, 20 from Central Bank of India and 32 from Jammu & Kashmir Bank Ltd. were selected.

FINDINGS

Customers' Satisfaction in State Bank of India

Table 1 shows that a total of 33 customers of the bank have reported overall highest satisfaction which is evident from the mean score of 3.20, which is above average, out of these, service class is the highest satisfied lot (3.47) while Businessmen have but up lowest mean score (2.86). It falls in the below average range. Practitioners seems to be moderately satisfied as they are showing mean score of 3.27. Table further show that mean scores of service class, customers and practitioners are higher than the overall average for the bank.

TABLE I

Customers' Satisfaction in Banks

State Bank of India

Number of Customers	*Category-I Attitude towards Banks*	*Category-II Attitude towards employees*	*Category-III Attitude towards Management*	*Average Mean Score*
$*B_{11}$	2.91	2.61	3.04	2.86
$*S_{11}$	3.43	3.55	3.42	3.47
$*P_{11}$	3.26	3.35	3.19	3.27
Average Mean Score	3.20	3.17	3.22	3.20

Comparative analysis of the three variables for the study shows that customers are showing good response towards management of the bank which is clear from the mean score of 3.22, which is higher than those of other two variables, i.e. attitude towards Banks (3.20) and attitude towards employees (3.17). Thus, table makes it clear that staff of the bank has not been able to make customers happy with their behaviour. That is why table is showing the lowest mean score (3.17) regarding this particular variable.

A comparative analysis of overall mean score and actual means of different variables shows that regarding customers' attitude towards Banks, mean score of service class (3.43) and practitioners (3.26) are higher than overall average. Mean score of 2.91 is lower than the overall mean score.

B – Businessmen.

S – Service Class.

P – Practitioners.

As regards second variable, i.e. attitude towards employees except businessmen (2.62), other two categories of customers have responded well, i.e. 3.55 of service class and 3.35 of practitioners.

Customers have reported highest mean score of 3.22 towards management of the bank. It is also higher than actual average of businessmen (3.04) and practitioners (3.19). To sum up, it may be stated businessmen, through having the frequent dealings with the bank, still is the least satisfied lot especially with the behaviour of employees.

Customers' Satisfaction in Punjab National Bank:

An overall mean score of 3.06 has been shown in the Table 2 comparing the actual responses with the overall response, it may be stated that it is lower than those of service class customers (3.33) and practitioners (3.14). Conversely, it is higher than that of businessmen (2.71).

TABLE 2

Customers' Satisfaction in Punjab National Bank

Number of Customers	*Category-I Attitude towards Banks*	*Category-II Attitude towards employees*	*Category-III Attitude towards Management*	*Average Mean Score*
B_8	2.60	2.52	3.01	2.71
S_9	3.32	3.45	3.21	3.33
P_9	3.23	3.05	3.16	3.14
Average Mean Score	3.05	3.00	3.13	3.06

A comparative analysis of overall average with the actual mean score, table under reference shows that overall mean score (3.05) is lower than averages for service class (3.22) and practitioners (3.23) but higher than that of business class (2.60), so far as attitude towards banks is concerned.

Customers do not seem happy with the staff of the bank which is clear from the mean score of 3.00, which is just average. It is again higher than that of businessmen responses (2.52) while it is lower than those than those of service class (3.45) and practitioners (3.04). Highest degree of satisfaction has been shown towards management of the bank (3.13) which is lower than those of service class and practitioners, i.e. 3.21 and 3.16 respectively, but higher than of businessmen (3.01).

Customers' Satisfaction in Central Bank of India

Table 3 signifies almost similar position as that of two banks discussed above. Though Central Bank is a small bank, still its customers are most satisfied, i.e. 3.30. This compared to actual averages shown by different categories of customers, it becomes clear from the table under reference that it is higher than those of businessmen responses (3.04) and practitioners (3.9) but lower than that of service class (3.67).

TABLE 3

Customers' Satisfaction in Central Bank of India

Number of Customers	*Category-I Attitude towards Banks*	*Category-II Attitude towards employees*	*Category-III Attitude towards Management*	*Average Mean Score*
B^8	3.18	3.01	2.92	3.04
S^6	3.51	3.55	3.96	3.67
P^6	3.25	3.28	3.04	3.19
Average Mean Score	3.31	3.28	3.31	3.23

Table further shows that comparing the variable-wise position of the bank regarding 1st variable, i.e. attitude toward banks, overall average (3.31) is higher than that of businessmen and practitioners (3.25) while lower than service class as lower (3.51). It clarifies the fact that businessmen and practitioners are more satisfied with the bank. Same is the position regarding variable attitude towards management, i.e. (3.31).

As regards third category i.e attitude towards management overall mean is lower than that of service class customers (3.06) and higher than is case of practitioners (3.04) very poor response has been shown by businessmen as clear from the below average mean score of 2.92 second variable i.e attitude towards employees shown lowest mean score of 3.28 which is lower than of service class customers (3.55) but higher than that of business men (3.01). It is equal to that of practitioners (3.28).

Customer Satisfaction in Jammu and Kashmir Banks Ltd.

The position of Jammu & Kashmir Banks Ltd. is different from other banks so far as mean score of practitioners is concerned. Their mean score has been estimated at 2.90 which is below average range. Lowest overall mean score of 3.01 has been worked out against which mean score of service class customers is the highest (3.41) as compared to Businessmen (2.71) and Practitioners (2.90) as shown in Table 4.

TABLE 4

Customers' Satisfaction in Jammu & Kashmir Banks Ltd.

Number of Customers	*Category-I Attitude towards Banks*	*Category-II Attitude towards employees*	*Category-III Attitude towards Management*	*Average Mean Score*
B[12]	2.93	2.75	2.44	2.71
S[10]	3.45	3.01	3.78	3.41
P[10]	3.11	2.87	2.71	2.90
Average Mean Score	3.16	2.88	3.03	3.01

Comparative analysis of the actual mean score of customers with the overall mean scores, variable wise, reveals that the highest degree of satisfaction lies in case of their attitude towards Banks (3.16). As against this overall average, mean scores of business call customers (2.93) and Practitioners (2.71). Customers responses regarding employees are very poor as is evident from the mean score of 2.88 which is lower than that of majority of service class customers (3.01) employees show dissatisfaction towards Bank.

Above table shows the customer's satisfaction in all the four commercial banks through judging their attitude towards three categories, viz. attitude towards banks, attitude towards employees attitude towards customers.

Comparative analysis of the level of customers satisfaction if four banks shows that overall average mean score has been estimated at 3.14 which is little above average. This shows that the customers re not fully satisfied with the services offered by the banks. On comparing it with the actual mean scores of four Banks, table shows that customers of Central Banks of India have shown a highest level of satisfaction i.e 3.20 whereas State Bank of India has been ranked second in order of satisfaction amongst customers. Customers of Punjab National Bank and Jammu and Kashmir Banks are showing almost same mean score i.e 3.06

TABLE 5

Customers' Satisfaction in Four Commercial Banks: A Comparative Analysis Between 3 Categories

Bank	*Category-I Attitude towards Banks*	*Category-II Attitude towards employees*	*Category-III Attitude towards Management*	*Average Mean Score*
SBI	3.20	3.17	3.22	3.20
PNB	3.05	3.00	3.13	3.06
CBI	3.31	3.28	3.31	3.30
J&K	3.16	2.88	3.03	3.01
Average Mean Score	3.18	3.08	3.17	3.14

and 3.0 respectively which is lower than the overall average mean score. It shows the dissatisfaction is prevailing in these two banks. Though Central Bank of India being a bank with few numbers of branches is having good relations with its customers. Likewise other three banks should also try to raise the level of customers satisfaction by providing maximum facilities to them.

SUGGESTIONS

To improve the level of customers satisfaction following suggestions are made:

1. Bank personal should be wiling and ready to serve the customers.
2. They should be quick and prompt in offering the services.
3. Personnel should be knowledgeable and Skillful and the same should be reflected in their behaviour.
4. Service should have consistent quality.
5. All the risks and side-effects, it any, should be conveyed to customers.
6. Banks should be conveniently located.
7. All dealings with customers must be kept confidential.
8. Employees should be properly equipped with communication facilities for easy access.
9. Physical should be polite, respectful, flexible, friendly and accommodating.

Notes and References

1. Ambastha, Manoj Kumar and Kumari Sunita, "Customer Satisfaction in Banking Industry", Marketing strategy and challenges. ***Indian Journal of Commerce,*** Vol. 52, p. 60, Oct.-Dec. 1999.
2. Yadapadithaya, P., "Relationship Marketing Emerging Issues and Challenges in the New Millennium", ***The Indian Journal of Commerce***, Vol. 52, p. 23, Oct.-Dec. 1999.
3. Vaishnav, B.L., Service Marketing: Global Perspectives", ***The Indian Journal of Commerce***, Vol. 53, July-September 2000, p. 3.

CHAPTER

22

Waste Control in Industries and Role of HR: A Comparative Study in Communication View Point

DR. V.L. DHARURKAR AND DR. MEENA CHANDAWARKAR

I. INTRODUCTION

Modern Industries are facing various problems related to waste control. We have to take special care regarding waste control in order to develop productivity on large scale. The Human Resources Development wings in Industries can play a vital role in this regard. HRD in the age of globalization is playing multipurpose role in planning, control, coordination and reshaping of products and services. The evolution of human resource management must be studied both in theoretical as well as professional aspects. The emerging HRM functions are strengthening regional as well as local industries and they are focusing on different aspects such as reshaping human behaviour, controlling the wastage and improving the productivity process. It is interesting to observe that HR is making a profound impact on industrial organizations and its work force. Every industry is developing human resource information system, which is changing the profile of modern industries. HR in the new age is correlating personnel management, organizational behaviour and it is also helping to control industrial relations. Towards more smoother line of action the modern HR bureau are trying to strengthen job analysis and job performance, new techniques for improving skills and abilities, and they are trying to develop multiple assessment techniques. All these are helping the modern industries to elevate job related stimulations at one hand and they are also trying to control waste on the other hand. In order to judge skills and abilities assessors are properly trained and evaluated. In waste control ToT (Training of Trainers) is also a significant phenomenon. In waste control process systematic procedures should be used to record observations at every level.[1] The present paper is a unique experiment which has contributed to the field of improving productivity in modern industries.

II. FIELD STRATEGY

Any study related to industrial development and HR must be based on field work. It has been pointed that "Human resource development is a broader function and has larger contours. Most literature has depicted HRD as a corporate concept."[2] The present paper is based on field work conducted in Marathwada region in Maharashtra State and North Karnataka region in Karnataka State. The study was conducted in 30 industries in both the regions and findings have been presented below. In such regional studies we have to observe various trends which benefit industries due to their serious efforts through HR interventions. It has been observed that "The word human is global and the context is national."[3] But the present study is based on regional land scape. The immediate context is provided by organization where the individual works.[4] Thus, the study of regional industries is conducted here in the globalized context having more regional and local concern. Such a concern helps a great deal to understand the problem more correctly on appropriate lines. The waste control phenomenon has three fold dimensions. It begins right from the acquiring of human resources to the end creating the favorable working conditions for constant increase in production. The first aspect is based on HR planning in a constant and continuous manner. The second aspect depends on selection, training and development of skilled working class. The third aspect is mainly based on testing, evaluation, and improvement of productivity process. In the third stage, industrial waste control becomes a more relevant aspect, which is studied in this paper.

III. STRATEGIC HUMAN RESOURCE DEVELOPMENT

The strategic human resource management is a serious concern related to national as well as regional industries. It has been observed that "For HRD function of organization, it has now become imperative to emphasize on strategic focus, particularly when organizations need to survive competing with the global level."[5] The regional industries have to focus on reshaping human resources for gaining global status in the competitive age. Strategic Human Resources Management in industries is playing a key role in productivity acceleration because HR units are creating awareness regarding time waste, money waste and resource waste. It is true that "HRD strategy is the pattern of decisions concerning policies and practices associated with the HR system. Each organizational system encompasses all functional area of an organization and thereby helps in formulation of organization wide strategies."[6] The success of regional industries depends on which model they use for waste control in the systematic manner. These industries can effectively exercise waste control as a consequence of proper training and orientation. It has been observed that "Developing Human Resource Information System, aligning HRP with corporate strategies, skill and competency mapping are important aspects of strategic HRP.[7] If more effective training is given there can be a positive change in strategic HR planning. In the selected industries it has been brought to light that training and orientation has helped industries to face the problem of waste control on three lines, such as time waste, resource waste and manpower waste. In waste control supervisor or quality control officer is a backbone of the system. If he is properly strengthened, waste control policy can be carefully implemented. Due to his training organizational change is possible and elements of performance appraisal can be neatly searched out and developed in the process of productivity improvement. It has been observed that when we debate regarding change in behaviour, attitudinal change is implied.[8] In waste control also this kind of attitudinal change is very important and proper training can bring such change and it can revitalize the organization towards positive approach. Positivism is vitally needed for productivity appraisal and performance appraisal. In this regard in order to understand the correct focus of HR in waste control of industries, the following questions have been raised in this study:

(i) What should be the objective of HR in waste control?
(ii) What are the different methods of training used by the company for waste control?
(iii) How far performance counseling and waste control measures are useful for improving productivity?
(iv) How to decide the success of HRM strategies with special reference to waste control?
(v) What are the issues in terms of criteria as well as human judgment and professional success in waste control?
(vi) Whether these criteria have been fulfilled?

All these questions were testified to 30 industries in the region by using oral tools of data collection and the trends were classified in nutshell. It is true that developing human resources is a key in waste control and it has two dimensions, one is related to proper training and development and the other is based on scientific perception and methodical communication.[9] Now it would be interesting to study current trends in the regional industries.

IV. CURRENT TRENDS IN HR IN REGIONAL INDUSTRIES

In the field of waste control the following current trends can be observed in the regional industries:

(A) Industries are aware of their current business status and they are improving performance by conducting internal and external audits. It has been observed that in corporate management growth is possible. It has been observed that "Organization expand production volumes by using their existing technology more effectively and thus, bringing unit costs down."[10]

(B) They are formulating new business strategies as per their mission statements and they are translating this mission in to strategic goals. Such goal setting helps them for changing corporate behaviour. It has been pointed that "One of the biggest behavioural barrier in corporate environment is to break the aspect of negativity."[11] This can be changed by HR training.

(C) Regional industries are formulating strategies to achieve these goals due to these steps. Corporate planning is a key factor for such success. "Planning activities represent an expenditure or investment of company resources. Such a expenditure can be justified only if the benefits attained sufficiently out weight the costs incurred."[12]

(D) Waste Control has been possible due to HR planning and education.

(E) In nutshell managing strategy for waste control is an ongoing process.

(F) Competitors introduce new products, technological innovations make production processes obsolete, and social trends reduce demand for some products or services while boosting demand for others.

(G) Strategic control keeps the company's strategy up-to-date.

V. WASTE CONTROL AND PROFIT

Here waste control helps to maintain profit neatly. Reduction in waste, sale of raw material as well as resources, controlling time and working hours waste increases productivity. Local knowledge and actual work of preparing plan are two important factors in productivity.[13] Further in the corporate industries intimate local knowledge is needed for realistic plan.[14] In the profit-making of a regional industry, local knowledge, planning of production as well as HR counseling are important factors. The success of a corporate industry in the modern age depends on positive working conditions. The shaping of working conditions in a righteous way helps to increase margin of profit. In a country like India the three aspects which adversely affect work behaviour are accidents, fatigue and absenteeism.[15] All these hurdles in profit can be overcome through systematic efforts of HR section. The companies which follow work ethics and CSR (corporate social responsibility) are gaining more profit in the age of globalization. It has been observed that "Professionals have a responsibility towards society because the approval of upright behaviour is finally endorsed by the society."[16] There are three basic aspects in profit-making. Sharu Ranganekar has observed that "There is three point plan: Sense of mission, sense of action, sense of loyalty."[17] Thus, value inculcation is the base of improving productivity. We can put an equation in this regard.

The reduction of Waste + Reduction of Cost + Value Addition = Profit

VI. BENEFITS OF HR TRAINING

On the basis of this field work, following benefits of training can be recorded in the process of waste control:

(A) Employ training and supervisors performance appraisal helps to limit wastes. It has been observed that "HRD is competence building, commitment building and culture building."[18] HRD managers can be benefitted by such training.

(B) Establishing quality control and product improvement strategies help each other for improving profit at every stage. It is true that every HR wing should be learning-oriented and should continuously learn renewing itself.[19]

(C) Continuous HR training empowers the staff by providing them advanced skills at working level. It is true that "HRD can develop an organizational culture in which supervisor, subordinate relationships, team work and collaboration well being, motivation and pride of the employees."[20]

(D) HR training benefits to improve the prestige of organization on national level. The first decade of new millennium is marked with beginning of knowledge, knowledge miracle. It is true that "Globalization entails new markets, new products, new mindsets, new competencies, and new ways of thinking about business."[21] Companies must understand these facts.

(E) HR brings competitiveness on global level. It has been observed that " Responsiveness includes innovation, faster decision-making, leading an industry in price or value, and effectively linking it with supplies and vendors to build a value chain for customers."[22] This customer oriented approach is bring a change in the corporate industries and continuous product improvement is an ongoing process.

VII. WASTE CONTROL AND TQM

Waste control must be a part of total quality management. It has been observed that "Management includes functions such as finance, human resource, materials and scheduling."[23] In such scheduling waste control is one of the factoRs. Any corporate agency that uses TQM system can have very well upper hand in productivity acceleration. Quality is essence of product improvement. The use and application of TQM as a movement can help company to overcome various difficulties in waste control properly. The TQM application can be benefitting on the following three grounds:

- TQM brings overall awareness about quality which leads to productivity.
- Product improvement can be possible in all respects of units assembling components and package.
- New trends can help to raise overall structures.

The TQM products along with quality circle as a movement must be implanted systematically. It is true that "Problem solving is an important process in quality management. If quality has to be improved then the existing problems are to be solved."[24] The quality circle is a Japanese movement which has been accepted world wide. The role relevance and effectivity of quality circle as a movement has manifold benefits. It has been observed that "implementing quality circle is a tough task. However, careful selection of motivational strategies and consistent projection of success stories can do the job effectively."[25]

The TQM can enhance product quality at one hand and it can create overall awareness regarding agriculture marketing of products. This application of TQM helps a great deal to overcome many difficulties.

VIII. SUGGESTIONS

Problems solving tools and preventive techniques must be properly used. Waste control must be given priority in the TQM road map. Evaluation of HR efforts in waste control is highly required. It has been observed that "Some organizations in some sectors may need to grow in order to remain viable players in a turbulent sector."[26]

The following suggestions can be made to improve waste control through HR management:

- The HR Department must undertake massive campaign for waste control.
- Special experts must be involved for arranging workshops on waste control.
- Both conceptual and applied areas must be conducted such as time waste, space waste as well as resource waste.
- Special research must be conducted to understand the problem and IT must be applied to such work culture.
- A WAN, (wide area net work) must be established to understand the problem and to solve it.

IX. CONTINUOUS EVALUATION

HR planning must go ahead because it is a part of the testing of industries regarding potentials and limitations in the field of productivity. Retaining high quality and internal as well as external mobility based on creative ideas can help for the compensation of wastage at many levels. Working conditions and high degree of responsibility can help control waste. Regarding evaluation strategies the R and D must prepare a plan innovative process must be developed to overcome these problems. It has been observed that "If the organizations pay attention to various problems, training can be of great use to the organizations in increasing their efficiency and productivity and thus, in building better HRD climate."[27]

RACE, (Research, Action, Communication and Evaluation) formula must be adopted. After conducting research e4valuation of performance must be made a systematic manner. A cycle of continuous evaluation must be prepared to overcome complications. The questionnaire prepared must be improved from time to time. A net work of research must be developed to overcome various problems. After each month a Evaluation Chart must be prepared and record must be maintained. This can help to improve productivity at all levels.

The evaluation report must present a model plan keeping in to view the next five years forecasting in a systematic manner.

X. SUMMARY

The HRD system must be made aware regarding waste control because quality of work life is based on learning habits of organization. Work culture must be developed with special reference to the program of action. To control waste, understanding the message of quality and waste control must be routed through every unit of organization so that performance can be improved. Motivation perspective must be developed based on human communication. Communication dynamics must be focused for waste control education. Proper language message treatment and fine relationship, removal of tension are important factors while developing communication strategy for waste control. A case study based on quality management can help to control waste and improve performance. In corporate industries we have to develop an optimum waste control strategy. A learning organization has to focus on increasing productivity at all levels. This can be possible by reexamination of waste control methods. We have to develop a sub system based on role and objectives to control wastage in production process. This study was conducted with special reference to the industries in backward areas. This approach can help for improving industries. We have to search for a new perspective based on cost saving, value addition and waste control. The adoption of high performance work practices also known as high commitment or high involvement work practices are intended to achieve better individual, team and organizational performance by increasing commitment and involvement of the staff. This can help to control waste at many levels and has proved that those companies who adopt HR strategies can be able to established leadership in management. Companies which use advanced communication tools can be more effective than those who do not use. Decision-making can be effective on the basis of communication problems in management can be solved on the basis of practical approach. The emergence of the managerial role in waste control is based on innovation new tools and techniques, result and activities with profitability and productivity.

Thus, the novel ideas evolved in this paper are based on field work and they can be applied to any company to plan waste control strategies. Especially, this can be very well used for improving strategies on regional level.

Notes and References

1. Saiyadain Mirza, S., "Human Resources Management", Tata McGraw Hill, New Delhi, 2003, p. 5.
2. Rao, T.V., "Future of HRD", Macmillan Publishers, New Delhi, 2008, p. 159.
3. *Ibid.*, p. 159.
4. *Ibid.*
5. Bhattacharya, "Human Resource Development", Himalaya Publishing House, New Delhi, 2009, p. 275.
6. *Ibid.*, p. 276.
7. *Ibid.*, p. 279.
8. *Ibid.*, p. 177.
9. *Ibid.*, p. 235.
10. Chandra, Ramesh, "Corporate Management", Kalpaz Publishers, New Delhi, 2002, p. 35.
11. *Ibid.*, p. 158.
12. *Ibid.*, p. 270.
13. *Ibid.*, p. 279.
14. *Ibid.*
15. Saiyadain Mirza, S., *op cit.*, pp. 415-30.
16. *Ibid.*, p. 430.
17. Ranganekar Sharu, "In the World of Corporate Managers", Vikas Publishing House, New Delhi, 2009, p. 226.
18. Rao, T.V., *op cit.*, p. 96.
19. *Ibid.*, p. 98.
20. *Ibid.*, p. 131.
21. *Ibid.*, p. 175.
22. *Ibid.*, p. 176.
23. Suganthi, L. and Samual, A.A., "Total Quality Management", Prentice Hall Publishers, New Delhi, 2008, p. 17.
24. *Ibid.*, p. 115.
25. *Ibid.*, p. 237.
26. Chandra, Ramesh, *op cit.*, p. 35.
27. Jain, V.K., "HRD Practices in Indian Industries", Anmol Publishers, New Delhi, 1996, p. 158.

CHAPTER

23

Impact of Information and Communication Technologies on Living Standard of Socially Backward Young Women in Kozhikode District

DINESH, M.P.

INTRODUCTION

The autonomous or empowerment of women is one of the central issues in the process of development of countries all over the world. Women empowerment studies have been concerned with the question of women's participation or non- participation in paid work, social work and employment. The fact is that waged labour does open up new doors, new ways of being and an opportunity to connect with other women and form bonds that are not familial. All these would undeniably have an impact on how women perceive themselves and others, it is clear that the ability to work outside the home enables women to recognize, challenge or confront structures of hierarchy and women perceive possibilities of transformation in identities and perceptions through employment. Further aspects that shape the way women and society respond to wage work have to do with the nature of the work itself.

While it is known that a large part of employment opportunities available to women consists of work that is considered either 'feminine' or subsidiary/inferior. However, money earned by poor women is more likely to be spent on the basic needs of life than that by men and that these realizations would bring women as the focus of development efforts.

WOMEN EMPOWERMENT

Empowerment is a multi-faceted, multi-dimensional and multi-layered concept. Women empowerment is a process in which women gain greater share of control over resources—material, human and intellectual like knowledge, information, ideas and financial resources like money and access to money and control over decision making in the home, community, society and nation and to gain power (UNO).

According to the Government of India, "empowerment means moving from a position of enforced powerlessness to one of power". The World Bank has suggested that empowerment of women should be a key aspect of all social development programmes (World Bank, 2001). For women in India, need to be empowered in several realms: personal, familial, economical, social and political.

INFORMATION AND COMMUNICATION TECHNOLOGIES (ICTs)

Information is the key to democracy. ICTs have the potential to improve interaction between government and citizens, fostering transparency and accountability in governance.

Information and communication technology comprises a complex and heterogeneous set of goods, applications and services used to produce, process, distribute and transform information. ICTs are a diverse set of technological tools and resources to create, disseminate, store, bring value-addition and manage information.

The ICT sector consists of segments as diverse as Telecommunications, Television and Radio broadcasting, computer hardware, software and services and electronic media, for eg.: the internet and electronic mail (United Nations, 2005).

WOMEN AND ICT

In our society, the freedom of women to seek employment outside of the family is a major issue. This freedom is denied in many cultures and this attitude in itself is a serious violation of women's liberty and gender equality, the absence of this freedom militate the economic empowerment of women creating deleterious consequences.

Knowledge liberates people from poverty and empowers them. It is also now well understand that any attempt to improve the quality of life of people in developing countries would be incomplete without the progress of their women folk. ICTs emerge as a powerful tool for gender empowerment in a developing country like India. According to the World Bank, teledensity in India had reached 3.8% of the population by 2001. The number of internet accounts is growing at a rate of 50% per annum. The IT and ITES sector is projected to grow 18% in the next five years to become an industry of Rs. 4.58 lakhs crores by 2011, according to an IDC[1] release.

The UNESCO reports on "Gender issues in the informal society" states the capability of women to effectively use information obtained through ICT is clearly dependent on many social factors, including literacy and education, geographical location, mobility and social class. Hence, the clear guidance on good practice for ICT project interventions from the micro level of participatory assessment of needs to the macro level of gender sensitized policies on infrastructure (Jorge, 2002, Huyer and Sisoska, 2002) is required.

SCOPE OF THE STUDY

It is generally agreed that female participation in production process, would result in increasing their economic and social status, consequently empowering them.

Kerala is the only state in India where the female labour force participation rate has declined continuously. The female work participation rate in Kerala has declined from 16.9% in 1991 to 15.3% in 2001 (Censes, 2001).

At the same time, Kerala economy has number of special features in social development. Kerala exhibits high sex ratio (1082/1000); high female literacy rate (88%), first Rank in Human Development Index (HDI), Gender Development Index (GDI) and Gender Empowerment Index (GEI). The study of Mridul and Praveena (2005) shows that plus two qualified unemployed female is higher compared to same qualified male candidate in Kerala. Another feature is the absorption of women employment in the non-agricultural sector and white-color job trend are comparatively high in Kerala.

On the contrary, Kerala economy faces number of hindrances for female participation in Production Process. Religions and caste factors are found to be significant influence of low female work participation rate in Kerala (Nirmala and Bhat, 2000). Qualified female generation (Mridul and Praveena, 2005), white-color preference of the Malayali society (Mohan Pillai, 2004), existing marriage conditions and customs and middle class peoples' negative attitude towards the female work participation are other major factors responsible for continuously diminishing female work participation rate in Kerala. Above all, domestic responsibilities also hinder the women to be at their work place.

The present study examines the job opportunities in the information and communication technology (ICTs) sector and how we have to utilize these opportunities for female participation and their economic empowerment in Kozhikode, Kerala. The paper presents findings from Kudumbashree ICTs Units, showing significant impacts on Women's employment, income, social roles and empowerment of poor women. Gender focused ICTs initiative, Kudumbashree ICTs units, involving significant state intervention has brought about positive changes to livelihood outcomes and empowerment of women. It shows how ICT-based initiatives empower women.

OBJECTIVES OF THE STUDY

1. To assess the impact of economic and social background of women workforce in information and communication Technologies (ICTs).
2. To study how ICTs are helpful for the enhancement of living standard of deprived women.
3. To asses ICTs infrastructure in rural areas for women empowerment.
4. To suggest the policy implications for the effective implementation of further ICTs-based programmes under the objectives of women empowerment.

Research Methodology

The data collection and analysis is based on the data collected from various Kudumbashree ICTs Units located in the three regions of Kozhikode district, Kerala—Kozhikode, Koilandy and Vatakara—through a systematic stratified simple random sampling survey using a structured questionnaire. The regions selected for the study was taken keeping in mind that their location and access to ICTs. Here, we use Percentage method, Testing of Hypothesis and ÷2 test for Analysis.

Review of Literature

Jawaharlal Nehru said, "to awaken the people, it is women who must be awakened; Once she is on the move, the family moves, the village moves and the nation moves." Education is one of the most important means of empowering women and giving knowledge, skills and self confidence necessary to participate in the developmental process.

Beijing Declaration and platform for action (1995), states: women should be empowered by enhancing their skills, knowledge and access that information technology. This will strengthen their ability to combat negative portrayals of women internationally and to challenge instances of abuse of the power of an increasingly important industry. With this statement, the international community was able to recognize, for the first time, the importance of the ICTs for economic and social empowerment of women.

United Nations (1999) observed that women's paid employment could change the perception of women's value and motivate them in the investment in their girl child's education and health. So UN underlines that women who work outside their house caters empowerment.

Huyer and Mitter (2003) find out that the women who can access and use ICTs. ICTs offer them income, education, health, access to information and awareness on public and private right and improve their well being.

Huyer and Sikoska (2003) observed that the income and knowledge from ICT sector enables women to perceive themselves as capable of undertaking decisions and making choices about their lives, which in turn requires sufficient levels of self-confidence and assertiveness. So, the empowerment directs the structural changes in all aspects of her life—social structures and relationships moral, cultural values and norms, institutions and power structures.

Hafkin and Huyer (2006) observed that information and communication technology (ICT) is a tool that could create greater impact on day today life of women. ICTs directly influence the empowerment process of women. That is, by using information and communication technologies, we have to direct the women into economic empowerment, social empowerment and political empowerment. At the same time, he gave the warning against 'gender divide' through ICTs. Since the participation of women in the ICT sector is lower in number compared to men there is a possibility of male domination in this sector.

The study of D.A. Patil, M. Dhere and C.B. Pawar (2009) finds that some key ICT-based initiatives in Asia in general and India in particular which are specifically designed for the empowerment of rural and deprived women; It shows how ICT-based initiatives empower rural and deprived women; it tries to clarify a theoretical dilemma between meaning and measurement of empowerment and provides a holistic definition of empowerment.

Choudhary (1996), in her study stressed the need for sharpening women's empowering strategies to make them effective and result oriented. She pointed out that money earned by poor women is more likely to be spent on the basic needs of life than that by men and that these realizations would bring women as the focus of development efforts.

Bhat and Rajan's (1990) study points out that more women are entering the labour force not only because of increase in female population in the working age groups but also because they can do so due to the dramatic decline in the number of births per woman since the 1970's. However, female work participation rate reveals a declining trend, which is primarily due to inadequate growth in employment opportunities, more pronounced in the case of educated women.

Nagaraj (1999) observed that wage discrimination shown to women is mainly for discouraging them from entering the labour force in Kerala. Over time, this can result in discouraging women from entering labour force preferring to remain primarily housewives. In Kerala, female work seekers may not be willing to accept employment at the wages being offered.

Lakshmi Devi's (2002), study reveals that educated female unemployment arises because they had not been able to find jobs of their preference. In a recent study of women's education, employment and job preferences, nearly three-fourths of the unemployed women reported that they were unemployed and full-time housewives because they had not been able to find jobs of their preferences.

Balakrishnan (2002), maintains that certainly there is high hope that ICT projects can play an important role in reducing gender inequalities. Since ICTs projects are gender blind, it gives more opportunities with high salary for women.

Pradeep and Dayasindhu (2004), find that Kerala has caught on to the coattails of India's software boom and can be investigated to see how demand for software labour has impacted women's paid employment in the mainstream ICT sector.

Shanta and Mohanan Pillai (2008) argue that poor women through Kudumbashree have made an entry into the lowest spectrum of IT enabled jobs. This ensures a minimum level of income adequate to meet the calorie requirement to keep them above the poverty line. If the full potential of this endeavor is attained it can provide much higher incomes and larger employment to the poor women. Over and above this in the specific context of Kerala with a high level of educated unemployment and a low female work participation rate, the ICT sector seems to have met the aspirations of many poor women for a white collared job.

Thus, the review of available literature reveals those numbers of studies have been made on women empowerment through female labour force participation. However, the specific studies of impact of ICTs on women empowerment are limited in the international, national and

Kerala literatures. Scarcity of literature on Women empowerment through female work participation, especially through ICTs, is visible. Hence, the present study makes an attempt to fill this research gap of the role of ICTs for increment female work participation, living standard and consequent empowerment of women in Kerala context. This paper presents findings from Kudumbashree ICTs Units in Kozhikode.

Kudumbashree

Kudumbashree, an innovative woman based participatory programme, aims at eradicating absolute poverty under the leadership of Local Self Governments formed and empowered by the 73rd and 74th Amendments of the Constitution of India. The Mission launched by the State Government with the active support of Government of India and NABARD has adopted a different methodology in wiping out absolute poverty by organizing the poor into community-based organizations. The Mission follows a process approach rather than a project approach.

Kudumbashree is envisaged as a Mission, a process, a project and a delivery mechanism for the poor. Identification of Kudumbashree can only be made through the various facets. The following facets are visualized for the Kudumbashree Programme:

- Identification of poor families using a non-monitory poverty index.
- Organizing the poor to a 3-tier Community Based Organization (CBO).
- Empowerment of women through Community Based Organizations.
- Formation of Informal Bank of poor women operating round the clock throughout the year, starting from thrift and credit operations.
- Formation of micro-enterprises.
- Convergent Community Action.
- As a further step to decentralization process.
- Establishment of rural marketing network throughout the state.

Kudumbashree ICTs Units-Study Findings

Economic uplift of the poor could be achieved only through sustainable micro enterprises. With this view, we searched out for inevitable enterprises, which can provide sustainable income. On the basis of this, among various Kudumbashree micro-enterprises like Ethnic Delicacies, Agriculture Nurseries, Soap Making Units, Direct Marketing, Food Processing, Dairy Products, Bio-Technology, Integrated Coconut Processing. Our study area is Kudumbashree ICTs Units.

There are 228 Kudumbashree ICTs units are now operational in Kerala (Kudumbashree, 2008). All these units are co-operative enterprises, which owned, managed and operated by women from poor families and, among these units, there are three types of ICTs-based enterprise.

* IT training units which provide IT training to Schools.
* Data entry and digitization units which manly create local digital content for public (and to a lesser extent private) sector organizations, but which may also do some other IT work and
* Hardware assembling and maintenance units, basic software and Web development work.

Example for a Kudumbashree ICTs Unit

Techno World III, near civil station—Kozhikode, **is** a Kudumbashree ICTs Unit initiated on 18/03/2004 by ten women belongs to below-poverty-line rural families. Total capital of the Project was Rs: 3,15,000, which includes;

Beneficiaries' share	=	Rs. 15000
Bank Loan	=	Rs. 175000
Subsidy from Local Government	=	Rs.125000
Total	=	Rs. 315000

Techno World III, near civil station—Kozhikode

They started with 8 computers and a printer. At present they have Rs.40,00,000 (Forty lakhs rupees only) turn over set up and 65 additional employees including one male on contract basis.

The Unit mainly undertakes Data entry work for Government departments like Transport office, Hajj office, Land records of various Revenue offices, Corporations birth and death records, voters' Lists, etc. There are 10 Kiosks at various centers under this Unit.

Source: Techno World III, IT Publication.

Analysis of Collected Data from Kudumbashree ICTs Units

The analysis based on the data collected from various Kudumbashree ICTs Units located in the three regions of Kozhikode—Kozhikode, Koilandy and Vatakara—through a systematic stratified simple random sampling survey using a structured questionnaire. The regions selected for the study was taken keeping in mind that their location—south, middle and north of Kozhikode—and access to ICTs. Here, we use Percentage method, Testing of Hypothesis and ÷2 test for Analysis.

TABLE I

Analysis of Data from Kudumbashree ICTs Units

Sl. No	*Empowerment Parameters*	*Opinion*				*Total (%)*
		No change	*Less Improved*	*Improved*	*Highly Improved*	
1.	Self-confidence	4 (3.70)	30 (27.78)	41 (37.96)	33 (30.56)	108 (100)
2.	Provides increased Economic prospects	8 (7.41)	18 (16.67)	40 (37.04)	42 (38.88)	108 (100)
3.	Decision-making power	19 (17.59)	31 (28.70)	30 (27.78)	28 (25.93)	108 (100)
4.	Change the status in the family and community	2 (1.85)	22 (20.37)	35 (32.41)	49 (45.37)	108 (100)
5.	Support from the family for working in an ICTs unit	11 (10.19)	21 (19.44)	36 (33.33)	40 (37.04)	108 (100)
6.	Contribution to family income	10 (9.26)	28 (25.93)	31 (28.70)	39 (36.11)	108 (100)
7.	Freedom in household Expenditure	6 (5.56)	23 (21.29)	50 (46.30)	29 (26.85)	108 (100)
8.	Participation in Political Activity	12 (11.11)	42 (38.89)	28 (25.93)	26 (24.07)	108 (100)
9.	Increased flexibility and skill development	20 (18.52)	43 (39.82)	26 (24.07)	19 (17.59)	108 (100)
10.	Sharing the household Responsibilities	17 (15.74)	39 (36.11)	32 (29.63)	20 (18.52)	108 (100)

In the three study areas, a total of 108 women respondents were interviewed by the researcher. All the respondents are women who are working in the Kudumbashree ICTs units. We have to explain the empowerment through employment in the Kudumbashree ICTs units with 10 parameters. They are:

1. Self-confidence (C).
2. Provides Increased Economic Prospects (E).
3. Decision-making Power (P).
4. Change the status in the family and community (S).
5. Support from the family for working in the ICTs units (F).
6. Her contribution to family income (I).
7. Freedom in household expenditure (H).
8. Increased Flexibility and Skill development (D).

9. Support from the family for sharing household responsibilities (R).
10. Participation in Political Activity (A).

With the help of notations (symbols), we have to explain the functional relationship between Women Empowerment through Employment in the ICTs sector, W, with its parameters in the following way:

W = f [C, E, P, S, F, I, H, D, R, A], where
W = Empowerment through employment in the ICTs units.

We have to explain the role of ICTs for empowerment with the help of the Table 1.

1. Self-confidence (C)

One of the very important factors concerning on women empowerment is Psychological development. In Kerala, the family gives more protection to girls than boys. Up to marriage, the girls live under the safe protection wings of parents. For their travel and all other activities accompanied by guardians.

At present the situation visibly comments that most of the skills the girls have, are lest unexploited, when compared to boys. The study of K.R. Mallika among college students in Kozhikode finds out that while more than 70% of boys are engaged in the physical supporting jobs like construction work, painting and catering jobs, only less than 15% of girls got engaged in works, that too, which required less amount of physical or intellectual skills such as doing primary tuitions and tailoring. (K.R. Mallika, [8th March, 2011], *Mathrubhumi Daily*).

For empowerment the Psychological development, Self-confidence, is necessary. In the study, we examine the confidence level of female workers. Out of 108 respondents, 41(37.96%) employees confess that their Self-confidence level improved with the ICTs jobs. Only 3.70% of (4 out of 108) employees comment that there is no change their Self-confidence with ICTs jobs. At the same time, 30.56% of respondents are happy with very good Self-confidence level. Majority of the respondents' (104 out of 108) view is that jobs in the ICTs sector have helped them to improve their thinking level and confidence in life. [Refer, Table 1].

2. Provides Increased Economic Prospects

Another important dimension of empowerment is Economic empowerment. Employment in Kudumbashree ICTs units brought considerable economic benefits on basic monthly remunerations for employees. Generally, this remuneration package is higher than jobs in the non- ICTs units in the private sector for women.

From the data collected, it is clear that majority of the respondents stated that their economic background sufficiently improved after getting a job in the ICTs parks. Out of 108 respondents, 42 (38.88%) employees accepted that their economic prospects high improved with ICT sector. In the same way, 40 (37.04%) employees said that improved their economic status and living standard also nurtured with ICTs employment. Only 8 (7.41%) employees are not satisfied with the salary in the ICTs sector. [Refer, Table 1].

3. Decision-making Power

All the theories relating to empowerment argue that Decision Making Power is the threshold for empowerment. It is clear that the decision-making power in the home, community and state are necessary for empowerment. The response of the respondents shows that more steps have to be initiated for the establishment of power for women in the decision-making process in their home and society, 19 (17.59%) respondents out of 108 answered that they are unhappy in this parameter. That is 'no change' in their status even after joined ICTs sector jobs. At the same time, 31 (28.70%) respondents have belief that they could slightly shift the power structure positively. However 30 (27.78%) respondents said improved and 28 (25.93%) employees claimed remarkable change in their authority at home and society.

4. Change in the Status in the Family and Community

We know that status is directly related to money, income and employment. Financial capacity and standard job provides good personality status in the family and society. Majority of our respondents feel happy with the status component. Out of 108 respondents, majority of female employees 49 (45.37%) confess that they experienced good reputation in the family and society after joining the ICTs jobs. In the same way, 35 (32.41%) respondents answered improved and 22 (20.37%) female employees feel that their status in the family and society is improving but slowly.

5. Support from the Family for Working in a ICTs Units

Majority of the employees stress that the support from family members is inevitable for their own development. Out of 108 employees, 19.44% (21) workers said that the support from inside and outside of the family improving slowly. On the contrary, 11 (10.19%) female workers suffer from the lack of support they should get from their family. At the same time, 36 (33.33%) respondents comment improved support from family and 40 (37.04%) female workers are very happy with highly improved support from family and community for their work participation in the ICTs units.

6. Contribution to Family Income

For the well being of the family and growth and structural change of the society, the work participation of youth is essential irrespective of their gender. To meet high cost of living of family and participation in national develop, the utilization of educated females' talent with male is inevitable in a state like Kerala. Out of 108 respondents, 98 employees have satisfied with their financial contribution to family income. An interesting point is the female employees from BPL families got financial safety after their participation in the ICTs works.

7. Freedom in Household Expenditure

In Kerala, the traditional culture is male dominated society. Day-today expenditure and other expenses related to family is absolutely managed by the male; either husband or father or uncle like that. In the survey, the researcher had to observe in detail whether there is any improvement in the freedom of household expenditure of female employees in the Kudumbashree ICTs Units. The Table 1 shows that 46.30% employees (50 out of 108) argued that they got more freedom for day-today household expenditure purpose. In the same way, 29 employees (26.85%) feel very happy for their highly improved freedom in household expenditure matters. Only 6 employees (5.56%) are unhappy with this parameter.

8. Participation in Political Activities

In Kerala, 50% seats in Local Bodies reserved for females. In additions to this; the seats of President, Vice-President and standing committees are also reserved up to 50% for females. An interesting point is, in last local bodies election majority of female candidates came from Kudumbashree and related self-help groups.

The interaction of Kudumbashree members with Local bodies and government officials lead way to political platform and decision making power. The collected data reveal that Kudumbashree ICTs units' members participate in the political activities frequently. 24.07% of sample (26 out of 108) respondents have highly improved their political dimension after joining ICTs jobs. Similarly, 28 respondents said improved and 42 respondents feel they are slowly improving in getting developed politically. Only 12 (11.11%) respondents feel unhappy with this parameter.

9. Increased Flexibility and Skill Development

Increased flexibility and skill development is another parameter for empowerment. In the given sample, 26 respondents (24.07%) admit that improved flexibility and skill development offers ICTs jobs. In the same sense, 17.59% respondents (19 workers out of 108) said that highly improved their flexibility and skill development due to involvement in this work and 43 female employees (39.82%) comment that slowly improved. However, 20 employees (18.52%) are unhappy in this parameter.

10. Sharing the Household Responsibilities

For women burdened still with traditional expectations of particular roles in family and in society. In the given sample, 17 female employees out of 108 (15.74%) felt that domestic responsibilities are affected by the longer working hours. Respondents reported that, when children or husband or relatives to be sick at home or hospital, it is women who typically take-off from work.

At the same time, 29.63% respondents (32 out of 108) said that the sharing mentality of husband and relatives in the household responsibilities improved and 20 respondents (18.52%) feel that it has improved remarkably. Like that, 39 respondents expected to improve the sharing mentality of family members.

For examining the goodness of fit between empowerment and its given parameters, based on collected data, we have to use test.

$$\chi^2 = (O_i - E_i)^2 / E_i,$$

where O_i=Observed frequency

E_i = Expected frequency and i=1, 2, 3....9, 10.

For analysis, here, we include the opinions that less improved, improved and highly improved in to one group. It is clear that less improved shows gradual progress from traditional expectation of women's life .That is tremendous change happens in the overall side of women who are employed in the ICTs field with various growth rate - less improved, improved and highly improved.

On the other hand, 'no change' sample in the given parameters show that some remedial measures should implement for the improvement of employees' living standard . The Table 1 shows that three factors for empowerment—1. Increased flexibility and skill development, 2. Decision-making power, and 3. Household responsibilities—become highest in the 'no change' opinion group. 20 respondents out of 108 (18.52%) argued that suitable measure should implement for improvement of flexibility and skill development parameter. 17.59% (19/108) respondents feel unhappy with the decision making parameter. Their decisions are absolutely controlled by guardians. 17 respondents out of 108 (15.74%) face tiresome of complete family responsibility with work in the ICTs units. To implement suitable measure for female workers to done their natural duties like pregnancy and delivery and child care without affecting their 'flexibility and skill development' in the work place, especially ICTs sector.

TABLE 2

Goodness of Fit-Kudumbashree ICTs Units

Empowerment Parameters	*Observed Frequency,* O_i	*Expected frequency,* E_i	$O_i - E_i = X$	X^2	X^2/E_i
Self-confidence	104	108	-4	16	0.1481481
Provides increased Economic prospects	100	108	-8	64	0.5925926
Decision-making power	89	108	-19	361	3.3425926
Change in status in the family and community	106	108	-2	4	0.037037
Support from the family for working in an ICTs unit	97	108	-11	121	1.1203704
Contribution to family income	98	108	-10	100	0.9259259
Freedom in household expenditure	102	108	-6	36	0..3333333
Participation in Political activity	96	108	-12	144	1..3333333
Increased Flexibility and Skill Development	88	108	-20	400	3.7037037
For Sharing the Household Responsibilities	91	108	-17	289	2.6759259
Degree of Freedom= n-1= 10-1= 9			Calculated Value		14.212963

Analysis of Collected Data from Kudumbashree ICTs Units

Hypothesis

H_0: The Women's employment in the ICTs units are significantly supporting for empowerment.

H_1: The Women's employment in the ICTs units are not significantly supporting for empowerment.

(Refer Table 2 given on next page)

Inference

Level of Significance, $\alpha = 0.05$

Table value, $\chi^2_\alpha = 16.9$

Calculated v. $\chi^2 = 14.212963$

Since the calculated value is less than the table value, we accept the H_0 and reject H_1.

Result

The Women's employment in the Kudumbashree ICTs units are significantly supporting for women empowerment.

CONCLUSION

The Role of Information and Communication Technologies (ICTs) as a tool for development has attracted the sustained attention of the government and NGOs. The millennium declaration adopted in 2000 underscored the urgency of ensuring that the benefits of new technologies, especially ICTs are made available to all. ICTs can be a powerful catalyst for political and social empowerment of women and the promotion of gender equality. There is growing agreement, however, that the impact of ICTs in developing countries is not gender neutral, necessitating an engendered approach to ICT-based projects. The paper presents findings from Kudumbashree ICTs Units, showing significant impacts on Women's employment, income, social roles and empowerment of poor women. Kudumbashree ICTs Units, a gender focused ICTs initiative involving significant state intervention has brought about positive changes to livelihood outcomes and empowerment of women. This paper concludes that ICT as a form of new technology are socially deterministic, with varied implications for women in terms of employment and empowerment dependent on the context within which the ICTs are utilized. It shows how ICT-based initiatives empower poor young women.

REFERENCES

Arun, S. and Arun, T. (2002), 'ICTs, gender and development: women in software production in Kerala', *Journal of International Development,* 14(1), 39-50.

Balakrishnan, R. (2002). Harnessing ICTs for the advancement of rural women . Paper presented at UN meeting on ICTs and their Impact on ICTs and their Impact on and use as an Instrument for the Advancement and Empowerment of Women. Seoul, Korea, 11-14 November.

Batnagar, S. and Schware, R. (2000). Information and Communication Technology in Development Cases from India, Sage Publications, New Delhi.

Beijing Declaration and Platform for Action (1995). Available: www.umn. Edu/humanrts/ Instree.

Bhat, P.N.M. and S., Irudaya Rajan (1990). Demographic Transition in Kerala, Revisited. *Economic and Political Weekly,* Vol. 20, No. 11, pp. 1547-60.

Castells, M. (2001). The Internet Galaxy, Oxford Press, Oxford.

Cecchini, S. (2002). Can ICT Applications Contribute to Poverty Reduction? Lessons from rural India, Poverty Reduction Group Working Paper, World Bank, Washington, DC.

Choudhary, M.A. (1996). Empowering strategies for rural women in India. *Kurukshetra,* 44(3): pp. 18-22.

D.A. Patil, A.M. Dhere, C.B. Pawar (2009). "ICT and empowerment of rural and Deprived Women in Asia." *Asia Pacific Journal of Rural Development,* Vol. XIX, No. 1, July 2009.

Dayasindhu, N. and Pradeep, G. (2003). 'Nurturing a software industry in Kerala', in: *The Digital Challenge.*

Dr. Agarwal, Ranjana (2004). Women's Careers in Information Technology.

Eapen, M. and Kodoth (2002): Family Structure, Women's Education and Work: Re-examining the High Status of Women in Kerala.

GOK (2009). *Economic Review*, Kerala.

Hafkin, N. and Huyer, S. (2006). Cyberella or Cinderella? Empowering Women in the Knowledge Society. New Hampton: Kumarian Press.

Huyer, S. and Mitter, S. (2003). ICTs, Globalization and Poverty Reduction: Gender Dimensions of the Knowledge society, Part 1. Poverty Reduction, Gender Equality and the Knowledge Society: Digital Exclusion or Digital Opportunity?

Huyer, S. and Sikoska, T. (2003). Overcoming the Gender Digital Divide: Understanding the ICTs and their Potential for the Employment of Women. INSTRAW Virtual Seminar Series. Santo Domingo: INSTRAW.

Huyers, S. and Sikuska, T. (2003). Overcoming the Gender Digital Divide: understanding ICTs and their Potential for the Empowerment of Women. United Nations—INSTRAW virtual seminar series on Gender and ICTs, No. 1.

Kudumbashree (2008). www.kudumbashree.org

Lakshmi Devi (2002). Education, Employment and Job Preferences of Women in Kerala: A Micro Level Study, Discussion Paper No. 42, KRPLLD, Trivandrum .

Mohanan Pillai, P. and Shanta, N. (2008). ICT and Employment Promotion Among Poor Women : How Can We Make It Happen ?. Working Paper 398.

Mridul Eapen and Praveena Kodoth (2005). Looking Beyond Gender Parity. *EPW*, Mumbai.

Nagaraj, K. (1999). Labour Market Characteristics and Employment Generation Programmes in India, in (eds.) Barbara Harriss, White and S., Subramanian, Ill-fare in India, Sage Publication New Delhi.

Nirmala, V., Bhat and Sham, K. (2000). Gender-wise Determinants of Higher Occupational Scales in Service Sector. *The Indian Journal of Labour Economics*, Vol. 43, No. 4, pp. 883-96.

Pradeep Kumar Panda (1999). Poverty and young women's Employment: Linkage in Kerala-CDS.

Technopark (2010). www.technopark.org/index

CHAPTER

24

Work Life Balance: Challenges and Solutions

ROHINI GUPTA SURI, DR. AMRIK SINGH SUDAN AND DR. SANTOSH GUPTA

1. INTRODUCTION

In organization and on the home front, the challenge of work/ life balance is rising to the top of many employers and employees' consciousness. In today's fast-paced society, human resource professionals seek options to positively impact the bottom line of their companies, improve employee morale, retain employees with valuable company knowledge, and keep pace with workplace trends. This article provides human resource professionals with an historical perspective, data and possible solutions-for organizations and employees alike-to work life balance. Three factors—global competition, personal lives/ family values, and an again workforce—present challenges that exacerbate work life balance. This article offers the perspective that human resource professionals can assist their companies to capitalize on these factors by using work life initiatives to gain a competitive advantage in the marketplace.

In a society filled with conflicting responsibilities and commitments, work life balance has become a predominant issues in the workplace. Three major factors contribute to the interest in, and the importance of, serious consideration of work life balance: (1) global competition; (2) renewed interest in personal lives/family value; (3) an aging workforce, Research suggests that forward-thinking human resource professional seeking innovative ways to augment their organization's competitive advantage in the marketplace may find that work life balance challenges offer a win-win solution.

The main purpose/objective of this paper is to discuss in detail the following issues of work life balance:

1. Introduction;
2. The Genesis of Work Life Balance;
3. Defining the Work Life Balance;
4. Personal Lives and Family values to the Forefront;
5. The changing face of Family;
6. A pivotal study;
7. Stress and the consequences for Employer and Employee;
8. The Employer's perspective: Return on Investment (ROI);
9. Diversity and Work Life Balance;
10. Is your organization culture—Ready for Work Life Initiatives?;
11. Communication is key;
12. Can Work Life Initiatives be Measured?;
13. Employee Time Saved;
14. Employee Retention;
15. Increased Motivation and productivity;
16. Absenteeism;
17. Increased Health Care costs and stress-related illnesses;
18. Family-friendly Benefits;
19. Work Life Balance Trends;
20. New Research;
21. Elder Care;
22. Work Life Balance in the Relief World;
23. Total Life Planning;
24. Is Work Life Balance Achievable?;
25. The Modern Work Life Power Balance; and
26. Conclusion.

2. THE GENESIS OF WORK LIFE BALANCE

Work Life Balance

A state of equilibrium in which the demands of both a person's job and personal life are equal. Phrases and words serve as cultural signposts to explain where we are and where we are going. The term "Work Life balance" was coined in 1986, although its usage in everyday language was sporadic for a number of years. Interestingly, work life programs existed as early as the 1930s. Before World War II, the W.K. Kellogg Company created four six-hour shifts to replace the traditional three daily eight-hour shifts, and the new shifts resulted in increased employee morale and efficiency.

Rosabeth Moss Kanter's seminal book (1977), Work and Family in the United States: A Critical Review and Agenda for Research and Policy, brought the issue of work life balance to the forefront or research and organizations. In the 1980s, companies began to offer work life programs. While the first wave of these programs were primarily to support women with children, today's work life programs are less gender-specific and recognize other commitments as well as those of the family.

Work life balance initiatives are not only a United States (U.S.) phenomenon. Employees in global communities also want flexibility and control over their work and personal lives. However, for the purpose of this article, the research and surveys presented focus on work life balance in the united States.

3. DEFINING WORK LIFE BALANCE

The idea of work life balance is based on a simple assumption: that work and life are somehow not related. This assumption itself to typical responses when the question of work-life balance comes up. If these responses were useful the question would still be asked as often as it does.

Let us examine the assumption: that work and life are in separate spheres. Is this true? If it is true, does it have to be true? Are the other approaches?

In reality, our society is set-up so that work life are separate. Typically you go to work, often doing whatever will get you paid, and then you go home and have your life.

Let's break this world down, at least for weekdays. For at least 9-11 hours you are at work or going to work or coming home from work. For 8 hours you should be sleeping. This leaves less than only 5-7 hours per day to actually live.

How sadly and miserably unsatisfying!

This is how it is, but if it has not always been this way does it have to remain this way forever? It has not always been this way.

In the early 1800's Thomas Jefferson witnessed the emergence of modern day capitalism and he was none too pleased. He spoke of what he called "wage slavery" where men should be selling their time instead of their wares and he waned that such a society of wage slaves was ill equipped to run a free nation. How could people whose only commodity was time ever have enough time to do the reflection and introspection it takes to make decisions for the whole nations?

In the past people didn't sell their time, they sold their wares. Life and work were the same thing; they did the work within their family and community and as they were producing their wares, for sale, they were raising their children, spending time with their wives, meeting and talking with people in their community, and debating the issues of the day.

In ancient Greece the people came to an "Agora" where they sold the wares they had produced or purchased and their most of the debates and discussions about politics and philosophy took place.

This idea we work and live at home is relatively new and if it is relatively new it is not necessarily the only alternative.

Now consider a different question: is the way you earn a living for your family in harmony with your life, is it part your life?

Getting to that spot, where work and life are in harmony, so that we live as we work and work as we live, is going to take some doing: it will require a change in attitude, approach, goals, and methods.

The subject of this article is the balance between work and life and we have concluded that instead of balancing these two we should work towards making them one. How can we do this?

Here are some thoughts to get your started:

1. work towards being self-employed using your passions and talents to create a family business,
2. work towards actually enjoying your relationships with your coworkers by sharing in community with them (this may take telling them about this concept),
3. works towards making sure your job or business is something you really like doing,
4. where possible, involve your family in your business/work by telling them about it in a positive way, teaching your children what do, or urging your superiors to have more family type events like an open house or special days for children and the like,
5. use technology, try and keep in touch with you wife or husband during the day, schedule times when you can both take a break and spend that time talking on the phone,
6. don't over extend yourself at work, make sure your schedule time for the family and down time for yourself,
7. look for work, in business for yourself or working for others, that is more family-friendly.

These are just a few ideas.

The goal should not so much be balance as harmony: you want to be able to LIVE while you are working and you want to be able to work while you live.

Life is a balancing act, and in American society, it is safe to say that almost everyone is seeking work life balance. But what exactly is work life balance? We have all heard the term, and many of us complain that we don't have enough of it in our lives. Among men and women alike, the frustrating search for work life balance is a frequent topic of conversation, usually translated into not enough time and/or support to do, to handle, to manage ... our work commitments or personal responsibilities.

> "Juggling competing demands is tiring if not stressful and brings lower productivity, sickness, and absenteeism, so work life balance is an issue for all employees and all organizations".

The meanings of work life balance has chameleon characteristics. It means different things to different groups, and the meaning often depends on the context of the conversation and the speaker's viewpoint. The following are working definitions of terms used regarding work life balance; some definition overlap and some are continuing to evolve.

Work family: A term more frequently used in the past than today. The current trend is to use titles that include the phrase work life, giving a broader work live connotation or labeling referring to specific areas of support (e.g, quality of life, flexible work options, life balance, etc.).

Work family conflict: The push and pull between work and family responsibilities.

Work life balance from the employee viewpoint: The dilemma of managing work obligations and personal/family responsibilities.

Work life balance from the employer viewpoint: The challenge of creating a supportive company culture where employees can focus on their jobs while at work.

Family-friendly benefits: Benefits that offer employees the latitude to address their personal and family commitments, while at the same time not compromising their work responsibilities.

Work life programs: Programs (often financial or time-related) established by an employer that offer employees options to address work and personal responsibilities.

Work life initiatives: Policies and procedures established by an organization with the goal to enable employees to get their jobs done and at the same time provide flexibility to handle personal/family concerns.

Work life culture: The extent to which an organization's culture acknowledges and respects the family responsibilities and obligations of its employees and encourages management and employees to work together to meet their personal and work needs.

4. PERSONAL LIVES AND FAMILY VALUES TO THE FOREFRONT

The American work ethic remains intact, yet in recent personal and family lives have become critical values that Americans are less willing to put on hold, put aside, or ignore, for the sake of work. Over time, the American workforce has begun to change course from being willing to spend every hour working to learning to manage the complexities of modern living. In addition, the impact of the terrorist attacks of September 11 has led many people to reevaluate their lives and consider the meaning of work.

Consequently, Americans are looking for options that allow for both a personal and family life, and many seek ways to have it all. As a result, the United States (U.S.) management philosophy that expects employees to put work first, ahead of personal lives and family commitments, is becoming less accepted.

- In a 2001 survey conducted by the Radcliff Public Policy Center, 82% of men and 85% of women ages 20 to 39 placed family time at the top of their work life priorities.
- In a 2001 study by Rutgers University and the University of Connecticut, 90% of working adults said they are concerned they do not spend enough time with their families.

5. THE CHANGING FACE OF FAMILY

With the growing diversity of family structures represented in the workforce in the new millennium, it is important that human resource professionals better understand the interface of work and family relationships and the resulting impact in the workplace.

Research by Parasuraman and Greenhaus (2002) documented that segments of the workforce may be subject to unique work/family pressures, yet often have few sources of support. The under-representation of these groups of individuals with potentially difficult types of work/family pressures represents a major gap in work/family research and employers' understanding of their needs. Typically, studies have focused on employed men and women who are married or living with a partner or those with children. Omitted from research are single-earner mothers and fathers, single and childless employees with extensive responsibility for eldercare, blended families with children from both partners' prior marriages, families with shared custody of children, and grandparents raising their grandchildren.

- From both the employer and employee viewpoint, the changing nature of what constitutes family is one of the complications of today's society.
- As human resource professionals design policies and programs to address employee retention, job satisfaction, employee morale, and productivity, this research warrants serious consideration.

6. A PIVOTAL STUDY

In their highly acclaimed book, Work and Family-Allies or Enemies, Friedman and Greenhaus (2000), two leaders in work life balance, bring forth new evidence to help us understand choices we make as employers and individuals regarding work and family. This pioneering study of more than 800 business professionals considered values, work, and family lives and found that "work and family, the dominant life roles for most employed women and men in contemporary society, can either help or hurt each other."

To handle work life balance, Friedman and Greenhaus emphasize that working adults learn to build networks of support at home, at work, and in the community. Conflict between work and family has real consequences and significantly affects quality or family life and career attainment of both men and women. The consequences for women may include serious constraints on career choices. Limited opportunity for career advancement and success in their work role, and the need to choose between two apparent opposites—an active and satisfying career or marriage and children. Many men have to trade off personal and career values while they search for ways to make dual-career families work, often requiring them to embrace family roles that are far different, and more egalitarian than those they learned as children.

This research reveals a compensatory effect between two forms of psychological interference: work-to-family and family-to-work. Specifically, support from two domains (partner and employer) has a significant impact on one another. The impact of partner support is greater when business professionals feel their employers are unsupportive of their lives beyond work. Conversely, for employees with relatively unsupportive partners, the employer family-friendliness reduces role conflicts more than partners. Thus, one source of support compensates for the lack of the other.

Looking at behavioural interference of work on family, the picture changes. In this case, the whole is greater than the sum of its parts: the combined impact of employer and partner support leads to a greater reduction in conflict than does independent employer or partner support.

7. STRESS AND THE CONSEQUENCES FOR EMPLOYER AND EMPLOYEE

We live in stressful times, and each of us deals with stress every day. In the past three years, an increasing number of employees surveyed indicate they are struggling with work life balance.

- A work life balance survey conducted in 2002 by True Careers states that 70% of more than 1,500 respondents said they don't have a healthy balance between their personal and work lives.
- "Holding a job, Having a life: Strategies for change", 2001 study by the Work Institute of America points out that employee-driven solutions help reduce overtime, stress, and workloads, and increase flexibility and family and leisure time.

Scientist agree that in moderate amounts stress can be benign, even beneficial, and most people are equipped to deal with it. However, increasing levels of stress can rapidly lead to low employee morale, poor productivity, and decreasing job satisfaction. Some of the specific symptoms that relate directly to productivity in the work environment are abuse of sick time, cheating, chronic absenteeism, distrust, embezzlement, organizational sabotage, tardiness, task avoidance, and violence in the workplace. Other serious repercussions are depression, alcohol and drug abuse, marital and financial problems, compulsive eating disorders, and employee burnout. Dr. Bruce S. McEwen, director of the neuroendocrinology laboratory at Rockefeller University, has been studying stress for more than three decades. As he notes "blaring car alarms, controlling bosses, two-career marriages, six-mile traffic jams, and rude salesclerks were simply not part of the plan.

Employee Assistance Programs (EAPs), offered by many employers, are an excellent resource for employees under stress. Employee Assistance Programs (EAPs) provide a myriad of services, form *drug and alcohol* abuse counseling to addressing family and marriage problems, financial and legal difficulties, and stress-related problems. In addition, in line with the times and the increasing stress levels in our society, a new profession has emerged: work life professionals. The concept of work life professionals originally develop as an extension of wellness programs (established as early as 1933) and Employee Assistance Programs (EAPs) (created in the 1940s). The Alliance for Work Life Progress (*http://www.awlp.org*), the national association for work life professionals, leads and promotes work life initiatives in business, academia, and the public sector to support a healthier work life balance between work, personal, and family life. Work life programs represent a largely untapped workplace solution that have the potential to significantly address stressful work environments.

8. THE EMPLOYER'S PERSPECTIVE: RETURN ON INVESTMENT (ROI)

An employer's commitment to work life initiatives is influenced by the perception of whether or not such initiatives have a positive return on investment. In recent years, employers increasingly realize that the quality of an employee's personal and family life impacts work quality and that there are concrete business reasons to promote work and family integration.

9. DIVERSITY AND WORK LIFE BALANCE

While some companies may view diversity and work life balance as separate functions, the business case for managing diversity is, in large part, the same for work life balance. Both diversity and work life initiatives promote employee commitment, improve productivity, lower turnover, result in fewer employee relations challenges, and decrease the likelihood or unethical business practices.

Diversity and work life initiatives can be found at the core of the new social contract being negotiated between employees. "The basic outline of the social contract, as it has emerged during the past several years, calls for workers to commit their best contributions and greatest energies to the job in return for interesting work, respectful treatment, developmental opportunities, and an environment that responds to individual needs. Where those provisions conflict (e.g., the degree of commitment and energy expected by employers *versus* the flexibility required by employees), the expertise of both diversity management and work life professionals ill be critical to find win-win solutions."

10. IS YOUR ORGANIZATION CULTURE READY FOR WORK LIFE INITIATIVES?

"A common thread that links the reasons work life benefits go unused in organizational culture". Before establishment work life initiatives, it is important to know if the organization's culture is open and ready to support work life programs. The path to determining culture readiness may be as formal as using an employee survey assessment or as simple as a thoughtful judgment made by the organization. The following provides food for thought regarding whether an organization is ready to begin work life initiatives.

As with most change initiatives, work life programs require support from senior management. In addition, for the work environment to be ready for work life benefits, it is helpful to have a "corporate culture that encourages employees to look at business in an entirely different way and supports and accepts employees as individuals with priorities beyond the workplace."

"Life cycles are another consideration. People need different things at different times of their lives," explains Sandra Burud, Ph.D., principal of Bright Horizons Family Solutions, discussing the business strategy of work life initiatives, "In the factory days, everything was standardized and synchronized—that's when the original benefits package was designed. It doesn't fit anymore," states Burud. "Furthermore," she continues, "Employers are realizing that work should be intrinsically interesting and satisfying to employees, and these are the folks who produce the best work. The manager's job is to get out of the way. The move from extrinsic rewards to intrinsic rewards has an impact on work life initiatives."

One of the challenges of work/life initiatives—from both the employer and the employee viewpoint—is equitability, which has been cited as a major concern regarding work life initiatives. When organizations are establishing work life programs, it is important to consider the purpose of the programs and whom they serve. For example, do the work life programs serve all employees or are they aimed toward employees who are parents or who are dealing with their elderly parents?

Another aspect of judging organizational readiness for work life programs is the employees' view of perceived support of the organization. The 1999 study by Thompson, Beauvais, and Lyness at the City University of New York and University of Rhode Island considered the links between an organization's work/family culture, the extent to which employees used work life benefits, the extent of work family conflict, and the employees' intention to stay with their company. To determine the culture readiness of an organization for work life initiatives, the researchers developed a series of questions to measure supportive work life culture, addressing perceived managerial support, negative career consequences for devoting time to family concerns, and organizational time demands and expectations that interfere with family responsibilities. The study revealed that more work/family benefits translated to greater commitment, less work/family conflict, and less intention to leave. Interestingly, the study result confirmed anecdotal evidence that a supportive work/family culture is closely related to work attitudes and perceived managerial support linked with less intentions to leave the organization.

11. COMMUNICATION IS KEY

Communication about work life programs is essential. Although an organization may offer a rich menu work life of benefits, the desired effect—yielding proactive business results—is unlikely to occur if employees do not know about the programs or understand them. Human resource professionals should consider four critical questions: (1) does the company culture truly support work/life benefits; (2) does the management philosophy, starting with senior management, sincerely endorse work life benefits; (3) do managers an supervisors understand the impact the work life balance has on their workforce; and (4) are employees aware of and do they understand the company's work life programs?

If a company already offers work/life benefits, the next step may be to repackage and recommunicate them so employees see how these benefits offer ways to manages work/family conflict. In addition, developing a human resources strategy that is clearly integrated with the company's mission will demonstrate how committed the organization is to employee needs. The following are suggestions to promote work/life programs:

- Review the human resources strategy to see if it supports the company's mission.
- Through questionnaires or focus groups, find out what employees feel about work life balance.
- Align work life initiatives with HR strategy (e.g., employer of choice).
- Create a work life award program using non cash incentives aligned with business objectives.

12. CAN WORK LIFE INITIATIVES BE MEASURED?

Work life initiatives create positive employer branding, promote being an employer of choice, foster organizational citizenship, and support diversity initiatives. Building a strategic business case for work life initiatives requires hard data documenting positive results on the balance sheet. Toward this end, there have been a number of studies dedicated to the financial impact of work life programs in the last decade.

One such study (1997) was conducted by WFD (formerly Work/ Family Directions) with research focused on work/life programs with a number of clients (e.g., DuPont, Johnson and Johnson, Hoechst Celanese, IBM, and others). At DuPont, for example, "employees who used the company's work/life programs were 45% more likely to agree strongly that they will 'go the extra mile' and are least likely to feel overwhelmed or burned out." Data from Hoechst Celanese documented that "some 60% of those surveyed reported the ability to balance work with personal and family responsibilities was of great importance in their decision to remain with the company".

There are several key factors to be considered in measuring Return on Investment (ROI) of work life programs: the audience, the culture, designing the data collection process, integration findings and making projections, and identifying the company's primary goal. For many organizations, however, quantifying the data presents the greatest challenge. The best place to start is to consider five key areas: (1) employee time saved; (2) employee retention; (3) increased motivation and productivity; (4) absenteeism; and (5) decreased health care costs and stress-related illnesses.

13. EMPLOYEE TIME SAVED

One of the most direct and measurable financial benefits of work/life initiatives is the amount of time saved by employees by using assistance, such as an EAP, to address and solve personal problems. Work Family Direction (WFD) conducted research with clients across industries with 300 employees and documented that an average of 17 hours per year were saved when employees used a consulting service that provided counseling, referrals, and research in areas such as parenting, education, childcare, adoption, and eldercare. According to this research, a minimum of 80% of this saved time would have involved employees working on their personal problems during the day, since most assistance organizations are only open during regular business hours.

14. EMPLOYEE RETENTION

The cost of employee turnover and accompanying loss of valuable company knowledge can be significant. Work life programs offer a solution to retention woes. A prime example is IBM'S 1992 workforce survey that documented "the highest performers are the most likely to consider their ability to balance work and personal responsibilities in a decision to stay with the company. Employees overall rated work-balance issues as sixth of16 factors that keep them with the company."

15. INCREASED MOTIVATION AND PRODUCTIVITY

Research indicates that company commitment to work life initiatives is closely aligned with employee motivation and productivity. A study focusing on work, well-being and stress illustrates this link, finding that "45% of men and 50% of women would turn down a promotion if the new position would leave them with less time for their personal or family life."

16. ABSENTEEISM

Research has documented that work life programs can reduce absenteeism. Johnson & Johnson "found that there was a 50% decline in absenteeism among employees who used flexible work options and family leave policies".

17. DECREASED HEALTH CARE COSTS AND STRESS-RELATED ILLNESSES

With increasing company focus on the high cost of health care, work life programs are becoming an intelligent choice to help lower the number of health care claims. According to the American Institute on Stress, one million workers are absent due to stress-related complaints, and American firms lose more than 5 million workdays annually due to illness, more than half stress-related. A 1992 study by North-western National Life Insurance Co. found that "72% of all workers surveyed experienced three or more stress-related illnesses 'somewhat or very often.'"

The entire story regarding work life programs, however, cannot be told strictly by quantitative measurements. To present a solid argument for work life policies and programs, the positive yet hard-to-measure factors of work life initiatives—corporate reputation, public relations, improved community relation, increased employee loyalty, and enhanced recruitment—should also be considered.

18. FAMILY-FRIENDLY BENEFITS

"Family-friendly firms have a significant impact on the lives and careers of business professionals who work in them." A review of more than 30 surveys regarding work life balance published from 1997 to 2003 reveals that the number of employers who offer family-friendly benefits has dramatically increased. According to the SHRM[R] 2003 Benefits Survey by the Society of Human Resource Management (www.shrm.org/surveys—available late June 2003), the percentage of employers offering family-friendly benefits continues to increase. "The survey documents that the top five family-friendly benefits offered are:

Dependent care flexible spending accounts (71% of respondents).

- Flextime (55% of respondents).

- Family leave above required leave of the federal Family and Medical Leave Act (FMLA) (39% of respondents).
- Telecommuting on a part-time basis (34% of respondents).
- Compressed workweeks (31% of respondents).

In 2002, the Work Life Today Survey conducted by the Alliance for Work/Life Progress revealed that an average of 56% of employees used work life benefits when offered. This survey indicated that the top two benefits provided were EAPs and flexible schedules. It should be noted, however, that each survey is organized differently, whether it be family-friendly or work life. For example, the SHRM[R] 2003 Benefits Survey EAPSs under the health section.

With the ever-increasing number of surveys and studies available that focus on family-friendly benefits, human resource professional who are considering offering work life programs, or expanding current work life initiatives, have substantial and convincing data to support their positions.

19. WORK LIFE BALANCE TRENDS

The challenge of work life balance in our society is unlikely to disappear. Identified and discussed as follows are four work life balance trends. Awareness of these trends will place the human resource professional in a position to better educate management and work closely with employees.

20. NEW RESEARCH

The topic of work life balance is getting a great deal of attention in the academic and corporate worlds, and new research is continually being conducted. The following is an example of new research in the growing of work life balance.

A 2003 study reveals that employees are often preoccupied with work when not working, and when in the company of family and loved ones, experience an inability to be meaningfully engaged in no work spheres. As researcher Ezzedeen explains, "modern work has become knowledge based, fluid, and intellectual; overworked people think about work all of the time. For many people, work has become cognitively intrusive.

To understand work life balance, Professor Swiercz and Doctoral Candidate Ezzedeen of The George Washington University developed and tested a cognitive approach, the "Cognitive Intrusion of Work." In simple terms, this means that work life balance is not just about finding "physical time" to do all that needs to be done. Instead, and more importantly, it is about the "cognitive space" necessary to process, organize, and respond to the thinking demands of life within a complex society.

Ezzedeen and Swiercz found that the cognitive intrusion of work results in lower job satisfaction, less happiness, a greater incidence of work life conflict, and more frequent burnout. Furthermore, their findings challenge long-held assumptions about the role of personality, gender, family status and age with regard to work life balance. They found that the experience of intrusion transcends demographics and personality and is rooted instead in the design of the job and the organizational culture of the employer. Commenting on the potential impact of this research on organization policies ad practices, Professor Swiercz states, "this research not only give employers a new benchmarking tool, it also a new theoretical basis for understanding this important social issue from a research perspectives."

21. ELDERCARE

One of the most significant trends in work life balance is the increasing focus on eldercare. Researchers point out that work life professionals anticipate eldercare will become a major issue in the coming years. According to census data, 13% of Americans are age 65 or older, and by 2030, 20% of Americans (about 70 million) will be over age 665. The population age 85 and older is the fastest-growing segment of the older population, growing by 274% over the past 25 years. In addition, the Labour Project for Working Families states that 40% of people caring for elders also have childcare responsibilities. These trends have led to the baby boomer generation being known as the "sandwich generation."

A growing number of companies offer work/life programs addressing eldercare. According to the SHRM[R] 2003 Benefits Survey, while not as mainstream as childcare programs, some organizations offer various forms of eldercare. Eldercare programs include eldercare referral service (20% of respondents), emergency eldercare (3% of respondents), subsidy of eldercare cost (2% of respondents), and paid eldercare (1% of respondents). The 2001 study by Hewitt Associates points out that some employers offer dependent care programs that include help with childcare expenses and assistance with eldercare.

Employee Assistance Programs can play an important role in a organization's eldercare program. Some of the lesser-known benefits of an EAP are referrals to community programs and contact regarding eldercare issues. Human resource professionals could significantly increase EAP usage by learning more about EAP eldercare support and communicating this to employees. Through these actions, HR professionals can emphasize the importance of EAPs to management and at the same time leverage EAP benefits to employees.

22. WORK LIFE BALANCE IN THE RELIEF WORLD

In the "relief world," comprised of organizations with employees and volunteers that provide service and care to communities in need locally and worldwide, the demands of an aging population in the coming decade are increasing the current strong competition for qualified individuals upon which relief organizations depend. Not only will there be fewer young, keen and free-to-travel individuals who will want to be convinced that agencies are caring, 'best-practice' employers, but more skills and experience will be possessed by older staff likely to have families and other commitments and thus, different priorities for their work life balance." Therefore, organizations that provides relief services may experience increasing difficulty staffing and retaining employees due to the pull of family commitments at home.

Furthermore, the challenges of work life balance will no doubt impact recruitment, retention and willingness to serve in hardship locations. In view of these factors, voluntary organizations and aid agencies whose missions and services are critical in many parts of the world may well have an even greater need for work life programs to attracts and retain staff.

23. TOTAL LIFE PLANNING

Total life planning is a new and innovative approach to work life benefits and helps employees examine important aspects of their professional and personal lives and understand how they relate. Their goal is "to encourage employees to look at their lives as a whole and assess relationships, emotional and physical well-being, careers, spirituality, and their personal financial situation. From these programs, employees can assess their available choices to improve balance in their lives and develop and individualized life plan the most successful programs set a goal-oriented environment with a meaningful and transformational component for each individual."

The concept of total life planning represents a paradigm shift in our society. One of the major benefits is renewed employee energy, enthusiasm for work, and increased productivity. Total life planning programs may be ordered in conjunction with benefits such as health, life, and disability insurance, or on a standalone basis. However, not all organizations may wish to consider total life planning, as it brings topics into the workplace that have traditionally been considered private.

24. IS WORK LIFE BALANCE ACHIEVABLE?

Work life balance in important for better productivity and a harmonious is work life, but maintaining such balance is the biggest dilemma of the corporate world. So, how do you accomplish that? We have discussed work life balance many a times. But do we know really think this is achievable? Yes, say more than 60% respondents on a shine poll. A lot of people even believe that **women are better** at this job. While we all understand that work life balance is important for better productivity and a harmonious life, striking a balance between your personal and professional lives is the biggest dilemma of the modern day work world and a necessity as well. Otherwise, you will be left puffing, juggling between deadlines and socializing or between business travels and vacations. When you don't have enough time for yourself, take damage control measures. But there's no one-size-fits-all approach to bring work/life balance. These 5 tips can seem useful when you try to restore equilibrium between your professional and personal commitments:

Work on the available option: Most organizations understand the importance of employee-oriented policies. If you like job but would like to manage time better, talk to the management about a flexible work routine. You can seek **telecommuting** or choose a part-time option. When you talk about the option you want, stress on the fact that it will improve your productivity and efficiency and thus, make it counter-productive for the company. Some professions are more demanding in nature. If you feel you can't maintain work/life balance; you can mull a change in career. You can do something within the same field but considerably less stressful. Don't forget that you can risk giving up other things like good money, more responsibilities, etc.

Remain grounded by not going overboard: You are good at work and hence, handle a lot of responsibilities. But don't try to accomplish too many things in a short span of time. That is simply not feasible; rather find means to simplify work. Priorities work and you will be able to manage time. Try to accomplish work in a human avatar and stop believing that you're a superman/woman. Try avoid stress situations.

Share and delegate responsibilities: Without checking into your efficiency meter, it can still be safely said that you can do only as much as your body and mind permit. You and your boss are both aware of this fact. So, if you're in a supervisory role, **delegate work** to people giving them independent responsibility. Thus, you mentor others to be future leaders and get to de-load some work too. Also share work among colleagues, so that it's not just you, who is working always.

Make work life balance a target: Work life balance is a tough call and one has to pursuer it with perseverance to reach there. Work on it as a target as it is dependent on your goals and responsibilities at the end of the day. Try and work on the ones that top your strategy agenda.

Slow down and mediate: Devote sometime to enjoy the simple pleasures and joys of life. Remember, child psychology, David L. Weatherford's words: **"When you run so fast to get somewhere, you miss half the fun of getting there".** Meditate as it makes you feel fresh. Keep time aside for yourself to rejuvenate. Don't get bogged down with work. Keep weekends and vacations for relaxing with family, detaching yourself from work.

It's very much possible to reach your professional milestones without affecting your personal life. And the key to success is to smart. Prioritise the important things in both your personal and work life and try to reach the goals in a systematic manner, without getting stressed or sacrificing personal life at the professional's altar.

25. THE MODERN WORK-LIFE POWER BALANCE

The demand for work-life power balance solutions by employees and managers is expanding at an unprecedented rate. As a result, work-life power balance is an increasingly hot subject in boardrooms and government halls today. In the future this will be the most important topic to discuss for executives and HR professionals.

In the following paragraphs you may locate very important information in regards to the work-life power balance and tips on how to carry out and manage it within your business. We'll discuss the crucial points of why this really is so essential for you along with your team. In following report we'll chat about the main elements of profitable work-life-power balance implementation.

Because of the business and personal links between these critical issues, executive now recognize that organizational objectives and individual work-life gorals are not "either/or" options. Leaders came to the conclusion that it's most all about how their staff perform but far more about that they also have a personal life. This is one major task of HR and the senior management to make this happen in a way the people can work a lot more efficient and client focused. There happen to be quite a few initiatives on promoting the new work-life power balance scheme but most of them have failed and not been applied correctly into the organizations. Besides that there have also been some positive predictions and examples where it has been implemented. You can find examples of errors that have already been made public so you do not need to do the identical mistakes all more than again. Let's start with learning from the past.

Doing something and getting it done is not the same. Even though some HR professionals are carrying out a good deal to apply a work-life power balance, they fail. Even if you do a lot, you won't get things done sometimes.

You have to start immediately if you want to get something done that produces great output.

Try to get your first actions running as soon as possible.

Whether you are just introducing a work-life plan or generating an already excellent one far better, you might be substantially improving your organization's bottom line benefits and changing individuals' lives for the far better. This is some of the most important work you are able to do. All the participants will be enjoying a difference in their private and professional life.

26. CONCLUSION

Work life programs have the potential to significantly improve employee morale, reduce absenteeism, and retain organizational knowledge, particularly during difficult economic times. In today's global marketplace, as companies aim to reduce costs, it falls to the human resource professional to understand the critical issues of work life balance and champion work life programs. Be it employees whose family members and/or friends are called to serve their country, single mothers who are trying to raise their children and make a living, Generation X and Y employees who value their personal time, couples struggling to manage dual-career marriages, or companies losing critical knowledge when employees leave for other opportunities, work-life programs offer a win-win solution/situations for employers and employees.

CHAPTER

25

Knowledge Management

ROHINI GUPTA SURI, DR. AMRIK SINGH SUDAN AND LEKHA NAGAR

1. INTRODUCTION

Knowledge Management is one of the hottest topics today in both the industry would and information research world. In our daily life, we deal with huge amount of data and information. Data and information is not knowledge, until we know how to dig the value out of it. This is the reason we need knowledge management. Unfortunately there is no universal definition of knowledge management, just as there is no agreement as to what constitute knowledge in the first place, we choose the following definition for knowledge management for its simplicity and broad context.

Knowledge Management (KM) refers to a multi-disciplined approach to achieving organizational objectives by making the best use of knowledge. Knowledge Management (KM) focuses on processes such as acquiring, creating and sharing knowledge and the cultural and technical foundations that support them.

Knowledge Management may be viewed in terms of:

- **People**—How do you increase the ability of an individual in the organization to influence others with the knowledge.
- **Processes**—Its approach varies from organization to orgnisation. There is no limit on the number of processes.

- **Technology**—It needs to be choosen only after all the requirements of a Knowledge Management initiative have established.
- **Culture**—The biggest enabler of successful knowledge-driven organsations is the establishment of a knowledge focused culture.
- **Structure**—The business processes and organizational structures that facilitate knowledge sharing.
- **Technology**—A crucial enabler rather than the solution.

Knowledge Management (KM) comprises a range of practices used in an organization to identify, create, represent, distribute and enable adoption of insights and experiences. Such insights an experience comprise knowledge, either embodied in individuals or embedded in organizational processes or practice. An established discipline since 1995. Knowledge Management includes courses taught in the fields of business administration, information systems, management and library and information sciences. More recently other fields, to include those focused on information and media, computer science, public health and public policy also have started contributing to Knowledge Management (KM) research. Many large companies and non-project organizations have resources dedicated to internal Knowledge Management (KM) efforts, often as a part of their 'Business Strategy', 'Information Technology', 'Human Resource Management' Departments. Several consulting companies also exists that provide strategy and advise regarding Knowledge Management (KM) to these organizations.

Knowledge Management (KM) efforts typically focus on organizational objectives such as improved performance, competitive advantage, innovation, the sharing of lessons learned and continuous improvements of the organization. Knowledge Management (KM) efforts overlap with organizational learning and may be distinguished from by a greater focus on the management of knowledge as a strategic asset and a focus on encouraging the exchange of Knowledge. Knowledge Management (KM) can help individuals and groups to share valuable organizational insights to reduce redundant work, to avoid reinventing the wheel *per se*, to reduce training time for new employees, to retain intellectual capital as employees turnover in an organization and to adopt to changing environments and markets.

This paper seeks to discuss in detail the following issues of knowledge Management:

1. Introduction of Knowledge Management;
2. What is Knowledge Management Related To;

3. History of Knowledge Management;
4. The Value of Knowledge Management;
5. Knowledge Management Today;
6. Knowledge Management Drivers;
7. Knowledge Management Objectives;
8. Knowledge Management Return on Investment (ROI);
9. Technologies—That Support Knowledge Management;
10. Present and Future State of Knowledge Management;
11. The Future of Knowledge Management;
12. The Effect of Knowledge Management (KM) Databases;
13. The Implications of Knowledge Management;
14. Knowledge Management Dimensions;
15. Knowledge Management Strategies;
16. Knowledge Management Motivations;
17. Knowledge Management Technologies;
18. Knowledge Management and implication in India;
19. The Challenges of Knowledge Management; and
20. Summary and Conclusion.

For thousands of years, humans have been discussing the meaning of knowledge, what it to know something, and how people can generate and share new knowledge.

—Rudy L. Ruggles, Knowledge Management Tools, 1997

Simply put, knowledge management is about capturing knowledge gained by individuals and spreading it to others on the organization is order to manage it better.

Although a number of management theorists have contributed to the evolution of knowledge management, such as Peter Drucker and Peter Senge/Ikuhiro Nonaka makes knowledge management an official discipline when he is appointed as the first distinguished professor of knowledge at the University of California in 1997.

Knowledge Management can be views from two perspectives:

- **Knowledge=Object**: relies upon concepts from "Information Theory" in the understanding of knowledge. These researches and practitioners are normally involved in the construction of information management systems, Al, reengineering, etc. This group builds knowledge systems, while the next group changes the way we use knowledge, which ultimately changes human behaviour.
- **Knowledge=Process**: concepts from philosophy, psychology, and sociology. These researchers and practitioners are normally involved in education,

philosophy, psychology, sociology, etc. and are primarily involved in assessing, changing and improving human individual skills and behaviour.

Knowledge Management, as we know it today, is generally considered to have begun in the 1950's when Alfred Sloan divisionalized General Motors. It sent a message of the techniques necessary for large-scale business management. Although that view has somewhat changed today with human talent as being viewed as the primary competitive differentiator.

With knowledge management, the unmeasurable must be measured.

"Every organization—not just businesses—needs one core competence: innovation. And every organization needs a way to record and appraise its innovative performance."

—Peter F. Drucker, *Harvard Business Review*, 1995

There are normally thought to be two kinds of knowledge:

- **Explicit Knowledge**—It can be expressed in words and numbers and shared in the form of data, scientific formulae, product specifications, manuals, universal principles, etc. This kind of knowledge can be readily transmitted across individuals formally and systematically. Also, it can easily be processed by a computer, transmitted electronically, or stored in databases.
- **Tacit Knowledge**—This is highly personal and hard to formalize thus making it difficult to communicate of share with others. Subjective insights, intuitions and hunches fall into category of knowledge. Furthermore, tacit knowledge is deeply rooted in each individuals' actions and experiences, as well as in the ideals, values and emotions that they embrace. The subjective and intuitive nature of tacit knowledge makes it difficult to process or transmit the acquired knowledge in any systematic or logical manner. For tacit knowledge to be communicated, it must be converted into words, models, or numbers that anyone can understand. Also, there are two types of tacit knowledge:
  - o The "technical" dimension—This encompasses the kind of informal and hard-to-pin-down skills or crafts often captured in the term "know-how". For example, master craftsmen develop a wealth of expertise at their fingertips, after years of experience. But they often have difficulty articulating the technical or scientific principles behind

what they know. Highly subjective and personal insights, intuitions, hunches and inspirations derived from bodily experience fall into this dimension.

- o The "cognitive" dimension—It consists of beliefs, perceptions, ideals, values, emotions and mental models so ingrained in us that we take them for granted. Though they cannot be articulated very easily, this dimension of tract knowledge shapes the way we perceive the world around us.

Robert Sutton, a professor of organizational behaviour at Stanford's University School of Engineering, says companies have wasted hundreds of millions on worthless knowledge management systems [Computerworld (January 3, 2000, p. 28)]:

- o The most valuable employees often have the greatest disdain for knowledge management. Curators badger these employees to enter what they know into the system, even though few people will ever use the information.
- o The managers of these systems know a lot about technology, but little about how people actually use knowledge on the job.
- o Tacit knowledge is extremely difficult to capture into these systems, yet it is more critical to task performance than explicit knowledge.
- o Knowledge is of little use unless it is turned into products, services, innovations, or process improvement.

Knowledge management systems work best when the people who generate the knowledge are the same people who store it, explain it to others, and coach them as they try to implement it. These systems must be managed by the people who are implementing what is known, not those who understand information technology.

2. WHAT IS KNOWLEDGE MANAGEMENT RELATED TO?

Knowledge Management draws from a wide range of disciplines and technologies:

- Cognitive science
- Expert systems, artificial intelligence a knowledge-base management systems (KBMS)
- Computer-supported collaborative work (groupware)
- Library and Information Science
- Technical writing

- Document management
- Decision support systems
- Semantic networks
- Relational and object databases
- Simulation
- Organizational science
- Object-oriented information modeling
- Electronic publishing technology, hypertext, and the World Wide Web; help-desk technology
- Full-text search and retrieval
- Performance support systems

Although around 20 kinds of disciplines and study areas were listed above, there is no way to include all of the related subjects to knowledge management.

3. HISTORY OF KNOWLEDGE MANAGEMENT

Knowledge Management (KM) efforts have a long history, to include on-the-job discussions, formal apprenticeship, discussion forums, corporate libraries, professional training and mentoring programs. More recently, with increased use of computers in the second half of the 20th century, specific adaptations of technologies such as knowledge bases, expert systems, knowledge repositories, group decision support systems, and computer supported cooperative work have been introduced to further enhance such efforts.

A broad range of thoughts on the Knowledge Management (KM) discipline exists with no unanimous agreement; approaches vary by author and school. As the discipline matures, academic debates have increased regarding both the theory and practice of Knowledge Management (KM), to include the following perspectives:

- Techno-centric with a focus on technology, ideally those that enhance knowledge sharing and creation.
- Organisational with a focus on how an organization can be designed to facilitate knowledge processes best.
- Ecological with a focus on the interaction of people, identity, knowledge, and environmental factors as a complex adaptive system akin to a natural ecosystem.

Regardless of the school of thought, core components of Knowledge Management (KM) include People, Processes, Technology (or) Culture, Structure, Technology, depending on the specific perspective. Different Knowledge Management (KM) schools of thought include various lenses through which Knowledge Management (KM) can be viewed and explained, to include:

- Community of practice
- Social network analysis
- Intellectual capital
- Information theory
- Complexity science
- Constructivism

1. During 70's

A number of management theorists have contributed to the evolution of Knowledge Management:

- o *Peter Drucker*: Information a Knowledge as organizational resources.
- o *Peter Senge*: "Learning Organisation".
- o *Leonard Barton*: Well-known case study of "Chaparral Steel", a company having knowledge management strategy.

2. During 80's

- o Knowledge (and its expression in professional competence) as a competitive asset was apparent.
- o Managing knowledge that relied on work done in artificial intelligence and expert systems.
- o Knowledge management-related articles began appearing in journals and books.

3. During 90's until now

- o A number of management consulting firms had begun in-house knowledge management programs.
- o Knowledge management was introduced in the popular press, the most widely read work to date is Ikijiro Nonaka's and Hirotaka Takeuchi's. The Knowledge-Creating Company: How Japanese Companies Create the Dynamics of Innovation (1995).
- o The International Knowledge Management Network (IKMN) went online in 1994.
- o Knowledge management has become big business for such major international consulting firms as Ernst and Young, Arthur Andersen, and Booz-Allen and Hamiltion.

4. THE VALUE OF KNOWLEDGE MANAGEMENT

Some benefits of Knowledge Management (KM) correlate directly to bottom-line savings, while others are more difficult to quantify. In today's information-driven economy, companies uncover the most opportunities—and ultimately derive the most value—from intellectual rather than physical assets. To get the most value from a company's intellectual assets, Knowledge Management (KM) practitioners maintain that knowledge must be shared and serve as the foundation for collaboration. Yet better collaboration is not an end in itself; without an overarching business context, Knowledge Management (KM) is meaningless at best and harmful at worst. Consequently, an effective Knowledge Management (KM) program should help a company do one or more of the following:

- Foster innovation by encouraging the free flow of ideas.
- Improve decision-making.
- Improve customer service by streamlining response time.
- Boost revenues by getting products and services to market faster.
- Enhance employee retention rates by recognizing the value of employees' knowledge and rewarding them for it.
- Streamline operations and reduce costs by eliminating redundant or unnecessary processes.

These are the most prevalent examples. A creative approach to Knowledge Management (KM) can result in improved efficiency, higher productivity and increased revenues in practically any business function.

5. KNOWLEDGE MANAGEMENT TODAY

According to a recent IDC report, knowledge management is in a state of high growth, especially among the business and legal services industries. As the performance metrics of early adopters are documenting the substantial benefit of knowledge management, more organizations are recognized the value of leveraging organizational knowledge. As a result, knowledge management consulting services and technologies are in high demand, and knowledge management software is rapidly evolving.

6. KNOWLEDGE MANAGEMENT DRIVERS

The main drivers behind knowledge management efforts are:

- **Knowledge Attrition:** Despite the economic slowdown,

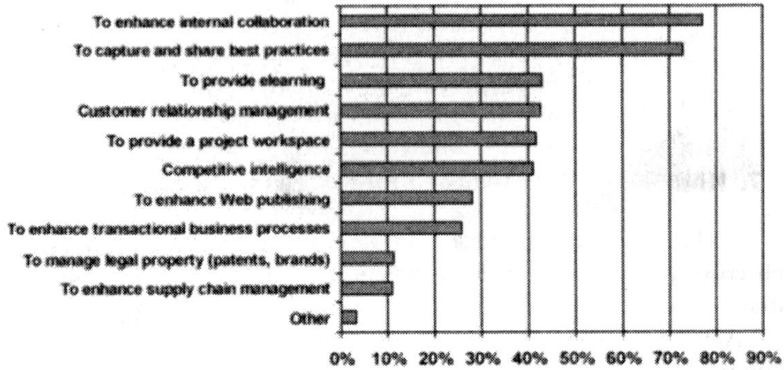

Source: IDC, 2002.

voluntary employee turnover remains high. A recent survey by the global consulting firm Drake Beam Morin revealed an average voluntary employee turnover rate of 20 percent with 81 percent of oraganisations citing employee turnover as a critical issue. Estimated annual costs of employee turnover was a triggering $129 million per organization. Much of this cost is due to knowledge attrition, which can be effectively minimized using knowledge management techniques.

- **Knowledge Merging:** Since 1980, the annual value of mergers has risen 100 fold reaching a cumulative $15 trillion in 1989. Over 32,000 deals were announced, triple the number of 10 years earlier and more than 30 times as many as in 1981. The recent frenzy of corporate mergers coupled with the increased need to integrated global corporate communications requires the merging of disparate and often conflicting knowledge models.
- **Content Management:** The explosion of digitally stored business-critical data is widely documented. Forester Research estimates that online storage for Global 2,500 companies will grow from an average of 15,000 gigabytes per company in 1999 to 153,000 gigabytes by 2003, representing a compound annual growth rate of 78%. As the volume of digital information expands, the need for its logical organization is critical for purposes of information retrieval, sharing and reuse.
- **E-Learning:** As the economy becomes more global and the use of PCs more pervasive, there has been a dramatic

increase in e-learning, also known as computer-based training. E-learning is closely linked to and overlapping with, but not equal to knowledge management, E-Learning can be an effective medium for knowledge management deliverables.

7. KNOWLEDGE MANAGEMENT (KM) OBJECTIVES

The graph above shows the results of a recent IDC study in which corporations cited various objectives for knowledge management efforts:

Activities related to these objectives include: creating knowledge sharing network that facilitate a corporate knowledge culture, developing knowledge leaders, optimizing intellectual capital by producing knowledge management solutions such as codification strategies and knowledge bases, and estimating revenue and efficiency gains resulting form knowledge management in terms of return on investment (ROI).

8. KNOWLEDGE MANAGEMENT (KM) RETURN ON INVESTMENT (ROI)

Although 65% of organizations that are currently implementing Knowledge Management (KM) initiatives have not measured the impact of their performance, large revenge gains and efficiency improvements have been recorded by numerous major corporations. For instance: Ford Motor Company accelerated its concept-to-production time from 36 months to 24 months. The flow on value of this has been estimated at US $1.25 billion, The Dow Chemical Company saved $40 million a year in the re-use of patents, chase Manhattan, one of the largest banks in the US, used Customer Relationship Management. Knowledge Management

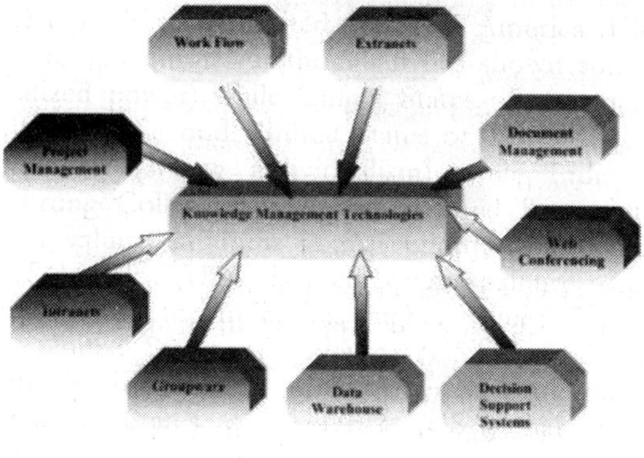

(KM) initiatives to increase its annual revenue by 15%, and Pfizer credit Knowledge Management (KM) practices for discovering the hidden benefits of the Viagra drug.

9. TECHNOLOGIES THAT SUPPORT KNOWLEDGE MANAGEMENT

The following diagram reflects the main technologies that currently support knowledge management systems.

These technologies roughly correlate to four stages of the Knowledge Management (KM) life cycle:

1. Knowledge is acquired or captured using intranets, extranets, groupware, web conferencing and document management systems.
2. An organizational memory is formed by refining, organizing, and storing knowledge using structured repositories such as data warehouses.
3. Knowledge is distributed through education, training programs, automated knowledge-based systems, experts networks.
4. Knowledge is Applied or leveraged for further learning and innovation via mining of the organizational memory and the application of expert systems such as decision support systems.

PRESENT AND FUTURE STATE OF KM

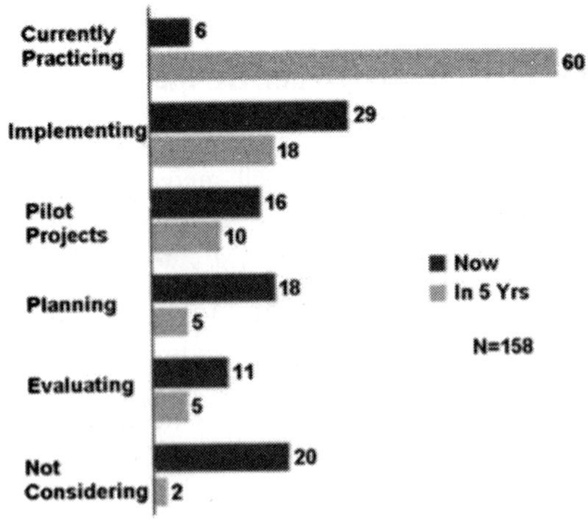

All of these stages are enhanced by effective workflow and project management.

10. PRESENT AND FUTURE STATE OF KNOWLEDGE MANAGEMENT (KM)

Currently, communities of practice such as the Knowledge Management (KM) Network and the development of standards and best practices are in a mature stage of development. Knowledge Management (KM) curricula such as certification. Corporate training and university graduate certificate programs are on the rise. Techniques such as data mining and text mining that use Knowledge Management (KM) for competitive intelligence and innovation are in the early stages of development. Finally, organizations are investing heavily in *ad-hoc* Knowledge Management (KM) software that facilities organizational knowledge. The chart below estimates the state of their current and future Knowledge Management Activities

11. THE FUTURE OF KNOWLEDGE MANAGEMENT

In the next several years ad-hoc software will develop into comprehensive, knowledge aware enterprise management systems. Knowledge Management (KM) and E-learning will converge into knowledge collaboration portals that will efficiently transfer knowledge in an interdisciplinary and cross functional environment. Information systems will evolve into artificial intelligence systems that use intelligent agents to customize and filter relevant information. New methods and tools will be developed for Knowledge Management (KM) driven E-intelligence and innovation.

12. THE EFFECT OF KNOWLEDGE MANAGEMENT (KM) ON DATABASES

Multiple corporate databases will merge into large, integrated, multidimensional knowledge bases that are designed to support competitive intelligence and organizational memory. These centralized knowledge repositories will optimize information collection, organization, and retrieval. They will offer knowledge enriching features that support the seamless interoperability and flow of information and knowledge. These features may include: the incorporation of video and audio clips. Links to external authoritative sources, content qualifiers in the form of source or reference metadata, annotation capabilities to capture tacit knowledge. Content will be in the form of small reusable learning objects and associated metadata that provides contextual information to assist Knowledge Management (KM) reasoning and delivery systems.

13. THE IMPLICATIONS OF KNOWLEDGE MANAGEMENT FOR ...

- **Database Users:** From business class users to the general public, database users will enjoy a new level of interaction with the Knowledge Management (KM) system including just-in-time knowledge that delivers precise relevant information on demand and in context. More complex, smart systems will translate to optimal usability and less time spent searching for relevant information. For example, data analysts will enjoy simplified access and more powerful tools for data exploitation. The use of knowledge bases can reduce customer service costs by providing customers with easy access to 24/7 self-service via smart systems that reduce the need to contact customer service or technical support staff. Database users may even create customized view of knowledge bases that support their needs.
- **Database Developers:** The design and development of knowledge-based systems will be considerably more complex than current database development methods. Developers must consider the overall technical architecture of the corporation to ensure seamless interoperability. The use of standardized metadata and methods will also facilitate both intra-platform choices will be equally more complex. Both knowledge-base developers and administrators must understand the role of the knowledge-base in the overall Knowledge Management (KM) system.
- **Database Administrators:** Database Administrators will evolve into Knowledge Managers. The Knowledge-base will store and maintain corporate memory and knowledge Managers will become the gatekeepers of corporate knowledge. The lines between technical roles such as Web Developer, Data Analyst or Systems Administrator will blur as there systems merge into and overlap with Knowledge Management (KM) systems. DBAs will need to have some knowledge about each of these disciplines.
- **General Public:** Even if they are not interacting directly with a knowledge base, the general public will benefits from the secondary efforts of improved customer service due to faster access to more accurate information by service providers.

14. KNOWLEDGE MANAGEMENT (KM) DIMENSIONS

Different frameworks for distinguishing between knowledge exist. One proposed framework for categorising the dimensions of knowledge distinguishes between tacit knowledge and explicit knowledge. Tacit knowledge represents internalized knowledge that an individual may not be consciously aware of how he or she accomplishes particular tasks. At the opposite end of the spectrum, explicit knowledge represents knowledge that the individual holds consciously in mental focus, in a form that can easily be communicated to others.

Early research suggested that a successful Knowledge Management (KM) effort needs to convert internalized tacit knowledge into explicit knowledge in order to share it, but the same effort must also permit individuals to iternalise and make personally meaningful any codified knowledge retrieved from the Knowledge Management (KM) effort. Subsequent research into Knowledge Management (KM) suggested that a distinction between tacit knowledge and explicit knowledge represented an oversimplification and at the notion of explicit knowledge is self-contradictory. Specifically, for knowledge to be made explicit, it must be translated into information (i.e., symbols outside of our heads).

A second proposed framework for categorising the dimensions of knowledge distinguishes between embedded knowledge of a system outside of a human individual (e.g., an information system may have knowledge embedded into its design) and embodies knowledge representing a learned capability of a human body's nervous and endocrine systems.

A third proposed framework for categorising the dimensions of knowledge distinguishes between the exploratory creation of "new knowledge" (i.e., innovation) *versus* the transfer or exploitation of "established knowledge" within a group, organization, or community. Collaborative environments such as communities of practice or the use of social computing tools can be used for both knowledge creation and transfer.

15. KNOWLEDGE MANAGEMENT STRATEGIES

Knowledge may be accessed at three stages: Before, during, or after Knowledge Management (KM)-related activities. Different organizations have tried various knowledge capture incentives, including making content submission mandatory and incorporating rewards into performance measurement plans. Considerable controversy exists over whether incentives work or not in this field and no consensus has emerged.

One strategy to Knowledge Management (KM) involves actively managing knowledge. In such an instance, individual strive to explicitly encode their knowledge into a shared knowledge repository, such as a database, as well as retrieving knowledge they need that other individual have provided to the repository. Another strategy to Knowledge Management (KM) involves individuals making knowledge requests of experts associated with a particular subject on an *ad-hoc* basis. In such an instance, expert individual(s) can provide their insights to the particular person or people needing this.

16. KNOWLEDGE MANAGEMENT MOTIVATIONS

A number of claims exist as to the motivations leading organizations to undertake a Knowledge Management (KM) effort 27.28. Typical considerations driving a Knowledge Management (KM) efforts include:

- Making available increased knowledge content in the development and provision of products and services.
- Achieving shorter new products development cycles.
- Facilitating and managing innovation and organizational learning.
- Leveraging the expertise of people across the organization.
- Increasing network connectivity between internal and external individuals.
- Managing business environments and allowing employees to obtain relevant insights and ideas appropriate to their work.
- Solving intractable or wicked problems.
- Managing intellectual capital and intellectual assets in the workforce (such as the expertise and know-how possessed by key individual).

Debate exists whether Knowledge Management (KM) is more than a passing fad, though increasing amount of research in this field may hopefully help to answer this question, as well as create consensus on what elements of Knowledge Management (KM) help determine the success or failure of such efforts.

17. KNOWLEDGE MANAGEMENT TECHNOLOGIES

Early Knowledge Management (KM) included online corporate yellow pages as expertise locators and document managements systems. Combined with the early development of collaborative technologies (in particular Lotus Notes), Knowledge Management (KM) technologies expanded in the mid-1990s. Subsequent Knowledge Management (KM) efforts leveraged semantic technologies for search and retrieval and the development of e-learning tools for communities of practice.

More recently, development of social computing tools (such as blogs and wikis) have allowed more unstructured. Self-governing or ecosystem approaches the transfer, capture and creation of knowledge, including the development of new forms of communities, networks, or matrixed organizations. However, such tools for the most part are still based on text and code, and thus, represent explicit knowledge transfer. These tools face challenges in distilling meaningful re-usable knowledge and ensuring that their content is transmissible through diverse channels.

As we approach the dawn of the 21st century we are humbled by the scale of challenges our world faces at the global scale. The unprecedented development and growth of knowledge during the 20th century not withstanding, the evolution of a peaceful 21st century will depend on our ability to address three interdependent global challenges of prosperity, security, sustainability. The new world order, though far from being fully defined and agreed to, is beginning to already point at some of the strategic threats and opportunities that would be determining factors in our ability to help shape a new century that we can look back as symbolic of the human sprit and its collective accomplishment at its best and most current stage in the evolution.

As it begins to dawn on us that for the first time in human history we may be able to see the impact of human decisions on the evolution of our society in our own lifetime, we become acutely aware of the unprecedented responsibility and the de-facto power we all posses to influence the path of our evolution. Traditional disciplinary knowledge is limited in its ability to support the challenging decisions that lie ahead. Global stability in the future will depend upon our ability as a society to simultaneously address the three fundamental issues o prosperity, security and sustainability. Today, well over three billion people across the globe earn less than two dollars a day. As they move toward achieving wealth, which has enormous consequences for the environment and global sustainability. At the same time, a vast number of citizens of the world remain illiterate, posing an enormous challenge for us in providing the necessary education and training to create global prosperity in the knowledge-based economy. We must find effective ways for capitalizing on emerging information technology to help the citizen of the world to gain access to education and opportunities for adapting and prospering in the next century.

We are entering into an era where the future will be essentially determined by our ability to wisely use knowledge, a precious global resource that is the embodiment of human intellectual capital and technology. As we begin to expand our understanding of knowledge as an essential asset we realize that in many ways our future is limited only by our imagination and ability to leverage the human mind. The future of the world lies in front of us unexplored and uncharted, and is full with great opportunities for expanding peace and prosperity in the next millennium through the responsible use of the knowledge.

As knowledge increasingly becomes the key strategic resource of the future our need to develop comprehensive understanding of knowledge processes for the creation, transfer and deployment of this unique asset are becoming critical. In the face of a globally expanding and highly competitive knowledge-based economy the traditional organizations are urgently seeking fundamental insights to help them nurture, harvest and mange the immense potential of their knowledge assets for capability to excel at the leading edge of innovation. Schools, Colleges, Universities, and training organizations (traditional suppliers of knowledge); and businesses and knowledge-based organizations in public sector (growing users of knowledge) are in need of an integrative discipline for studding, researching and learning about the knowledge assets—human intellectual capital and technology. This would be similar to the development of Operations Research as an integrative discipline during the Second World War. An international society of knowledge professionals can provide the necessary of our time on study, research and learning dedicated to the underlying disciplines and their integrative evolution in to the emergence of Knowledge Management as a new discipline.

The task of developing the Knowledge Management as a new discipline is a challenging endeavour. This new discipline must successfully respond to the diverse needs of Knowledge-based organizations and the knowledge professional in a timely fashion. The pioneers of must recognize that a great many of the organizations of today are primarily run largely on the basis of insights gained from the successes of the manufacturing based capital intensive industrial economy of the past. These organizations have fallen or are rapidly falling out of alignment with the evolutionary direction of the future as the economy transitions form the industrial economy to one that is rapidly becoming an intellectual capital and technology-based global knowledge economy.

The problems of the future will be open-ended, complex, global and adaptive in nature. As academics we must look for a new synthesis of knowledge, integrating hard and soft science, to create the knowledge assets necessary for addressing the challenges in a rapid evolutionary era.

The participation of experts from traditional academic disciplines (i.e., information technology, management, cognitive sciences, economics, finance, policy, law, social sciences), and business and government will be essential to the cohesion of an integrative body of knowledge leading to the formation of Knowledge Management discipline and a community of scholars, teachers and professionals associated with this new discipline. The professional society can serve as a home for enabling the development of the Knowledge Management discipline. However, it must be created as a hybrid, independent and entrepreneuring organization with strong global participation and ties to leaders in academia, industry and government.

In order to fulfil its mission the professional society need to foster three unction: academic education, research and advanced technology program. These three interconnected set of functions need to be pursued in a hybrid way by combing the strengths of the distinct institutional methods of academia, industry and think tanks. The professional society may become and incubator for implementing new 21st century models of operations for each of the three functions: an experiential learning based academic environment; a collaborative research community dedicated to life-long knowledge-based learning; and a multimedia and information technology-based knowledge-era tools development program for supporting the performance of the knowledge professionals and organizations.

The professional society needs to foster the developments in the three strategic areas through the establishment o founding partnerships with innovative universities, premier research universities/institutes and a group of leading edge industry and government collaborators from the United States and aboard. One of these founding partners needs to step up and become the founding home of the professional society, where the society's activity will be coordinated. Beyond the founding partners, opportunity for participations in the society should be open to all credible academic, government, NGOs and industrial organizations worldwide, including students, research fellows and professionals across a rang of disciplines.

The first strategic development focus for the professional society needs to be in the areas of education. There is need of a comprehensive executive education program in Knowledge Management for leaders, executives, knowledge professional and policy-makers. The program may offer an executive Master of Science degree and should be designed to include the state-of-the-art topics addressed through scholarly rigor-enabling it to earn support of faculty members at the leading universities as well s accreditation. This degree program, in collaboration with the leading edge developers of multimedia and information technology based knowledge-delivery-tools, can serve as an alpha site for testing and implementing state-of-the-art innovations in global instruction and learning delivered through the distance learning technology (including learning and instruction delivered locally in real life settings at a representative set of host organizations who are the leaders in applying Knowledge Management). The post-graduate educational commitment should be explored as a means for testing and implementing networking based tools for delivering just-in-time learning to the graduates of the maters degree programs throughout their professional careers. The knowledge network consisting of students, faculty and graduates across the globe can from the prototype of a knowledge community/university of the future where on-going learning, resident in the network, will be the true competitive advantage. The educational function, by convening annual real/virtual conferences addressing the topics at the frontiers of the Knowledge Management, by serving as a knowledge resource network for sharing the latest research insights, and by providing professional advice and mentoring to instill the career self-management philosophy, can serve as a basis for facilitating the formation of a network based global knowledge management professional society.

The second strategic development area of research also needs to be carried out collaboratively with the university, government and industry partners, and research fellows/interns, addressing the foundational topics in Knowledge Management through the basic and applied research work. These topic will aim to understand the processes and practices for generation, identification, assimilation, and distribution of knowledge as an asset for use by knowledge professional and organizations, a well as, to study the key knowledge sector policy innovations at the individual, organizational and societal levels. The economics of knowledge, addressing the accounting, valuation, depreciation of knowledge asset, should be considered as one of the core theme on the overall research agenda. The initial research will lead to the formulation of specific and significant Knowledge Management research topics, some of these will be topics leading to the M.Sc. dissertations and Ph.D. thesis. The results of research (in the form of best practices, case studies, papers, thesis and dissertations) should be published and can be relevant inclusions in the instructional material used in the degree programs under the educational strategic development area.

Some research topics should be considered to leverage the uniquely global knowledge network represented by the professional society. Two areas are outlined below for the purposes of illustration, further development require broader participation by the professional society members:

1. *A global-change knowledge network* as an experiment in the formation of global community of student, scholars and professionals to facilitate the co-emergence of stability, based on prosperity, security, and sustainability, in the twenty-first century.
2. *Research on global intellectual capital investment* to explore assistance based on knowledge, skills and technologies that flow from investments in learning and development as global aid imperative rather than aid based on charity or altruism.

In order to carry out these programs, the professional society should foster the creation of strategic nodes, with excellence in areas of complementary expertise, both within the United States and abroad to create a virtual knowledge network, working within a common vision, addressing problems locally but from a global perspective. The organization of the society must evolve to an inclusive shared stewardship and commitment.

The third strategic development area addresses the advanced technology. The professional society needs to carry out technology development projects to maintain the society's educational development efforts at the cutting edge and to cost effectively provide enriched learning experiences to a diverse group of learners across the world. Each of these projects needs to be pursued through partnerships with the researchers and technology developers in the multimedia and approaches including those drawn from industry, federal laboratories and intelligence community (such as artificial intelligence, data mining, knowledge representation, virtual reality, simulation, telepresence, teleoperations, Global Information Systems, distance learning and cognitive sciences) in an environment open to exploration and learning will lead to a unique knowledge-era-tools development program for supporting the needs of the knowledge professionals and organizations. In particular the professional society is in a unique position to foster the development of tools to support the strategic needs of the education function, a cost effective just-in-time knowledge delivery capability and the formation of a global knowledge Network.

The professional society will require a strong and sustaining commitment form the home institution for an initial incubation period, a visionary leadership, an enterprising faculty/scholars and dedicated staff, a strong advisory board and sponsors, and leading edge partners and collaborators in the industry. Since the Silicon Valley is recognized worldwide for its leadership in information technology based innovations the initial focus on the Silicon Valley may serve as an ideal test bed for the professional society's home. A Silicon Valley home node will be critical to the success of collaborations in all of three strategic development areas and therefore essential to the early launch and acceptance of the society. However, from its inception the professional society will be international in its outlook and scope.

The professional society will serve as a beacon to knowledge professionals and organizations in the emerging knowledge-based economy. In order for the professional society to accomplish this challenging mission it must be developed as a highly entrepreneuring, enterprising, innovative and integrative catalyst organization. It must bring together the important components of Knowledge Management education and research, and the development of learning technology to significantly increase the productivity of knowledge-based performances and innovation (of professionals and organizations), to enable significant advancements in the capabilities for just-in-time knowledge delivery and in knowledge sector policy innovations.

The understanding of knowledge processes and its management will provide valuable insights to the policy-makers, learning organizations, knowledge professionals, individual citizens to focus their investment in learning, unlearning and life-long learning. In short, the challenge is to help provide the decision-makers at all ages, at all levels and at all places with the necessary intellectual capital to be competent performers and proactive opportunity creators in the evolution of the global knowledge-based economy of the 21st century.

18. KNOWLEDGE MANAGEMENT AND IMPLICATIONS IN INDIA

Knowledge Management is still in its infancy in India. Knowledge Management is no longer a luxury for Indian companies. Unlike in the West, where titles of Chief Knowledge Officer of Chief Learning Officer have been in vogue for years, in India such dedicated people are few and far between. Few companies have thought it fit to create a person responsible for it. Implications in the case of India are enlisted as follows:

- Companies like Infosys have therefore experimented with a system of rewards and incentives for inducing knowledge sharing.
- A few companies have attempted to include knowledge

sharing as part of annual appraisals of employee performance.

- The hardware in terms of making Knowledge Management technology available is the easy part. But the software part of getting staff motivated to use it regularly is proving difficult.
- Most Indian companies are finding it easy to get their employees to generate and share knowledge using the technologies that are put at their disposal.
- Establishment of websites regarding knowledge management prospects.
- Knowledge Management is no longer a luxury for Indian companies. It is a necessary that can make all the difference between survival and an early demise. This scenario is being put as a challenge before all the Indian workers to cope with innovative 'Dimensions of Knowledge Management'.

19. THE CHALLENGES OF KNOWLEDGE MANAGEMENT

It is very important to note that Knowledge Management problems can typically not be solved by the development of a technology solution alone. The major challenges that an organization may face are discussed under:

1. **Getting Employees on Board:** The major problems that occur in Knowledge Management usually result because companies ignore the people and cultural issues. In an environment where an individual's knowledge is valued and rewarded, establishing a culture that recognizes tacit knowledge and encourages employees to share it, is critical. The need to sell the knowledge management concept to employees should not be under-estimated after all; in many cases employees are being asked to surrender their knowledge and experience, the very traits that make them valuable as individuals.
2. **Knowledge Management Requires Ongoing Maintenance:** As with my physical assets, the value of knowledge can erode over time. Since knowledge can get state fast the content in a knowledge management programs should be constantly updated, amended and deleted. The relevance of knowledge at any given time changes, as do the skills of employees. Therefore, there is no end point to a Knowledge Management programme. Like product

development, marketing and research and development, Knowledge Management is a constantly evolving business practice.

3. **Dealing with a Data Deluge:** Companies deligently need to be on the look out for information overload. Quantity rarely equals quality and Knowledge Management is no exception. Indeed, the point of Knowledge Management programs is to identify and disseminte knowledge gains from a sea of information.

20. SUMMARY AND CONCLUSION

Organisations are realizing that intellectual capital or corporate knowledge is a valuable asset that can be managed as effectively as physical assets in order to improve performance. The focus of Knowledge Management is connecting people, processes and technology for the purpose of leveraging corporate knowledge. The database professionals of today are the Knowledge Managers of the future, and they will play an integral role in making these connections possible. The emergence of rapidly expanding technologies for distribution and dissemination of information and knowledge has brought to focus the opportunities for development of knowledge-based networks, knowledge dissemination and knowledge management technologies and their potential application for enhancing productivity of knowledge work. The challenging and complex problems of the future can be best addressed by developing the knowledge management as a new discipline based on an integrative synthesis of hard and soft sciences. A Knowledge Management professional society can provide a framework for catalyzing the development of proposed synthesis as well serve as a focal point for coordination of professional activities in the strategic areas of education, research and technology development. Preliminary concepts for the development of Knowledge Management discipline and the professional society are explored. Within this context of Knowledge Management discipline and the professional society, potential opportunities for applications of information technologies for more effectively delivering or transferring information and knowledge (i.e. resulting from the NASA'a Mission to planet Earth) for the development of policy options in critical areas of national and global importance (i.e. policy decisions in economic and environmental areas) can be explored, particularly for those policy areas whereas global collaborative knowledge networks is likely to be critical to the acceptance of the policies.

REFERENCES

Bhatacharya, Pransanjit, "The Faster Growing Sector 'Information Technology", *The Economic Times*, New Delhi, 29th June 2001.

Drucker, Peter F., "Technology Management and Society", New York, Harper and Row, 1970.

Dudeja, V.D., "Management Information in the New Millennium: Challenges and Opportunities", Common Wealth Publisher, New Delhi, First Edition, 2000.

Elias, M. Awad and Ghoziri, Hassan, M., "Knowledge Management", Dorling Kindersley (India) Pvt. Ltd., First Edition, 2006.

Kim Anm Zimmerman, "Using Knowledge Management", Your Intellectual property", July 2001, Effective Executive.

Murthy, C.S.V., "E-Commerce", Himalya Publishing House, Mumbai, First Edition, 2002.

Shelda, Debowski, "Knowledge Management", Jhon Wilsey and Sons (ASIA) Pvt. Ltd., Singapore, First Edition, 2006.

The Knowledge Society, "The Richest Resource", *The Economic Times*, 2nd January, 2000.

Vasudevan Committee Report, Published by Reserve Bank of India (RBI). Published Reports of different organizations such as NASSCOM, RBI and other agencies dealing in the field of Information Technology (I.T).

CHAPTER

26

Activity-based Learning in Commerce Education: An Awakening Call

DR. ANIL GUPTA, DR. AMRIK SINGH SUDAN AND KULDEEEP SINGH MANHAS

ABSTRACT

Education is the imparting and acquiring of knowledge through teaching and learning. It has developed from the human struggle for survival and enlightenment. Talking of the education, one feels proud of saying that I am an educated person because the education process involved sharing information about gathering food and providing shelter, making weapons and other tools, learning language and acquiring the values of a given culture. Whatsoever we are getting educated day-by-day and what's good about is that it is a never ending process.

As is well known, India is a country of extremes, with the best of many things and worst of many others. As the time passed and rural and urban areas started filling with Technological advances, the focus shifted to the quality and innovation along the quantity. The context has been undergoing transformation adding a whole new dimension to education. With changing scenario, efforts are being made to make Commerce education an activity-based learning to accelerate the production of knowledge and its dissemination through sophisticated communication channels. In countries like Finland, Norway and Germany there is a tradition of helping youngsters to develop career lines without going to Universities. At present the education is aided with a variety of technology. Computers, projectors, internet and many more techniques. In the light of this an attempt is made as under to focus on activity-based learning in Commerce education.

BEYOND TRADITIONAL LEARNING

Today's global job market demands that a student should be comfortable with the complexities of ill-defined real world problems. The on-going technological advancements pose great challenges to both students and especially teachers. In the past, the black boards and books were the main aids for teaching where students imbibed knowledge through class lectures.

Today's generation with easy access to electronic gadgets like cell phones, Internet, Social networking sites, video capturing and editing devices, peer-to-peer file sharing, etc. needs different stimuli to learn. The technique of learning "How to learn" carries greater value than mere acquisition of facts. Today's generation prefer non-linear hyperlinked-logic of learning than the linear acquisition of knowledge.

In this age of technological advances, students have access to a vast reservoir of information. However, not all the information is useful or reliable yet the role of the teacher is to equip the students with requisite skills to enable them to distinguish between reliable from unreliable information. To make commerce education more effective and popular efforts are made to equip the students with psychomotor capacity to move, perceive, apply physical skills, and to bring a paradigm shift from remembering, understanding and applying to more important skills of analysing, evaluating and creating. So efforts have been made to improve commerce education but much still remains to be done. Despite impressive changes, the significant change has been the information technology. With the help of information technology, teaching can be effective, improving the quality and relevance of class room instructions. All this is possible with infrastructural and economic factors. We would like to suggest that there are three areas in which information technology can significantly alter the current landscape. IT as stated help in better teaching by enabling collaborating development of teaching material and making it available to all, improving the quality of instructions by enabling mass participation and creating supplementary materials. It may also help improving operational efficiency by enabling the collection of disaggregating data and the use of information-based methods to procure and distribute materials.

NEEDS PARADIGM SHIFT

The most important question is what should be done to rectify all the maladies that plague the Commerce education. It has come to the surface from interaction with students, industry and other related persons that the course curriculum or the syllabi which is being taught is incongruent to the needs of market demand. The course curriculum which is being taught needs to be upgraded as it is designed to just distribute degrees and not to impart knowledge.

The slant of the syllabi should be more towards analysis and practical orientation by laying more stress on case studies and live projects. For this, case-based teaching methods needs to be adopted to make the students understand a concept of the case.

As India marks towards skill-based manufacturing, the three most important aspects are going to be—How productive are the assets that are deployed, How efficiently they are put to use and how cost competitive they are when put to use? In all these three dimensions, Commerce students need to be imparted the knowledge of diverse technologies and processes by bring technical resources, application, state of art infrastructure and workshop training/knowledge. In today's business environment the knowledge of managing assets throughout their life cycles to deliver the real and on-going value with reduced total cost of ownership is very important which can only be imparted through qualified manpower and use of technology and also this can be facilitated through open access to literature, open software and by strengthening libraries and infrastructure.

IMPART PRACTICAL KNOWLEDGE

Today most of the college students are wandering without any purpose. Many of them just take admission for getting maintenance allowance or scholarships. They do not acquire the proper skills which suit their capabilities and intelligentia. Do you think that implementation of semester system, internal assessment system and grading system is sufficient for maintaining quality and requirements of the market.

The activities for the students include communication skills workshop where focus is on their language skills through stories, role play and presentation, entrepreneurship training workshops to impart knowledge on preparation of project reports, location of site, understanding the extrinsic factors obstructing growth of entrepreneurship in the state (local area), providing students ample opportunities to express themselves in the class rooms, etc. There is a question always in the mind why students opt MBA? The simple answer is that MBA helps to facilitate a track change in once job. For those who realise that they are stuck following a bad career choice, an MBA can open a window of opportunity. For example, a doctor can implement the management expertise gained from an MBA to run a well-organised hospital with a well-structured administration. An MBA is also on the way to establish a firm providing consultancy services and the like services. In the same manner if the practical knowledge to commerce students is imparted, they will be able to take up the job activities either as a private consultant or in any corporate sector of the economy.

There is a demand for good resources in the industry in order to give "tailor made" product to the business world. "Academic-Industry" partnership has to be forged to impart practical knowledge and it should form part of course curriculum otherwise there is no point in having scores of affiliated colleges and couple of lakh of students if the university cannot design, develop, conduct, monitor and ensure the job-oriented commerce education. Again it is stressed upon that problem-solving and communication skills should not be overlooked while designing the course curriculum as it is also very important for students to be able to comprehend complex scenario and to simplify them.

EMERGING TRENDS

The crux of Globalization in education sector is that the teaching, learning is no longer for "Building of Nations" but businesses for "Profit-making". Students, Professors, Institutional Organizers are the parties affected by the globalization of educational services. We have strengths in some areas as there are 602 Universities and deemed Universities, more than 20,000 Colleges. In addition to Central Universities there are 4 more types of Universities/Autonomous Institutions which are recognised by UGC and have powers to accord degrees to students. These are State Universities, deemed universities, private universities and autonomous institutions. As on today there are 42 Central Universities, 275 State Universities, 130 Deemed Universities, 90 Private Universities and 65 Agricultural Universities besides Indian Institute of Technology, National Institutes of Technology, Indian Institutes of Information Technologies, Indian Institutes of Scientific Education and Research, Indian Institutes of Management and one School of Planning and Architecture.

NEW INITIATIVE FOR INCLUSIVE GROWTH

"Inclusion" is the growth mantra for policy-makers today—not just in India but the world over. And for inclusive growth, what could be more crucial input than business education. Good and Wholesome Business Education empower one, opens up opportunities for individual and group entitlements, productive employment and participation in various areas of social life. In all business education catalyzes economic growth. If India, with its predominantly young population wants to reap demographic dividends, then India is to ensure quality of education, to promote the spirit of quest and to workout meaningful curricula aligned with job market. In the following sections, we discuss some of the challenges being experienced by business education as Business Education is imparted to bridge the gap between employable education and skill to young people so that they are able to meet the growing needs of a rapidly expanding economy.

COMMERCE EDUCATION IN NEW MILLENNIUM

In the new millennium the commerce education embraces information technology and E-commerce, Commerce Education, Human Resource Development, Corporate Governance, International Trade, Finance and Accounting, etc. for industries of all types, service sector and even for non-profit-making organizations. The area of education makes clear that Management Education and Commerce Education are the two branches of Business Education. Information Technology has revolutionized the trade, Commerce Education and all spheres of life. However, the digital gap exists between developed and developing countries as 6-7% business is transacted at Internet. Commerce Education in India has not been able to cater to the demand for the market. The commerce education must move beyond the class-room and outside the text books to make the learning job-oriented. Mushrooming growth of private institutes have been offering professional courses in Management and Commerce education across the country but these are not able to lead to employability of Commerce and management graduates due to weak base of + 2 level, lack of interaction with business houses, non-preference of commerce graduates and post-graduates in banks and insurance sector.

Time demands that commerce graduates and post-graduates should be exposed to global environment issues like WTO, FDI, ISO Quality Standard, Indian and International Accounting standards. So long as we are not able to re-structure the curriculum and to have interaction with industry, commerce education will not make a much head way. A paradigm shift in the mind set and thinking is essential for effective utilization of human resources in view of the dynamic competition posed as a result of globalization, liberalization and information technology.

CHALLENGES AND SUGGESTIONS

The Indian economy has opened up the Global Competition and commerce education as such has become a major instrument of this change. Still there are challenges before this sector. Some of the challenges the Commerce Education facing today are as under:

1. The slant of syllabi should be more towards analysis.
2. Interface of students with the mandarins of business world is very important to prepare the students for the market.

3. Seminars and presentations should form the regular fulcrum of the course curriculum.
4. Students should be given practical orientation focusing on problem-solving and communication skills.
5. The syllabi of management subjects be designed to lay emphasis on developmental aptitude, the spirit of team work, eloquence of their presentation, their participatory role and so on.
6. University should review and redesign the curriculum by involving representative of professional like Company's Secretaries, Chartered Accountants, Cost Accountants, and Executives, etc.
7. Modern business the worldover is influenced greatly by computer aided technology and India as a part of world trade organization cannot afford to remain uninfluenced by the cybernetics.
8. There is a need to bridge the gap that exists between the Indian Accounting Standards and International Accounting Standards and that of the world in order to ensure better harmonization of accounting practices required in view of the present global economic scenario.
9. Strong and viable rapport between academic and industry must be developed to expose the learner to the job situation.
10. Traditional teaching methodology be done away with and replaced by learner centered teaching methodology like case studies, role playing, project report, assignments, quiz programs, business games, simulations, etc. This will developed interest and total involvement of students in the teaching learning process.
11. Bringing the learners into the class room and sustaining their interest till the last day of intrusion is the toughest job requiring ingenuity of high order. The smallest hurt to the learner's ego or barest reflection on his ability to learn turns him away from class room. Therefore, in the 21st century teaching methods should be made high tech to create the interest of the learners.
12. Without trained teachers even the fanciest buildings, best textbooks and facilities will not do.
13. Comparing the number of Universities in India with other parts of the world, which are considered as the knowledge countries, India has a very small size of Universities and colleges to meet the emerging needs of advanced research.

This situation calls for serious remedial action and the creation of new era in the field of knowledge economy. Besides, the acute shortage of funds is another area posing a serious threat to knowledge economy.

14. No doubt, globalization concept has made an impact on minds of young generations and they are towards global business knowledge, it would be a sufficient if teacher so teach that the young minds will learn how to discover the methods of solving problems that are known or unknown. Still the young must be taught how to think when the methods to be used is not known and the problem to be solved has not been formulated as mere education about global business and degrees in commerce. On the other hand, teachers have to develop the intellectual capacity to see beyond what seems to be the obvious. Good, bad or indifferent teaching has become a common denominator. In the present times when the teachers' pay scales are much better, they are still identifying as discontented persons. Most of them look upon teaching as job for which the pay cheque is the only reward. We must appreciate that higher education in India is still a luxury. Unfortunately India's limited resources and manpower are being frittered away leading to glowing devaluation of standard in education and in academic integrity.

It is further suggested that the curriculum has to be such that students need to be educated as specialist and generalist, not exclusively as one or the other. Non-cognitive skills such as communication skills, negotiating skills, and interpersonal skills need special attention in the class room teaching but not at the cost of technical or cognitive knowledge. Some teaching of ethical and moral values be included in syllabus to inculcate the value of life as these values are eternal values of truthfulness, honesty, sincerity and integrity.

CONCLUSION

Commerce education at secondary and higher levels in various forms formal and informal is a measuring tool for economic changes. Development experts and planners have been reshaping the educational system to meet the development needs. The innovation in curriculum as well as techniques of teaching bringing positive changes in the quality of commerce education so that when a student goes out with the degree is well acquainted with the up-to-date knowledge of computers, accounting standards, knowledge of corporate world, etc. The very essence of commerce education is to make our policies and programs cost-effective and result-oriented to overcome socio-economic, political and moral problems.

Further the quality of Commerce Education can be improved largely by the optimum use of all resources available. Business houses need to take part in education reforms system by investing in development programs for teachers as it will give the opportunity to implement creative ideas which will in turn improve the commerce education.

In brief, at the end Commerce Education is not to produce machines out of pupil. Vladimir Nabokov, a U.S. critic, Poet and Novelist says, "discussion in class which mean letting twenty young block heads and two cocky neurotics discuss something that neither their teacher nor they know." Activity-based learning in commerce education will provide an opportunity to students to develop the cognitive capacity to think and to solve problems. Also the practical orientation would help students get experience which would boost the confidence once they enter the job market.

References

Marketing Services—S.M. Jha, 2004, Education, Himalayan Pub. House, New Delhi.

Yojana—A Development Monthly, Vol. 49, Sept. 2005, Sept. 2009, Jan. 2010 and June 2010.

Yojana Journal, Vol. 38, Jan. 26, 1994, Vol. 42, Aug. 1998.

The Indian Journal of Commerce, Vol. XLIX, Dec. 1996.

The Indian Journal of Commerce, Vol. 57, July/Sep. 2004.

The Economic Times, Feb. 4, 2005, p. 11, Candice Zachariahs.

The Tribune, August 24, 2010, p. 12; Nov. 16, 2010; Dec. 22, 2010; March 22, 2011; July 19, 2011.

Daily Excelsior, Aug. 12, 2011.

State Times, Sept. 7, 2011.

Kurukshetra—A Journal of Rural Development, Vol. 53, No. 8, July 2005, pp. 18-19.

CHAPTER

27

Extensiveness and Effectiveness of Corporate Governance Regulations in South-Eastern Europe

DR. AMRIK SINGH SUDAN, ANU CHANDHA AND DR. RADHA GUPTA

INTRODUCTION

Corporate governance has been a dominant policy issue in developed market economies for more than a decade. In the transition economies, it took some time for corporate governance to advance the ranking of policy priorities, but since the late 1990s it become one of the most intensely debated issues.

At least two sets of events are responsible for the heightened interest in corporate governance. During the wave of financial crises in 1998, in Russia, Asia and Brazil, deficiencies in corporate governance endangered the stability of the global financial system. Just three years later, confidence in the corporate sector was sapped by corporate governance scandals in the United States and Europe, which triggered some of the largest insolvencies in history. The scandals and crises, however, are just manifestations of a number of structural reasons why corporate governance has become more important for economic development and a significant policy issue,[1] especially for transition economies in South-Eastern Europe (SEE), that do not have the long-established (financial) institutional infrastructure to deal with corporate governance issues.

Privatization has raised corporate governance issues in sectors that were previously in the hands of the state. Firms have turned to markets to seek capital, and mutual enterprises and partnerships have converted themselves into listed corporations.

Privatization has raised corporate governance issue in sectors that were previously in the hands of the state. Firms have turned to markets to seek capital, and mutual enterprises and partnerships have converted themselves into listed corporations.

The private, market-based investment process is now becoming more substantial for most of these economies, being underpinned by better corporate governance. The role of institutional investors is growing in many of these countries, with economies moving away from "pay as you go" retirement systems. This increased delegation of investment has raised the need for corporate governance arrangements.

Also, due to technological progress, liberalization and opening up of financial markets, trade liberalization, and other structural reforms, the allocation within and across countries of capital among competing purposes has become more complex, as has monitoring of the use of capital. This has made good corporate governance more important, but also more difficult to achieve.

Furthermore, programs of deregulation and reform have reshaped not only the local South-Eastern Europe (SEE) financial landscape but also the European one. Long-standing institutional corporate governance arrangements are being replaced with new ones but, in the meantime, inconsistencies and gaps have emerged.

European financial integration has increased, together with trade and investment flows. This has led to numerous cross-border issues in corporate governance.

All these developments have enhanced the need for formulating corporate governance rules and have led to the adoption all over the world,[2] and by most of the South-Eastern Europe (SEE) countries, of corporate governance codes, as a measure for dealing with each country's specific governance problems (besides international organizations' efforts, such as the Organization for Economic Co-operation and Development (OECD) corporate governance principles). These initiatives have recently resulted in improvements of formal legal rules, as well as in the drafting of soft-law recommendations.

These *sets of rules and regulations*, whether international, national, or denominator, they aim at shaping comprehensive standards of good governance, which are mainly the protection of minority shareholders, the avoidance of conflicts of interests and the request for disclosure and transparency,[3] the constitution of the boards (the issue of independent directors and supervisory board members), smaller boards to secure better coordination, the formation of monitoring, compensation and nomination committees, as well as the claim for one-share-one-vote.[1]

Yet, *corporate governance practices* tend to differ quite substantially across countries and companies, especially among developed and transition economies in Europe, with some specific feature for South-Eastern Europe (SEE) economies, which explains the concerns regarding the effectiveness of corporate governance rules, especially in these countries.

The paper is structured as follows. Section 2 identifies the key problems of corporate governance in transition. To do so, it describes some current features of the institutional environment in the South-Eastern European countries and subsequently identifies and analyzes the common and specific characteristics of the corporate governance issues facing these countries. Section 3 addresses the issues of corporate governance measurement for South-Eastern Europe (SEE) countries, with an emphasis on Romania. To this end, its first part includes a review of the methodological framework for determining both the extensiveness and the effectiveness of corporate governance legislation, as defined by the European Bank for Reconstruction and Development (EBRD). The second part focuses South-Eastern Europe on the actual results of the assessment exercise, by supplementing the investigation with an analysis into the effectiveness of legal institutions as opposed to the law on the books, and attempts to provide a measure to the law on the books, and attempts to provide a measure of corporate governance in Romania and other South-Eastern Europe (SEE) emerging markets. Section 4 concludes by emphasizing the corporate governance enforcement gap in the South-Eastern Europe (SEE) region and by identifying research issues that require further study.

II. SPECIFIC CORPORATE GOVERNANCE ISSUES IN THE TRANSITION ECONOMIES ON SOUTH-EASTERN EUROPE (SEE)

A. Key Facts

Although transition countries in South-Eastern Europe (SEE) vary considerably in history and current institutional setup, they share certain important features. They all had, and some still have, quite a large sector of former state-owned enterprises, in the process of restructuring. In addition, these economies inherited a dysfunctional legal system, and in many cases they had to construct basic institutions from zero.

Consequently, a useful prerequisite for the analysis of corporate governance in transition economies is a short review of corporate control issues under central planning.

Enterprise structures under central planning were characterized by two distinctive features that have persistent influence until the present day.[4] First, budget constraints for socialist enterprises were soft: passive finance was provided under the central plan, so they did not have to worry about external financing. Hence, the concepts of financial discipline and accountability were essentially absent. Second, the state, as the owner of most assets, had a pervasive monitoring problem in trying to ensure that managers of socialist enterprise acted according to the targets set out by the central plan.[5]

The two problems were closely interrelated. Absent the sanction of enforcing financial discipline by cutting off supplies and ultimately forcing an enterprise to close down, the problem of corporate control could never be resolved.

When central planning was abolished, the lack of external financing became a serious constraint, on enterprises. The problem of substituting government finance with new sources of external finance in thus, very much at the heart of the issue of corporate governance and restructuring in transition.

Parallel to the reduction of state financing, economic reforms in transition countries also fundamentally altered the structure of ownership rights through privatization.

Starting in the mid-1990s, the corporate governance debate within transition economies revolved around specific privatization issues, whereas initial efforts in the move toward responsible corporate governance included legislative, judicial and corporate initiatives to provide investors with more disclosure and transparent information.[6]

However, in this endeavor governments were constrained by the power of incumbent managers, who had accumulated implicit control rights as a result of weak state monitoring under central planning. In many South-Eastern Europe (SEE) economies, privatization simply led to the explicit recognition of these control rights through the allocation of ownership titles to insiders. Furthermore, new outside owners were often dispersed and weak (particularly where voucher privatization prevailed). As a result, transition has, in some instances, created an extreme version of the two classical issues of corporate governance: on one hand, the control of managers by dispersed outside owners and, on the other hand, the protection of minority shareholders against strong block-holder interests.[7]

Against this background, external investors have been cautious in providing new capital, and restructuring efforts have been disappointing. Indeed, unchecked by owners and with little access to new funds to finance risky restructuring, managers faced incentives that were skewed towards asst stripping and expropriating minority shareholder.[8]

B. Major Findings

At the risk of simplification, the issue of corporate governance in the transition economies of South-Eastern Europe (SEE) may thus be summarized as follows: (1) the persistence of soft budget constraints (i.e. the refinancing external financing; (2) the remaining influence of the state over corporate decision-making through a nexus of subsidies, regulatory favours and tax arrears provided in exchange for residual control rights; (3) poor investor protection (especially minority), with an entrenched position of incumbent enterprise managers, who retain effective control rights even where privatization has shifted ownership to outsider; (4) concentration of ownership that also undermines the liquidity of equity markets-ownerships and control relatively closely held by identifiable and cohesive groups of "insiders" who have longer-term stable relationships with the company (i.e families, banks, and workers). Furthermore, there is a strong dependency on banks high debt/equity ratios and less developed capital markets.

All these problems are closely intertwined. Enterprises will be unable to access external sources of funds as long as they remain subject to extensive state intervention and/or insider control. Conversely, insider control will remain pervasive as long as potential investors are doubtful about the possible returns on their investments and refrain from acquiring substantial amounts of shares. And, as long as enterprises are unable to survival at least of key enterprises.

With external funds accessible at reasonable costs, the need for state support would be reduced, which would lead to a change in the ownership structure of firms. New emissions of change in the ownership structure of firms. New emissions of shares would over time crowd out insiders, who may also find it attractive to part with their current holdings, provided that outsiders are willing to offer a reasonable price.

Other specific corporate governance issues in the transition economies of South-Eastern Europe (SEE) involve weaker legal systems and corruption. Of courses, there are differences in the degree of these matters, depending on the state of transition.

Court delays, as a measure of contract enforceability are higher in Civil Law countries than in Common Law countries.[9] Most of the South-Eastern Europe (SEE) transition countries have adopted a Civil Law system. According to the World Bank,[10] South-Eastern Europe (SEE) economies are generally belonging to the group with very high court delays (especially Serbia, Bosnia and Herzegovina and Montenegro). Such delays, therefore, increase the costs of using courts for conflict resolution and deter foreign, as well as domestic investors (highest costs for Macedonia, Albania, and Bosnia and Herzegovina).

Furthermore, the transition economies of South-Eastern Europe

(SEE) have to deal with the problem of corruption. The "corruption perception index (CPI)" compiled at the University of Passau[11] relates to perceptions on the degree of corruption as seen by business people, academics and risk analysts. It ranges between 10 (highly clean) and 0 (highly corrupt). The 15 Members-States before the enlargement were ranked within the top 25 of this Survey (with 109 countries). With scores ranging from 2.4 to 3.4, the South-Eastern Europe (SEE) states are significantly more corrupt (or, at least, are seen as more corrupt by market participants) than their European partners (Romania—3, Bulgaria—4).

Consequently, the predominant corporate governance issue in the transition economies of South-Eastern Europe (SEE) could be summarized as *"frequent insider control, difficult outside finance"*. Whereas hardened budgetary constraints and improved investor protection could has an effect on investment decisions, the prevailing issue in most of these countries seems to be enforcement or the effectiveness of corporate governance regulations.

III. MEASURING CORPORATE GOVERNANCE IN SOUTH-EASTERN EUROPEAN (SEE) COUNTRIES

A. Conceptual Issue: The Research Method

Corporate governance codes are, in general, not mandatory regulation and, therefore it is optional for companies to adhere to them. The situation changes if codes are becoming listing requirements at stock markets or formal legal rules by the legislators, which we can observe in many developed European countries (such as the UK or Germany).

Most South-Eastern Europe (SEE) transition economies (with the exception of Romania, where a corporate governance code has been issued by the stock exchange in 2003) have adopted a corporate governance code. Whereas we can find differences within the scope of the codes, they have in common that they abide closely by the OECD Principles.

Given the characteristics of the corporate governance systems institutional surrounding in South-Eastern Europe (SEE) transition economies and the specific corporate governance issues in these countries, formal legal rules which may arise from such codes cannot rely only on a basis of broad minimum standards (that are general legal criteria which are unclear and fuzzy and therefore require judiciary decision-making and classification), as it is often the case in the developed economies, but on binding legislation (mainly directives, that are legal commands which differentiate legal from illegal behaviour in a simple and clear way, adopted in the harmonization process), that can at least partially reduce the existent shortcomings.

The level of compliance of specific legislation with international standards and best practices is defined by the EBRD as *"extensiveness"* ("Law on the books") and is estimated in respect of corporate governance regulations proclaimed into law.

In order to analyze the corporate governance related legislation of each country, the EBRD created a questionnaire based upon the OECD Principles of Corporate Governance: (1) rights of shareholders; (2) equitable treatment of shareholders; (3) role of stakeholders in corporate governance; (4) disclosure and transparency; and (5) responsibilities of the board. Based upon the assessment results for individual countries, a rating system has been developed to show how countries have progressed in the corporate governance area.

"Effectiveness" ("Law in action"), on the other hand, looks at how the legal regimes work in practice, as opposed to the quality of the law on the books.

Changes in the legislation say little about the effectiveness of the new laws; this depends on the voluntary compliance rate on the one hand, and on the effectiveness of legal institutions that are charged with enforcing the law, on the other. Both are mutually reinforcing.

We use an index of the effectiveness of corporate, banking and capital market law in transition economies, constructed by the EBRD, to measure the actual enforcement of regulations in South-Eastern Europe (SEE) economies.

The effectiveness index is taken from the EBRD Transition Reports,[12] which use survey data to rank countries according to the effectiveness of legal reforms (speed, simplicity, enforceability and the institutional environment have been used as measures for the effectiveness of disclosure and redress mechanisms).

B. The Research Results

1. The extensiveness of corporate governance legislation in South-Eastern Europe (SEE)—law on the books

Two comments can be made in relation to recent corporate governance (CG) developments, as described in Table 1.

First, whereas European Union (EU) accession countries might be performing better in terms of economic transition, they donot always have better "laws on the book" than other non-accession countries. This observation is supported by the 2005 country ratings calculated by EBRD[2] (high compliance – Macedonia; medium compliance—Albania, Bulgaria, Croatia, Serbia, and Montenegro; low compliance—Bosnia and Herzegovina, Romania).

One possible explanation is that the EU accession process itself focuses on the harmonization of national laws with the EU *acquis* and, accordingly, better corporate governance regimes can be considered as a

TABLE I

Corporate Governance Development in South-Eastern Europe (SEE) Countries

Country	*Set of rules that serve as a national code of CG*	*Extent of compliance with OECD principles*	*Existing code-mandatory or voluntary*
Albania	Company Law (1992)	It is not in the OECD principles	It is fully mandatory
Bosnia & Herzegovina	Law on Banks and Law on Business Companies	Law on Business Companies and Law on Banks are based on the OECD principles	Law on Business Companies and Law on Banks are Mandatory
Bulgaria	Law for Public Offering of Securities	The existing law is based on the OECD Principles	The existing code is mandatory.
Croatia	Croatian Company Law	Croatian Company Law closely resembles the German Law model on CG and EU directives	There are provisions in the Croatian Company Law which are mandatory. However, certain matters may be regulated by the company articles of association.
Macedonia	Company Code and the Rules of the Stock Exchange	OECD principles are highly respected	The corporate governance rules are to a great extent mandatory
Serbia & Montenegro	Law on Enterprises and Law on Securities and other financial instruments market	The future code will be in accordance with OECD principles	Both laws are mandatory
Romania	Such a national code does not exist. Corporate governance is observed according to provisions in the Companies Law, the Securities Law and NSC's and BSE's regulations and procedures	The existing Code of Corporate Governance issued by the BSE is, to a certain extent, inspired by the OECD principles	The Corporate Governance Code issued by the BSE is mandatory only for the companies listed as members of the "plus tier" of the BSE.

Source: Own compilation, based on EBRD Legal Sector Assessment, 2004.

side effect of harmonization (but not as a goal for EU accession purposes). The fact that the European Commission has recently decided to take a more direct and organized approach in taking corporate governance issue within the EU further supports this explanation and has a more direct consequence on the new acceding countries' legislative framework.

The second comment is related to the evolving EU legislation and, consequently, permanent harmonization process for the EU accession countries. Over the years, a dilemma constantly facing accession countries was that whilst they were endeavoring to establish a "EU compatible" regulatory framework at the national level, the relevant EU norms themselves were not standing still and the global economic environment in which countries were trying to thrive was also changing very fast. If we take Romania as an example, throughout the accession process it was not uncommon to see a piece of legislation adopted only in the previous year being amended to include new relevant EU regulations (e.g company law in December 2006 and securities law in July 2004).

Hence, for the countries which joined the EU in 2007, the high level of harmonization anxiety resulted in more recent efforts put into the legislative process in order to comply with the provisions of the EU Action Plan on corporate governance (e.g. the Romanian framework was improved in 2004, 2005 and 2006 with the enactment of new provisions aimed at harmonizing national laws with EU legislation; according to the last assessment, that considered the 2004 improvements, the ranking of Romania improved to "medium compliance"–62%; for the 2005 and 2006 amendments to the legislation, it is too early to evaluate whether these provisions have addressed the fallings identified in the assessment).

With respect to the five dimensions of Corporate Governance that, according to the EBRD, define its extensiveness, Table 2 confirms the empirical findings outlined above, meaning that major problems of the South-Eastern European (SEE) countries are evidenced in the "disclosure and transparency" sector and in the "role of stakeholder" sector, with specific reference to control arrangements.

TABLE 2

Level of Compliance with OECD Principles on Corporate Governance

	Central Europe and the Baltics	*South-Eastern Europe*
Rights of shareholders	70%	72%
Equitable treatment of shareholders	83%	78%
Role of Stakeholders	79%	63%
Disclosure & Transparency	62%	52%
Responsibilities of the board	70%	72%

Source: Own compilation, based on EBRD Legal Sector Assessment, 2004.

In Romania's case, the major fallings in the observance of the OECD corporate governance principles were identified in the areas of the responsibilities of the board and disclosure and transparency.

2. *The effectiveness of corporate governance legislation in South-Eastern Europe (SEE)—Law in action*

The results of the legal indicator survey conducted by the EBRD in 2005 (aimed at measuring the effectiveness of disclosure and redress mechanisms) indicate that:

- as far as disclosure is concerned, South-Eastern Europe (SEE) is generally characterized by the persistence of a complex legal framework, limited competence and experience of institutions and limited availability and use of case law (with an especially weak institutional environment in Albania, but relatively sound in Bulgaria, Croatia and, more recently, Romania), and with difficult enforcement of judicial decisions (the average time needed to obtain a court order varies from a few months in Bulgaria and Romania to three or more years in Bosnia and Herzegovina; a relatively effective framework for disclosure was reported in Bulgaria, Romania and Serbia and Montenegro); and

TABLE 3

Effectiveness of Corporate Governance Legislation in Romania (58%)

Disclosure	*Refers to a minority shareholder's ability to obtain information about their company*	*68.25%*
Speed		61%
Simplicity		78%
Enforceability		72%
Institutional environment	Refers to the capacity of a country's legal framework to effectively implement and enforce CG legislation	62%
Redress	Refers to the remedies available to a minority shareholder whose rights have been breached	47.8%
Speed		38%
Simplicity		48%
Enforceability		77%
Institutional environment		37%
Costs	Refer to estimated expenses a minority shareholder must pay to take legal action	39%

Source: Own compilation, based on EBRD Legal Sector Assessment, 2005.

- with respect to redress mechanisms, Romania and Serbia and Montenegro have the most effective legislation in the South-Eastern Europe (SEE) region. Major weaknesses consist in long periods of time needed for obtaining an executable judgment (the average time needed varies from 18 months in Romania, to two years in Bulgaria and more than five years in Serbia and Montenegro), complex legal proceedings and a weak institutional environment (Bulgaria offers only one course of legal redress, whereas in Romania and Serbia and Montenegro, minority shareholders can choose between several different procedures which are generally deemed clear and enforceable).

A clear picture on Romania's situation is presented in Table 3.

CONCLUSION

Taken together, the two assessments (i.e. of the extensiveness and of the effectiveness of the corporate governance (CG) regulations provide a two-dimensional view on the quality of corporate governance legislation and on the functioning of legal regimes in the South-Eastern European (SEE) countries. The conclusive results, under the form of the enforcement gap, are illustrated in Table 4.

TABLE 4

Enforcement Gap

	Extensiveness	*Effectiveness*	*Enforcement gap*
Romania	58%	58%	0
Croatia	65%	50%	15%
Serbia and Montenegro	71%	53%	18%
Bosnia and Herzegovina	60%	38%	22%
Bulgaria	72%	50%	22%
Albania	75%	42%	33%
Macedonia	83%	42%	41%

Source: Own compilation based on EBRD Transition Report, 2005..

Several general conclusions can be drawn from this assessment:

- first the quality of the legal framework on corporate governance is improving in all countries, but its implementation is lagging behind. Countries that have developed a solid institutional environment can generally offer an effective legal framework (e.g. Romania, Croatia),

Nevertheless, good laws on the books are not enough to guarantee the effectiveness of a system. The sound environment needs to be coupled with a corporate governance framework in line with coupled with a corporate governance framework in line with international standards and with an effective civil and/or administrative procedural framework;

- second, recent EU member states (Bulgaria, Romania) and candidate countries (Croatia, Macedonia), whilst displaying a better institutional environment, do not systematically outperform other transition countries with regard to the effectiveness of disclosure or redress mechanisms; and
- third, the existence of implementation gaps undermines the usefulness of legal provisions and diminishes the confidence of foreign investors in the legal system as a whole. Consequently, most South-Eastern Europe (SEE) countries, whereas still needing to upgrade their corporate, banking and stock market legislation, should also focus on implementing and understanding the utility of this legislation in practice, in order to provide an explicit signal for investors that are essential for the development of their financial markets.

The most important conclusion of this paper is that a key aspect of weak corporate governance in South-Eastern European (SEE) countries—namely the difficulty in attracting external finance—cannot be solved only by improvements, however radical, in the corporate governance legal framework. The extent of legal reform in these areas of the law has been impressive by any standard. In fact, many of the South-Eastern Europe (SEE) countries, which received foreign legal technical assistance, can today boast higher levels of investor rights protection on the books than some of the most developed market economies.[5] Yet, the development of the law has not been matched constraint on financial market development is the absence of effective legal institutions, or what we have termed "effectiveness". Improving the law on the books in such an environment is at best a partial solution, but will not be rewarded unless a commitment to rule-based governance of markets is made credible.

In their analysis of law and finance around the world, show that effective law enforcement is not a substitute for poor laws on the books. The situation of South-Eastern Europe (SEE) economies indicates that the reverse can also be considered as valid: the existence of laws cannot become a substitute for weak institutions.

Notes and References

1. Pistor, K., Rraiser, M., Gelfer, S., " Law and Finance in Transition Economies", *The Economics of Transition,* 8, 2000, pp. 325-68.
2. European Bank for Reconstruction and Development (EBRD), "Legal Indicator Survey", EBRD, London, 2005.
3. Puffer, S., McCarthy, D., "The emergence of corporate governance in Russia", *Journal of World Business*, Vol. 38, 2003, pp. 284-98.
4. Coffee, J.C., "Privatization and corporate governance: The lessons from securities market failure", *Journal of Corporation Law*, 25, 1999, pp. 1-39.
5. Kornai, J., "The Socialist System: The Political Economy of Communism", Princeton, N.J., Princeton University Press, 1992.
6. Djankov, S., Glaeser, E.L., Porta, R. La, Lopez-de-Silanes, R., and Shleifer, A., "The New Comparative Economics", World Bank Policy Research Working Paper No. 3054, 2003.
7. Dyck, A., "Privatization and Corporate Governance: Principles, Evidence, and Future Challenges", *World Bank Research Observer*, 1, 1999, pp. 59-84.
8. Black, B., Kraakman, B., Tarassova, A., "Russian Privatization and Corporate Governance: What Went Wrong?", *Stanford Law Review*, 52(5), 2000, pp, 1731-1808.
9. Clasessens, S., "Corporate governance and development", in Focus 1: *Corporate Governance and Development,* World Bank, Washington, D.C., 2003.
10. World Bank, "Doing Business Survey", 2006. <http://rru.worldbank.org/DoingBusiness/default .aspx>.
11. Transparency International, "Corruption Perception Indices", 2004, 2005, 2006.
12. European Bank for Reconstruction and Development (EBRD), "Transition Report. Business in Transition", EBRD, London, 2005.

References

Porta, La R., Lopez-de-Silanes, F., Shleifer, A., Vishny, R., LLSV, "Legal Determinants of External Finance", *Journal of Finance* 52, 1997, pp. 1131-50.

Khachaturyan, A., "The Economics of Corporate Governance Regulation in the EU", University of Siena Working Paper, 2004.

Bergolf, E., Pajuste, A., "Emerging Owners, Eclipsing Markets? Corporate Governance in Central and Eastern Europe", in Cornelius. P and Kogut. B, *Global Competitiveness and Corporate Governance*, Oxford University Press, 2003, pp. 267-302.

Bergolf, E., Pajuste, A., "What do firms disclose and why? Enforcing corporate governance and transparency in Central and Eastern Europe", *Oxford Review of Economic Policy*, 21(2), 2005, pp. 178-97.

Bergolf, E., Thadden, Von. E., "The Changing Corporate Governance Paradigm", in Pleskovic and Stigllitz (eds.), *World Development Conference*. The World Bank, 2000.

Bergolf, E., Claessens, S., "Enforcement and corporate governance", World Bank Policy Research Working Paper No. 3409, 2004.

European Bank for Reconstruction and Development (EBRD), "Legal Sector Assessment", EBDR, London, 2004.

Gregory, H.J., Simmelkjaer, R.T., "Comparative study of corporate governance codes relevant to the European Union and its member states", Weil, Gotshal & Manges LLP, 2002.

Becht, M., Bloton, A., Roell, P., "Corporate Governance and Control", in G. Constantinides, M. Harris, and R. Stulz (eds.), *Handbook of the Economics of Finance*, Amsterdam, 2003.

Porta, R. La, F. SilanesLopez-de, Shleifer. A., Vishny, R., LLSV, "Law and Finance", *Journal of Political Economy,* 106, 1998. pp. 1113-55.

Porta, R. La, F. SilanesLopez-de, Shleifer, A., Vishny, R., LLSV, "Investor Protection and Corporate Valuation", *Journal of Finance,* 57, 2002, pp. 1147-70.

Braendle, U.C., Noll, J., "The Societies European—A Step Towards Convergence of Corporate Governance Systems?", 2005, available at SSRN: http://ssrn.com.abstract=704881.

CHAPTER

28

Development of Handloom Industry in J&K State

DR. AMRIK SINGH SUDAN AND DR. RAIS AHMAD

ABSTRACT

INTRODUCTION

Jammu and Kashmir State occupies a significant place in the handloom industry of the country. It has been famous for its handicrafts right from very early times. A large number of people are engaged in the production and distribution of handloom products in the state. Handloom industry play a vital role in the economy of the State and this sector is job and export oriented. Handloom sector is backbone of the industrial activity of Jammu and Kashmir. It has been occupying the predominate position in industrial development on the one hand and overall development on the other. Handloom weaving being the oldest traditional craft of Jammu and Kashmir State has earned world wide reputation for higher quality products. This industry is spread throughout the valley of Kashmir and hilly and backward areas of the state. Kashmir, since times immemorial has enjoyed universal fame for novelty of her industrial manufacturers and the exquisite quality of her arts and crafts. In olden days, world looked to Kashmir for supply of finest quality of silk and woollen textiles.

Travellers and historians from earliest times have paid glowing tributes to skill of Kashmiri and to their acumen and enterprise. These unique and novel products of Kashmir caught the imagination of kings and queens, emperors and empresses in the past. Emperors like Akbar, Napoleon, Bonaparte and Queen Victoria offered and accepted rare and celebrated gifts which included Kashmiri shawls and other high prized products prepared by Kashmiri artisans.

Arts and crafts are an important part of Jammu and Kashmir's heritage. Hand weaving is one the earliest crafts in the State. Hand spinning was the basic occupation of women in olden times, while men did the weaving. During the process of spinning, the women recited anecdotes, legends and historical tales to their children. This scene has been changing now, and alongwith the handspinning of pashmina and some part of wool and cotton, mill spun yarn in wool, silk or cotton is progressively being used.

There are about 57,000 handlooms in the J&K State. Roughly 70 percent of the looms works on woollen yarn, 20 percent on cotton and 10 percent on silk. Most of the rural weavers are agriculturists, weaving being their subsidiary occupation. The handloom industry provides continuous employment to an estimated 2 lakh weavers/workers in the state.

The government of Jammu and Kashmir has also been patronizing this industry from very early times. After independence the, government have formulated a well thought programme for survival and growth of this industry. They have taken a number of steps from time to time to give it a proper shape. Jammu and Kashmir State Handloom Development Corporation Ltd. was set-up by the State government in the year 1981, so that the corporation could take all there measures to rectify the causes which are punctuating the pace of growth in J&K State. The corporation has addressed itself to task of modernizing the existing looms and also to supply yarn to weavers on subsidized basis, apart from solving the problems of the unemployment. Presently J&K State Handloom Development Corporation Ltd. is providing employment to one lakh weavers/workers.

The present paper has been prepared with the following objectives in mind:

(i) To discusses in detail the Development of Handloom Industry in J&K State;

(ii) To highlight the various problems faced by the Handloom Industry in J&K State; and

(iii) To suggest some suggestive measures to overcome these problems.

DEVELOPMENT OF HANDLOOM INDUSTRY DURING MUGHAL PERIOD

During the Mughal period (1586-1753) the handloom weaving industry got sufficient opportunity to develop and prosper. During the Mughal period shawl industry received a great stimulus. Akbar was very fond of shawls and took great pains to improve this industry. The finest of the weavers unleashed the artistic urges of the Mughals who contributed a great deal to the designing of fabrics in Kashmir. During Akbar's reign the functional use of textiles was expanded from shawls and chaddars to materials of various dimensions. Designs with names like param-naram and Narharma (like the river) were evolved during the reign of Akbar, when the ranks in the nobility were determined by the type of Kashmiri textiles awarded by the king. Around the time of the renaissance in the west, the artistic achievement in India inspired. European traders, scholars and generals to visit its great craft centres. The French generals Allard and Ventura serving in the Sikh army, established large trading houses in Kashmir for procurement and export of textiles ranging from embroidered shawls to plain suitings. Following this, the European developed a huge counterfeit shawl industry by initiating the Kashmiri art forms, which lasted for a century. This industry died but motifs are still in vogue.

Akbar the first Mughal emperor to conquer Kashmir in 1586 introduced and encouraged the new art of shawl making, carpet weaving in silk and finest wool with floral animal and bird designs. The Kashmiri shawls were the most important the exports of Kashmir during the Mughal. This tradition was continued by Jahangir and Shahjahan during whose periods the crafts of the state of Jammu and Kashmir gained popularity world-wide. Kashmiri shawls and carpets used to form conspicuous part of the presents sent by the Mughal and Afghan kings to foreign courts which gave futher impetus to the popularity of Kashmiri shawls and other articles in Iran, Afghanistan, Turkistan and Russia and thereby increasing the demand for Kashmiri crafts in these countries. According to William Moorecratft, "The whole value of shawl goods alone manufactured in Kashmir during the closing year of Afghan rule, was about 30 lakhs of rupees per year".

DEVELOPMENT OF HANDLOOM INDUSTRY DURING SIKH RULE

During the Skih rule, the shawl industry in Kashmir is said to have been in flourishing condition. Maharaja Ranjit Singh conquered Kashmir in 1819 and extended indirect patronage to the industry of the state through his governor. He also gave some rewards to the chief weavers of the state. But a terrible famine somewhere in 1832, gave a serious jolt to the handicraft industry in the state, which, however, revived soon. During the Sikh period, the annual production of shawls, woollen fabrics and other items was worth Rs. 75.4 lakhs and the output of shawls alone was estimated at Rs. 35 lakhs.

DEVELOPMENT OF HANDLOOM INDUSTRY UNDER THE DOGRAS

Handicrafts and Handloom Industry of the Jammu and Kashmir received a fresh impetus under the Dogras who while residing in the valley solved various problems of craftsmen and set the trade of these crafts on scientific line. Maharaja Gulab Singh, the founder of Dogra rule in Jammu and Kashmir was a great patron of arts and crafts. His course of encouraging industrial development of the state particularly in this sector was further encouraged by Maharaja Ranbir Singh. His successor Maharaja Partap Singh introduced the Kashmir Handicrafts to Britishers, "When he presented an exquisitely carved-wood door to them at the time of the famous Delhi Darbar". Maharaja Partap Singh was locally known one of the most well known patrons of Kashmir arts and crafts. He used to visit various manufacturing establishments/units, ordered the supply of their goods in bulk and then distributed these among the members of his personal staff or his nobles. The popularity of Kashmir fabrics in western countries were mostly due to efforts of Dogra rulers who used to send presents of shawls and carpets to British and European nobility. From Europe the taste of shawls also spread to other countries and the merchants from U.S.A and Europe came to Kashmir to buy shawls to meet the growing demand in their homelands.

During the reign of Maharaja Hari Singh, the last Dogra ruler of the State, the handicrafts received a major setback as his policy was to replace hand-made articles by machine made goods. He believed in the modernization of the state by introducing modern techniques of production, increased utilization of power and available resources. He liked machine made goods to replace cottage industry. His policies created competition between the indigenous producers and producers of machine-made goods. Thus, competition created dilemma for the handmade goods and the overall handicraft industry received a major setback.

The handloom industry of Jammu and Kashmir state was also threatened by the efforts of the East India Company to Copy Kashmiri shawls and carpets. Though these attempts couldn't thwart the Kashmiri handicrafts but received death blow when war broke out between Germany and France in 1870. In the war, France lost miserably and had to pay a large war indemnity which rendered France unable to spare money for importing Kashmiri fabrics. The famine of 1878-79 also adversely effected Kashmiri crafts. A large number of craftsmen left the valley for Amritsar, Ludhiana, Lahore and other parts of the country where they carried on trade.

The golden period of the Kashmiri craftsmen was interspraced with dark patches. Afghan and Sikh rulers imposed heavy taxes on the trade, the burden of which forced the poor weavers to migrate to the plains. The less fortunate ones had their thumbs cut when they were unable to pay their taxes and quite a few died of starvation.

From the middle of the 15th century and early fifties of the 19th century, Kashmir woollens, particularly shawls, held almost the same reputation as the Muslin of Dacca, brass-ware and brocade of Banaras and silk of Murshidabad. In fact, during 1600-1860, it was one of the most industrialized parts of India.

By 1850, the annual output of shawls alone was valued at Rs. 40-50 lakhs. The cottage industries and arts and crafts were at their glory in 1850's. The Kashmir shawl industry had found its way to European markets as early as 17th century. As a result of British policy towards Indian industries, there was alround decline and ruination of industries in India including Kashmir.

The decline in industry was checked during 1890-1914, as the demand for Kashmir arts and crafts revived on account of increasing influx of visitors to the valley of Kashmir and the Swadeshi Movement led by Mahatama Gandhi Ji. During the period of First World War (1914-19), the industry had sufficient demand and therefore, registered tremendous increase in production. But after the war, the industry experienced a challenge to its survival.

The State Government realized the potential of these art and crafts and initiated firm steps to organize the industries on sound footings. In this directions, the Government established the Directorate of Industries in 1923. This step saved the state industry from the ill effect of the great depression in 1930's. The question of formulating a sound and proper industrial policy was first discussed in 1924 in the State Council. A Cottage Industries Board was established in 1936, but it could not do any tangible work.

The Second World War (1939-45) proved a boon for industries in Kashmir. On account of the demand from defence forces and drastic reduction in imports, there was increasing industrial activity during the IInd World War as is evident from the facts that, "there was great increase in the production of all industries particularly greater in woollens, numdha and silk." The number of factory establishments increased from 71 in 1943 to 99 in 1946. The labour employed in them also increased from 10,000 to 17,000 during the same period. The production in cottage industries and crafts also increased quite significantly.

On account of increased demand, the production and exports of wool and woollens increased form Rs. 30.5 lakhs in 1938-39 to about Rs. 65.00 lakhs in 1945-46.

The partition of the country and the tribal raid aided by Pakistan gave a severe blow to the economy of the state. Even after the ceasefire in January, 1949, the State continued to suffer from unstable economic conditions for several years as most of the trade was conducted through Rawalpindi, Lahore and Sialkot. The state economy was completely shattered as a result of partition of the country. Some of the cotton crafts and hand-woven textiles industries suffered on account of stoppage of import of wool and cotton from Pakistan.

DEVELOPMENT OF HANDLOOM SECTOR AFTER INDEPENDENCE

One the eve of independence and even thereafter, industrial activities in the state were limited to Handicrafts and Sericulture. The state has industrially lagged behind as compared to the other more advantageously situated states. This has been due to inadequacy of infrastructural facilities such as power, network of means of communication, raw material, etc. But the beginning of 1970's sought to bring about a change from the past. The State Government offered a package of incentives for the rapid and smooth development of industries in the state.

State Government also introduced a scheme for providing special concession for promotion of industries in the backwards areas which included capital investment subsidy, etc.

There is also an urgent need to stimulate private investment from within the state and outside. In order to quicken the pace of development, steps must be taken to stimulate private investment and entrepreneurship within the state. There is also need to promote private investment from outside, which will bring in with it technology as well as good managerial practices. These problems will be examined in details in subsequent pages.

In mid 1970's, the State Government took an aggressive line of action in making a break through in the industrial field. A development review Committee comprising of various experts from various disciplines made a study of problems and prospects of Industrial Development in the State. Big industrial houses in the country were invited to make investment in the state.

DEVELOPMENT OF HANDLOOM SECTOR IN J&K STATE DURING SEVENTIES AND EIGHTIES

The origin of handloom weaving in Jammu and Kashmir is very old and this handicraft is believed to be known since the 15th century A.D. However, the industry has grown and developed for long, mainly on the strength of its economic significance to the weavers to remain within their houses and the availability of raw wool locally were the major factors contributing to the growth of the industry.

The handlooms which have a considerable potential for growth and employment generation, have been sadly neglected in the state. An integrated plan of action for the development of handloom sector ought to be formulated for immediate implementation. This, should aim at expansion of production capacity, improvements in technology and wider marketing. While drawing up such a programme, the organizational and administrative aspects deserve particular attention with a view to strengthening and streamlining existing arrangements. The development of handloom sector would be of great benefit to the spinning industry also. Adequate resources would have to be found for the further development of the handlooms. The Khadi and Village Industries Board which has already done considerable work in woollen handloom should also be actively associated.

In this context, the report of the Sivaraman Committee and the Handlooms Commissioner's organization at the centre has been consulted and the central financial assistance under the Twenty-Point programme has been tapped for the development of handloom projects.

The development of handloom sector would not only provide substantial employment but also be helpful in utilizing locally available raw materials. There is a great need for the development expenditure to be incurred on the training, design development and promotion. For this, adequate outlays in the state plans have to be made.

State Government accords high priority to the development of handicrafts and handloom sectors as they play a vital role in the economy of the state. These sectors are job and export-oriented. The first organized effort to bring about a planned development of handloom industry/sector was reportedly made by the State Government during the sixties by attempting to bring handloom weavers into the fold of primary weaver's cooperative societies and by extending financial assistance to them. Besides, an Apex Marketing Society was also set-up to extend marketing support to the primary Weaver's Societies. Notwithstanding these measures, the handloom industry in the state could not register significant growth/development due, mainly, to constraints affecting supplies of raw materials, innovative designs, dyeing and finishing, managerial skills, etc. The weavers' cooperative societies which were conceived to play a vital role in handloom development have, therefore, remained by and large dormant.

During the 70's a renowned trust was given to handloom development in the state with the setting up of the following intensive handloom development projects:

(i) Intensive Handloom Development Project, Pampore.
(ii) Intensive Handloom Development Project, Bandipore.
(iii) Intensive Handloom Development Project, Udhampur.
(iv) Export-Oriented Handloom Development Project, Samba.
(v) Kani Shawl Project, Kanihama, Srinagar.

Further, during the early eighties, the industry received a shot in the arm with the setting up the State Directorate of Handlooms and the State Handloom Development Corporation Ltd. These institutions were conceived to evolve promotional measures and to provide the industry with auxiliary facilities in the field of raw materials supplies, training and extension and marketing of the end products.

The handloom has its own antiquity and is the backbone of arts and crafts that Kashmir produces. Kashmir is famous for producing quality of silk and pashmina. The State Government has launched special programme of plantation of mulberry for raring silk worms in the whole state.

Even though the handloom industry in the state of Jammu and Kashmir has immense potential of becoming the second largest employment-oriented industry next only to agriculture, recognition of its importance came only as late as 1980-81, when the Government of Jammu and Kashmir decided to set-up a full-fledged Directorate of Handloom Development. It was a late beginning but a purposeful one. In the last thirty one years, the Directorate of Handloom Development has tried their best with their available resources and staff to activise nearly 15,000 looms, but the needed thrust was not forthcoming.

In order to give further boost to the 'Handloom Development' in the state, a separate J&K State Handloom Development Corporation Ltd. was set-up on 30th June, 1981 with the sole purpose of revitalizing as many handlooms as possible within a reasonable time frame. This become evident, the first discernable change in the attitude and approach of the Government towards the development of the handlooms. A budget outlay, though meager and inadequate was also provided on the basis of the past track record and the corporation's immediate capability of achieving targets.

Jammu and Kashmir State Handloom Development Corporation Ltd. has been set-up which is involved in modernizing the existing looms in the state and also supplies modern looms to the weavers on a subsidized basis to give encouragement to the handloom sector in terms of employment and production of handloom products. The Jammu and Kashmir State Handloom Development Corporation Ltd. also helps the handloom weavers in its fold by providing them raw material, etc. and other technical guidance for producing quality fabrics. The product range of the Corporation comprises mainly woollen products including tweed, blazer blankets, shawls, etc. The corporation has also set-up its own sale and marketing centres in the state and some major cities in the country.

In addition to J&K State Handloom Development Corporation Ltd. in the State, there are Handloom Weaving Industrial Cooperatives, Khadi and Village Industries Commission, Khadi and Village Industries Board, etc., who are engaged in giving a fillip to the development of handloom industry in the state.

In view of the importance of handloom sector, its contribution to the economy of the state, priority being given by the State Government to this sector which forms an integral part of the 20-Point Programme, employment potential next only to agriculture in the decentralized sector and government's policy of enlarging the cooperative coverage, etc. The J&K Government engaged the services of Agricultural Finance Corporation Ltd., Bombay for studying the socio-economic conditions of handloom weavers, to assess the coverage of looms/weavers and adequacy of existing facilities, to study the extent of availability of pre-loom and post-loom facilities and suggest necessary improvements. In addition, the Agricultural Finance Corporation Ltd. was asked to make a study/analysis of the functions, coverage and efficiency of J&K Handloom Development Corporation, Handloom Primary Weavers' Societies as well as Khadi Board in the sphere of raw material supply, service facilities, marketing support, financial accommodation, price support, etc. and based on its studies prepare a detailed project report.

Jammu and Kashmir state had been a traditional weaving area of handloom fabrics. Kashmir was well known particularly for Kashmir silk, tweeds, chadders, loies, pattoos and shawls. Samba was well known for cotton fabrics and block prints. But with the advent of proweloom textile in the market and lack of proper atmosphere for the growth of handloom sector suffered as serve blow.

On the eve of independence of the country and for quite sometime after industrial activity in the Jammu and Kashmir, State was limited to handicrafts and sericulture and the only industrial units worth the names were some forest-based units.

However, beginning of seventies, marked a change from the past. Jammu and Kashmir State Government formulated a set of comprehensive incentives for the development of industries in the state and these incentives provided a large number of concessions for promotion of industries in the backward areas which included capital investment, subsidy, tax concession, etc. At the same time, the small scale sector also made some progress but not at the desired speed.

In the mid of 70's the Government of Jammu and Kashmir took a more aggressive line in making a break-through in the industrial field. A Development Review Committee comprising of various experts from the various disciplines made a comprehensive study, *inter alia* problems and prospects of industrial development in Jammu and Kashmir state. In his way, at the same time, large scale industrial houses in the country were invited to make investment in the state.

To encourage handloom industry, several projects came up in the state. In the year 1975-76, the Government of India sanctioned several Intensive Handloom Development Projects (IHDPs) and Export-Oriented Handloom Development Project to be established in the state. Both these projects were centrally sponsored under the Twenty-Point programmes and accordingly the funds were provided by the Central Government on the basis of 25% grant and 50% loan for IHDP and 50% loan for EOHDP, Samba.

The Government of India insisted that projects could be implemented only through a corporation. As such the Intensive Handloom Development Projects (IHDPs) was transferred to Small Scale Industrial Development Corporation Ltd. w.e.f 1-4.1977 from the Directorate of Industries and Commerce who were initially vested with the execution of the projects. The Export Oriented Handloom Development Project (EOHDP) was sanctioned by Government of India in 1976-77 and was entrusted to Small Scale Industrial Development Corporation Ltd. for execution. Similarly around January, 1981, Government of India sanctioned two Intensive Handloom Development Projects based at Bandipora and Udhampur respectively and entrusted them to Small Scale Industrial Development Corporation Ltd. for execution.

DEVELOPMENT OF HANDLOOM SECTOR IN JAMMU AND KASHMIR STATE DURING FIVE YEAR PLANS

Developments upto Fifth Plan

The important of handloom sector during plan period was recognized in early sixties when efforts were made to bring handloom industry on proper footing. A number of handloom weavers societies were organized during 1960-70 period and the State Government also provided financial assistance to them. Besides, an Apex Marketing Society was also organized in 1962 to give marketing support to primary handloom societies. However, adequate attention was not paid to the systematic development of handlooms and the societies did not get regular supply of yarn, improved design, bank finance and support for marketing their produce. Most of the societies worked on their own and arranged their own marketing. The societies did not have full time paid managers nor provided much needed common facilities to members. As a result, the traditional handloom woollen products suffered a setback and all the handloom societies, except a few, became defunct by 1980. No serious attempts was made for the revival of such societies and the coverage of weavers by the handloom societies remained nominal.

With the publication of Sivaraman Committee Report in 1974, a number of State Governments established Handloom Development Corporations durings the Fifth Plan Period. Central Government also launches 25 Intensive Development Projects and 21 Export Oriented Production Projects in the state. However, in Jammu and Kashmir State, handlooms continued to be looked after by the Directorate of Handicrafts. Later, with the importance, assigned to handlooms in the 20-point programme, the State Government decided to set-up the Jammu and Kashmir State Handloom Development Corporation Ltd., and it came into being on 30th June, 1981 and started functioning from 1981-82. As the handloom cooperatives were very weak in the state, the emphasis was shifted towards the corporation and it became the main agency for implementing the intensive projects of the development of handlooms. On its creation, the following production units were transferred from Small Scale Industrial Development Corporation Ltd. (SICOP) to the Jammu and Kashmir State Handloom Development Corporation Ltd. (JKHDC).

1. Intensive Handloom Development Project, Pampore.
2. Intensive Handloom Development Project, Bandipore.
3. Intensive Handloom Development Project, Udhampur.
4. Export Oriented Handloom Development Project, Samba.
5. Kani-shawl project at Kanihama, Srinagar.

Subsequently, the handloom silk factory of Jammu and Kashmir Industries and Handloom Woollen Weaving Factory Shrinbagh of Jammu and Kashmir Industries were also transferred to the Corporation. The IHDPS/EOHDP which were started originally with Government of India's assistance were later on transferred and the State Government started providing necessary assistance.. In the year 1983-84, the corporation started another IHDP for Poonch-Rajouri with headquarter at Rajouri. In the year 1986-87, corporation also started 2 another IHDP's for Doda and Leh. In this manner, the corporation has covered about 2000 looms in the project areas and provided them with raw material pre-loom and post-loom processing facilities and the production is supervised. The finished products were procured and marketed by the corporation. In addition to J&K Handloom Development Corporation Ltd., in the state, there are Handloom Weaving Industrial Corporations, Khadi and Village Industries Commission, Khadi and Village Industries Board, etc., who are engaged in giving a fillip to the development of Handloom Industry in the state.

During 1980-81, the State Government took another important decision and a separate Directorate of Handloom Development was created to give boost to handloom activities. The Handloom Cooperative Societies as well as the corporation were brought under the jurisdiction of Director of Handlooms. Besides, the Director of Handlooms was appointed as Managing Director of the Corporation.

The Directorate started organizing new handloom societies. However, its efforts hampered for want of supporting field staff. Only two Assistant Registrars were provided for the entire state to look after the organization, promotion and supervision of new societies. The Director Handlooms also being the Managing Director of J&K State Handloom Development Corporation Ltd., devoted its attention mostly on the working of the corporation. Besides, the corporation which was expected to cover the weavers outside the cooperative fold, actually adopted the members of cooperative societies functioning in Poonch-Rajouri, Pampore and Bandipora project areas, without involving the societies in any way. As much, societies in these areas, became dormant. The Apex Society did not come to the rescue of the primary societies for the supply of yarn or marketing their finished products and concentrated its activities in dealing with master weavers. The Handloom Cooperative Association, Jammu (a Regional Society) confined its activities to a few societies in Jammu district as it could not find any marketing outlet. The Cooperative Banks in the state also did not provide adequate credit support to meet the working capital requirements of handloom societies. The State Government was also reported to have issued instructions to make its purchase from J&K State Handloom Development Corporation only. The above factor hampered the growth of handloom cooperatives. Though the Department organized 22 new societies during 1985-86 and proposed to organize another 25 societies during 1986-87, the coverage of weaving families by cooperatives remained miserably poor. As on 30-06-1986, there were 90 handloom societies in the state of which 39 were defunct for long.

On the other hand, the J&K State Handloom Development Corporation Ltd., expanded its activities and opened a few more Common Facilities Centres (CFCs). There were 23 CFCs in the project area covering about 2000 looms. Its production rose form 3.54 lakh metres in 1981-82 to 9.64 lakh metres in 1984-85. The corporation, however, could not provide full employment to the weavers adopted by it and on an average, the weavers got employment for about 130 days in a year. The main impediments faced by the corporation were non-availability of raw material in time, organizational difficulties and inability to mount a major sales campaign. It has yet to make an entry in the export market.

Development During Sixth Five Year Plan

During the Sixth Five Year Plan, an outlay of Rs. 300 crores (out of total outlay of Rs. 900 crores for all the sectors) was provided for the handloom sector, though the actual expenditure was Rs. 4.49 cores only. During the period there was expansion in the production of handloom cloth by the handloom cooperative sector as well as by J&K State Handloom Development Corporation Ltd. During the year 1984-85, the production was to the tune of 11.04 million metres, the value of which was Rs. 3.96 crores only. The main emphasis was of the (i) establishment of common facility centres in the areas of concentration including coverage for dyeing and finishing requirements of weavers, (ii) establishment of handloom development projects in Udhampur-Doda in Jammu Division and Bandipora (Baramula) in Kashmir Division on the Pattern of IHDP Pampore; and (iii) creation of new training centre for various crafts of the industry.

Seventh Five Year Plan

The outlay of Seventh Plan under handloom sector was marginally increased to Rs. 5.70 crores against actual expenditure of Rs. 4.49 crores in sixth plan period. The major trust was to strengthen the data base of handloom sector publicity, rebate on retail sale, rebate on wholesale to Jammu and Kashmir State Handloom Development Corporation, assistance to UNDP Wool Utilization Centre, Study Tour of Weavers and Training Centres. The thrust to encourage handloom cooperative sector is not visible. The approved Seventh Plan outlay for handloom sector under main heads is given below:

Sl. No.	*Main Head*	*Approved outlay (Rs. in Lakhs)*
1.	Handloom Directorate: State Sector	263.00
2.	Training programme: District Sector	184.50
3.	JK-HDC/IHDC	122.50
	Total	570.00

The approved seventh plan outlay for various sectors in the state was Rs. 1400 crores of which approved outlay for handloom was Rs. 570.00 lakhs forming 0.47% of the total outlay. The total plan outlay for handicrafts sector at Rs. 11.39 crores was double than that of handloom sector. Further, out of the total outlay under handloom sector, an amount of Rs. 1.00 crore (including rebate assistance of approximately Rs. 44 Lakhs) was earmarked for development of handloom cooperatives. It works out to 0.7% of the plan outlay. The outlay was thus very meager in view of the fact that the present coverage of handloom cooperatives is to be increased to 50% and the State Government may have to take a major policy decision to provide additional financial outlay.

During 6th and 7th Plans, the industry attained a reasonable level of growth particularly in the spheres of production, product quality, modernization and marketing. However, compared to other hilly states, particularly in the north-eastern region, the level of development attained by Jammu and Kashmir State in handloom development remains to be small and the industry continues to exist in the potential stage.

Handloom Strategy for Eighth Plan

Handloom industry in the state has an immense potential in production as well as a largest employment oriented industry. The outlay of Seventh Five Year Plan for this sector was approved for an amount of Rs. 447.50 lakhs against which an amount of Rs. 367.85 lakhs has been spent during the first four years. The anticipated expenditure for 1989-90 was Rs. 100 lakhs against the approved outlay of Rs. 108 lakhs. By the end of the Seventh Plan 5022 weavers shall be brought under corporate fold. The anticipated production at the end of Seventh Plan shall be 7.33 million metres valued at Rs. 33.96 crores.

The proposed outlay for the Eighth Plan is Rs. 713.00 lakhs. In addition to the existing scheme, some new programmes are also proposed for implementation. The approved 8th Plan outlay for handloom sector in J&K State under main heads is given below:

Sl. No.	*Main Head*	*Approved outlay (Rs. in Lakhs)*
1.	Directorate of Handloom: State Sector	458.00
2.	Training programmes: District Sector	255.00
	Total	713.00

Source: Draft Eighth Plan and Annual Plan, 1991-92, Planning and Development Department, J&K Government, Jammu, pp, 168-69.

PROBLEMS OF HANDLOOM INDUSTRY IN JAMMU AND KASHMIR STATE

The major problems of handloom industry in Jammu and Kashmir are discussed as under:

Problem of Raw Materials

The state handloom sector has to depend mainly on raw material supply from state spinning mills which is inadequate to meet the total demand. Woollen yarn accounting for about 82% of the total handloom production in the state is much in demand and its prices have been going up quite regularly. Present market price of woollen yarn is Rs. 210-240 per kg. which is about 50% higher than the 1987-88 prices levels. Even at this price, local availability of woollen yarn is inadequate and about 50% of the total woollen yarn requirements estimated at 2.50 lakhs kg. per month is being met by imports from Amritsar, Ludhiana, Panipat, New Delhi and Kanpur markets. The problems relates to specially to services constraints in the free supply of yarn of requisite counts at reasonable prices.

Pashmina and shahtoosh yarn are produced in Ladakh region of the state. Pashmina and shahtoosh products fetch very good prices in domestic as well as overseas markets. The demand for the same has been increasing at an estimated rate of 25 percent per annum. However, their production is estimated to be rising @ 105 per annum only, thus leaving a wide gap in demand for and supply of pashmina and shahtoosh yarn. To make matter worse for J&K State Handloom producers/weavers, traders/businessmen from Punjab and New Delhi buy pashmina and shahtoosh in bulk and resell the same to J&K State Handloom weavers, dealing in pashmina and shahtoosh products at increased prices.

During the field survey, it has been found that majority of the weavers is facing the problem of shortage of raw material. The yarn provided by the J&K State Handloom Development Corporation Ltd. and Handloom Cooperatives to their units is not sufficient enough to enable the weavers to continue their weaving for whole of the month. So the weavers are compelled to sit idle after 20 or 25 days for want of raw materials.

The survey also revealed that more than 95% weavers/ cooperatives societies are not even aware of National Handloom Development corporation (NHDC); the apex national level organization engaged in supply of quality raw materials at reasonable prices. Consequently, private cooperatives, production centres entirely depend on traders who sell yarn to them at much higher prices. Quality of yarn supplied by these traders is not reportedly uniform and constant, resulting in loss of productivity, quality and finish of the products.

Inadequate Financial Support

J&K State Handloom industry suffers seriously from inadequate financial from commercial banks, financial institutions and other development agencies like Khadi and Village Industries Board, Khadi and Village Industries Commission, etc. It has been revealed during the field survey that about sixty-five weavers depend entirely on their own funds for their raw materials and other working capital needs. The weavers need finance to purchase the yarn during the shortage of supply of yarn from the J&K State Handloom Development Corporation Ltd. and Handloom Cooperatives. But these weavers do not have any easily available finance to purchase yarn. There is no proper bank loans scheme for the weavers to purchase the yarn and other accessories of looms. Moreover, few weavers who have some knowledge of the existing schemes hesitate to approach the banks because of the red-tapsim and cumbersome formalities of the banks.

Total financial support to state handloom industry by government, central and state agencies was much less than the requirements of the industry as already pointed out earlier.

Marketing and Sales

Marketing and sales network supporting the state handloom industry is rather inadequate. J&K State Handloom Development Corporation Ltd. (JKHDC), J&K Industries Limited (JKI), J&K Sales and Exports Corporation (JKSEC), Gandhi Ashram (GA) are major state level organizations providing marketing and sales support to J&K state Handloom industry, Central Cottage Industries Emporium, Handloom and Handicrafts Exports Corporation, Apex Cooperative Marketing Societies of Handloom (ACASH) also coordinate and promote marketing in the handloom sector at state and country level. However, these organization/institutions are catering to units/weavers who reside/ operate from cities/towns. In the absence of proper development of the cooperative sector in the state, weavers/units in interiors have to depend on traders/middlemen for marketing and sales support. Government organisation/institution largely procure handloom products through these trader/middlemen without visiting the actual weavers/artisans. In this scenario, weavers/artisans have to remain satisfied with the prices offered by these traders which in most of the cases remain quite low. Sometimes in order to recover their rent and to meet their day-to-day expenses weavers have to make distress sale of their products below their production cost.

Since J&K State Handloom Development Corporation Ltd., Dastkar Anjuman, Gandhi Ashram, Khadi and Village Industries Board (KVIB) and Khadi and Village Industries Commission (KVIC), have their own production centres/units, these organizations prefer to provide marketing and sales support to units/weavers run/supported by them. Thus, a large number of individual weavers/units depend upon the private traders/agents/middlemen to sell their products.

Inadequate Marketing Organisation

The marketing of handloom products is in a very miserable condition. Handloom products are not only facing the problem of cut-throat competition from the mill sector but also have a poor market organization. Due to lack of efficient market information system and research studies, required to understand the current market trend and consumer's preference, the market horizons of handloom products have become limited.

Absence of Cooperative Sector

Handloom Cooperatives in the state are in a very poor condition. Cooperative movement is hardly visible in the state handloom industry. There are only 1.5 percent of total handloom units in the cooperative sector. Majority of the cooperative societies are either defunct or partially functional. In the absence of proper development of cooperative sector, state handloom industry does not get incentives/support from central government agencies. This is the main drawback of the state handloom industry and needs immediate attention.

Lack of Proper Dyeing and Finishing Facilities

Dyeing and finishing are two most important factors which contribute in a big way for the improvement in the quality of handloom products. But unfortunately these facilities are not available to the weavers of state handloom industry. So the quality of handloom products is bound to the poor.

Lack of Sufficient Common Facility Centres

Common facility centres have been established to provide preloom facilities in the shape of winding, wraping and twisting and technical guidance in respect of weaving, designing and quality to the weavers. The Directorate of Handloom Development Department has opened such centres which give training to the weavers. But these centres are mostly established in urban areas benefiting the few weavers.

Lack of Design Development Facilities

During the field survey, it has been observed that most of the handloom units in the state use designs of their own. There is no institution to guide the weavers to adopt the new and upto date designs which are preferred by the consumers in the market. So they continue to use the old and obsolete designs on their products and lose market share to organized mill sector.

Obsolete Production Practices/Tools

The use of obsolete production practices and tools is the main cause of low productivity level and poor quality of handloom products prevailing in the J&K State Handloom industry. During the field survey, it has been found that more than 60% looms are age old pitlooms. About 38% looms are pedal looms and frame looms mostly supplied by J&K State Handloom Development Corporation Ltd. Pitlooms are difficult to operate, time consuming, less productive and very strenuous to the weavers/artisans. Per loom yield of the handloom industry in the state is estimated at 4.7 metres/loom per day which is well below the national average of 5.2 metre/day yield per loom. Allowing workers to operate with their existing looms should may be ousted from the domestic and overseas markets. Although frame looms provided by the J&K State Handloom Development corporation Ltd. are of better quality than pit looms. But even the frame looms are not as efficient as power looms which most of the units wish to install.

Low Wage Rate

During the field survey, majority of the weavers expressed woefully that the wages paid to them by the J&K State Handloom Development corporation Ltd. and handloom cooperatives are very low and it is very difficult for them to maintain their families with the existing below subsistence wage rates.

Lack of Processing Facilities

Wool processing facilities in the state are inadequate and much below the requirements of state handloom industry. United Nation Development Programme (UNDP) aided wool processing/utilization projects, Nowshera, Srinagar has an installed capacity to process wool upto 50,000 kg. per annum. This is expected to increase upto 1.10 lakh kg. wool per annum by 1991 as efforts are on to modernize and increase the capacity of the project. However, even the increased capacity would be well short of total demand for the woollen yarn which is estimated at about 5.5 lakhs kg. per month. Therefore, weavers/wool producers have to go to Amritsar, Ludhiana, Patnitop and Kanpur for wool processing purposes. Samba processing project/centre caters to cotton processing needs of the state. However, Samba alone is not capable for meeting the requirements of weavers and services nearly 10 percent of the total demand for processing and dyeing of the J&K State Handloom industry.

State dyeing facilities are also limited to UNDP aided project at

Nowsher and Samba only. UNDP project takes up dyeing of yarn mainly on experimental basis, from small and tiny unit holders. However, it is not in a position to provide dyeing facilities to al the weavers and willy nilly either they have to go to outside sources or take up dyeing and processing work on their own. In case of later, quality of dyed yarn more often than nor suffers resulting in less value for the finished products. Thus, inadequate dyeing and processing facilities in the state is one of the major problems faced by the J&K State Handloom industry.

Market Intelligence/Design Development

Market intelligence and design development are weakest link of state handloom industry. J&K state handloom industry has remained isolated from the changes occurring in public taste, preferences in the field of designs, combination of raw materials used, colour combination, quality and finish of the products, etc. barring individual efforts by big businessmen/handloom dealers, there has not been adequate efforts from state level organizations to know the pulse of the buyers in domestic as well as overseas market. Result of this has been that weavers in the state have been producing handloom items without knowing consumers preferences leading to low demand and importantly lesser value addition to state handloom products.

Lack of Quality Testing and Inspection Establishments

The importance and need for strict quality control measures in an unorganized and diverse industry like handloom is immense. But unfortunately during field survey it has been found that handloom industry in the state does not have any strict and scientific measures to control the quality of its products which consequently had led to the deterioration in the quality of handloom products.

Lack of Adequate Incentives and Welfare Schemes

Although there are many incentives and welfare schemes for the weavers under the Directorate of Handloom Development Department, J&K State. These incentives and welfare schemes are as under:

Incentives

(a) Market development assistance.
(b) Subsidy on yarn to handloom weavers.
(c) Share capital assistance scheme.
(d) Loan for purchase/renovation or modernization of looms.
(e) Training-*cum*-study tour of weavers.
(f) Loan for purchase of sewing machines.
(g) Organisation of industrial weavers' cooperative societies.
(h) Government share participation.

Welfare Schemes

(a) Thrift Fund Scheme.
(b) Workshed-*cum*-Dwelling for Handloom Weavers.
(c) Group Insurance Scheme for Handloom Weavers.
(d) Medical Reimbursement Scheme.
(e) Education Scheme.

Inspite of all these schemes, very few weavers have been benefited out of these schemes and the socio-economic condition of the weavers remain almost same as it was before partition of the country.

Other Problems

- The cooperative coverage under handloom industry is hardly 4 percent compared to the national average of about 45%, even after four decades of efforts by the state government. Other states in the country have not only taken advantage of various schemes and incentives drawn up by the central government, but also have shown very good progress in the field of production of handloom cloth is a mere 3 lakh metres with turn over of about Rs. 4.5 crores annually.
- There is an urgent need to strengthen the coverage of handlooms under cooperative to give a fillip to this industry.
- The financial outlay of Rs. 7.13 crores for the Eighth Plan out of a total outlay of Rs. 3920.00 crores in the state gives an impression that handloom have been given a very low priority. There is an urgent need to increase the outlay during the remaining period of Eighth Plan and also in the formulation of the Ninth Plan.
- Due to lack of adequate attention and meager capital outlay in this employment and income generating sector of the state economy, a sizeable percentage of installed looms numbering about 25,400 remain dormant without any work programme. There is an imperative need to find whole time work for all the looms installed in the state through a well planned revival programme.
- The average earnings of the handloom weaver has been found to be very meager at nearly Rs. 11.34 per day compared to the average earnings of handloom weavers in other prosperous states. The number of working days at 150 per year for a weaver are also on the lower side. Therefore, there is need to

argument weaver's daily earnings and also to provide them work for optimum workable days in a year.

- The marketing and export potential of this industry has not been encouraging so far. Kashmir silk and shawls are world renowed but its marketing and export potential still remain untapped.
- The present situation calls for immediate revival measures, taking into account the vast potential of handlooms to provide better living conditions to thousands of weavers in the state. The Government of J&K feels convinced about the imperative and urgent need for the revival of the handloom industry in the state by giving it top priority in view of its contribution to the economy of the state with the least investment.

In addition to the above problems, handloom industry also faced problems in regard to availability of skilled manpower, non-availability of quality chemicals, dyes a colours for processing and colours of the yarn, lack of coordination between various developmental agencies and handloom weavers, etc., which in cumulative effect seriously hamper the functioning of the handloom industry.

SUGGESTIONS REGARDING THE J&K STATE HANDLOOM INDUSTRY

The suggestions/measures to overcome the problems faced by the handloom industry in the J&K State are discussed as under:

1. Ensuring continuing and uninterrupted supply of yarn to weavers

One of the major reasons for the stagnation of the J&K State Handloom industry is the erratic supply of yarn to the weavers. Frequent shortage of yarn coupled with high prices have been noticed as the factor in the under-utilization of the looms as well as to the weavers remaining idle without job. In order to overcome this and to ensure the uninterrupted supply of yarn at more or less steady prices, a yarn bank is recommended to be established in the state.

Proper arrangements between raw procuring, processing and weaving agencies as well as between filature units and consumer agencies need be established for steady flow of the raw material according to the requirements of the consuming agencies. Permanent pashmina purchase centres be established at suitable places in Ladakh/Kargil districts by the State Wool Board.

2. Adoption of an effective development strategy for J&K State Handloom Industry

It has become imperative to define a development strategy for handlooms for the J&K State. This strategy must give top priority to handloom sector. This can be achieved only by treating handloom at par with tourism. Only then the needed thrust to the developmental activities will be forthcoming. This will involve upgradation of the office of the Directorate of Handloom Development to the position of the Commissioner, Handloom Development, provision of adequate outlay of finance, staff and strengthening the implementing agencies.

3. Data Bank

Lack of data has been the main bottleneck in evolving an effective development plan. Presently, to obtain data pertaining to any sector/component of the handloom industry, whether it is marketing, production programme, export, quality control, a long drawn out procedure and paper work involving enormous time and labour are to be compiled with one of the most important pre-requisites for the successful implementation of the project is the availability of accurate/correct data in time. A computer centre is recommended to be established under the J&K State Handloom Cooperative Marketing Federation or Directorate of Handloom Development. The computer centre shall cover marketing product development, handloom census, updating socio-economic survey, finance and accounts control quality and production control.

4. New Marketing Organisation

In order to achieve a sales turn over Rs. 80 crores in five years, it is recommended to set-up a new marketing organisaation exclusively for marketing of handloom products of J&K State. The organization proposed shall be the J&K Handloom Cooperative Marketing Federation Ltd. This organization will be equipped with the needed capital-base, staff and convincing promotional budget to achieve the target.

5. Modernisation of Looms

All existing fly shuttle frame looms should be fitted with two roller long length weaving device and take up let-off motion.

All existing fly shuttle frame looms, other than those engaged in weaving blankets be fitted with fly wheel attachment controlling the sway of the sley.

Handloom both in cooperative and corporate sector should be modernized by modification and replacement. Expansion in weaving capacity will be achieved by installing improved handlooms.

6. Post-looms processing and finishing facilities

In order to add value to the women fabrics and to reduce the dependence on outside agencies, the following additional facilities are recommended:

(a) Expansion and addition to the existing UNDP Nowshera Wool Processing Plant.
(b) Expansion and addition to cotton processing plant/house Samba to include facilities for polyester cloth and blends.
(c) An upto date and well-equipped pure silk printing, processing and finishing unit.
(d) Setting up a garment factory to support export efforts.

7. Quality Control and Product Development

Quality control laboratory is to be established directly under the Directorate of Handloom Development Department along with a Research and Development Wing. A Weaver's Service Centre is recommended to be established with the help of the Development Commissioner for Handlooms, Government of India, New Delhi, for product development and design services, in cooperation with J&K State Handloom Development Corporation Ltd. The Wool Board may be requested to develop mechanical dehairing preparatory to spinning. It is suggested that spinning of pashmina on hand driven mechanical charkhas be promoted. The Wool Board should sell raw material to worsted spinners in the form of wool-tops instead of raw wool.

8. Strengthening the Cooperative Movement in the State

Only viable handloom societies may be organized covering 60 looms and 80 weavers. They should be organized within the first three years of the project period. All the societies will be served by Common Facility Centre. For the work-shed type activities, work-shed of 30 looms is recommended along with common facility centres for 60 loom-bolloms. The Society should be managed by well-trained paid secretaries/managers, besides the technical staff.

9. Financing of the Societies

The Apex Bank/Jammu Central Cooperative Bank may create a separate cell in their banks exclusively to attend to the work of development and financing weaver societies. In case, the Central Cooperative Banks are not in a position to finance handloom cooperative societies, the Apex Bank may finance such societies directly.

In order to encourage the banks to create a cell, the Director of Handloom Development Department may be the initial stages, place one inspector each at the disposal of the banks for a period till the cell becomes functional.

Handloom societies lying defunct for the last 10 to 15 years may be wound up. Some societies have not yet installed the looms given by the department. Such cases may be reviewed immediately for appropriate action. Rebate claims of handloom societies remain pending for a number of years. The budgetary allocation for paying rebate should be suitably increased. The handloom weavers who are members of agricultural credit societies may be provided finance from Cooperative Banks by suitably devising a credit limit application form.

10. Training Institutes/Centres

Training institutes are to be reorganized throughout the J&K State for the purpose of imparting training to the exiting weavers on weaving more on paying styles and also for giving training to new entrants.

11. Need for Weavers' Service Centres

The All India Handloom Board has been setting up weavers' service centres in different parts of the country. There is no such centre established in Jammu and Kashmir so far. At present, there seems to be no institution in Kashmir where improved handloom equipment improved methods of processing including dyeing and printing are demonstrated. The weaver service centre would, therefore, come a model centre where all available improved equipment used in silk and wool weaving could be displayed. Centre should have a full preparatory unit including vertical and horizontal warping unit.

The weavers' service centre in the country are concerned with servicing, research and training and cover various fields of design dyeing and finishing. Unfortunately, the state has not been able to find a suitable place for the centre in Kashmir itself.

12. Need for Finishing Plant

In view of the cottage industry of the handloom industry, individual weavers are not in a position to do proper finishing of the woollen goods produced by them. It is, therefore, suggested that small processing plants may be set-up in the vicinity of those areas where concentration of handlooms is sizeable.

14. Setting-up of Handloom Advisory Board

In order to evolve an integrated policy for the development of handloom industry working on cotton, wool and silk, there are Handloom Advisory Boards/Handloom Development Corporation in various states of India. It appears that in Jammu and Kashmir, this industry has been neglected and therefore, there is an immediate need to set-up an Handloom Advisory Board at the state level. The composition of the Board should be such as to ensure the representation of all the concerned interests including, Government, cooperatives and private sector of the industry. The major functions of the Board shall include formulation of an integrated policy for future development of handlooms in the state which have considerable employment of potential as well as export possibilities in foreign countries. The Board may also lay down policy for coordination in different departments looking after handloom industry in Jammu and Kashmir. Provisions for adequate outlays may be considered in the annual plans of the state.

15. Development of Woollen Khadi

Khadi sector has evinced a gradual growth in the state. However, to make it provide gainful employment to a larger section of population, it is felt that certain developmental measures discussed in ensuring paragraphs deserves due consideration by the concerned authorities.

16. Finishing Facility in Jammu

In the absence of washing and finishing facility for woollen in the lower region of the state, goods/products have necessarily to be sent to Amritsar for processing. Gandhi Sewa Sadan—Sole Khadi Institution in Jammu region is in dilemma because sending of goods by road to Amritsar and back not only involves heavy transportation cost but sometimes due to heavy rush, goods have to wait for about two months before being processed. This *inter-alia* affects the supply schedule of the unit.

Processing charges which are being paid by this institution presently are reproduced in Table 1.

TABLE 1

Processing Charges for Various Items

S. No.	*Items*	*Rate*
1.	Barrack Blankets	Rs. 17.00 per piece
2.	Tweed cloth	Rs. 12.50 per metre
3.	Shawl	Rs. 15.00 per piece
4.	Chadder	Rs. 16.00 per metre
5.	Civilian Blankets	Rs. 16.00 per piece

Total expenses which are being incurred by Gandhi Sewa Sadan towards this head are estimated to be around Rs. 6.00 lakh per annum. In view of such a colossal expenditure on processing alone, it is considered ideal for the Sadan to have washing and finishing of its own.

Establishment of a washing and finishing unit will go a long way in the development of woollen textiles in the lower region of the state. Proposed unit which is to be controlled by the Gandhi Sewa Sadan will not only meet the demand of its promoter but a large number of unorganized weaver who have hither to been selling their product unprocessed will be in a position to make use of this facility and get higher returns for their finished product.

17. Spinning

The Jammu and Kashmir State has got inadequate spinning capacity of woollen yarns and the handloom industry is to largely depend on supplies from outside the state. There is presently no programme for expansion of spinning capacity on woolen systems. This needs immediate attention.

18. Scouring and Dyeing

The facility available in the UNDP Wool Utilisation Project, Nowshera, Srinagar is grossly inadequate. They can handle only 50,000 kgs. of wool provided they do not dye and yarn. Adequate capacity to handle dyeing of at least 4 to 5 lakh kgs. of wool and yarns will have to be developed in the state.

19. Dyes and Chemical Facilities

The state is to depend on supply of dyes and chemicals from firms outside the state from whom it is not possible to obtain quality dyes and chemicals as per requirement from time to time. A dyes and chemicals bank needs to be created with the Directorate of Handloom Development Department/J&K State Handloom Development Corporation Ltd., and these facilities should be made available at every common facility centres and training institutes/centres of the Directorate of Handloom Development in the state.

20. Hand-Spinning Facility

Hand-spinning needs be promoted as a craft in the primary schools. This can be a subsidiary means of income for most of the rural population and the demand of hand-spun yarn, particularly of woollen fabrics is practically unlimited.

21. Designing Facility

The need to create a center on the lines of school of designs in the handicraft unit is indicated for handloom sector as well. Some work has been done by the J&K State Handloom Development Corporation Ltd. under the guidance of Mr. Martin Hardingham of CFTC. A little work is also being done in the UNDP Wool Utilisation Project,

Nowshera. Some work will also be done in the weaver service centre, Srinagar which is proposed to be set-up by government of India. However, all these efforts will not give adequate returns unless a central design centre is made available in due course of time to look after the needs of handloom industry.

22. Printing Facility

The printing industry in J&K State is practically non-existent. A few units scattered here and they are not in a position to produce quality printing work nor they can meet even the local requirements. Practically, all the silk fabric is to be sent outside the state for printing and processing. The Export Oriented Handloom Development Project, Samba under the J&K State Handloom Development Corporation Ltd. is introducing screen printing. However, if the work is to be centralized in the interest of proliferation of various schools of printing and designs, setting up of various training units particularly for silk will have to be encouraged.

23. Processing Facilities

The facility available for processing of handloom fabrics in woollen, cotton and silk sector is practically non-existent. The UNDP, Nowsehra, Srinagar can only touch woollen fabrics. There are no facilities available for carbonization and quality raising. If the exports are to be promoted, at least half dozen processing plants/units will be required to be established in different parts of the state for woollen, silk and cotton processing.

24. Readymade Garments Units/Knitting Units

The concept of mechanized cutting and tailoring for exports has not caught on in our state as yet. At least two such units for cotton and woollen need to be established urgently in the state.

25. Increase the Number of Common Facility Centres

The number of common facility centres of the Directorate of Handloom Development Department, which are established to impart the weavers pre-weaving facilities should be increased and made more effective and efficient in rural areas.

26. Import of Machinery

The import of machinery for finishing, processing, sampling, testing and garment-making needs to be liberalized for handloom industry.

27. Import of Dyes and Chemical and Accessories

For export purpose, the import of dyes and chemical processing ingredients as well as accessories like press buttons and hooks, etc. also need to be liberalized.

28. Regularize the Supply Position of Yarn to Weavers

The government of Jammu and Kashmir should take steps to make yarn available to the weavers at the nearest place without any delay in the supply. The Directorate of Handloom Development, the J&K State Handloom Development Corporation Ltd. and Handloom Cooperative Societies should put every effort to regularize the supply of yarn during the shortage period so that the weavers are able to continue the work throughout the month or year.

29. Marketing Research

Marketing of handloom products is in a doldrums situation when compared to the marketing of mills products. So marketing information system of the handloom products should be overhauled. Marketing research should be initiated from time to time by the Directorate of Handloom Development Department, the J&K State Handloom Development Corporation Ltd., and Apex Cooperative Societies to understand the current market trends and consumer's preferences and only those products should be produced which have a ready and easy market.

30. Procedures

Procedures for sending sample for exports as well as the actual exports need be simplified and the export should be permitted directly from Jammu and Srinagar which are both connected by air.

31. Tariff Concession

The air tariff on carriage of handloom goods/products within the country provides for certain concessions. These concessions should be further improved and also extended for an international handloom cargo.

In order to give a new meaning and content to the revival programme for handlooms in the state, the state government may adopt the development goals which are enunciated below:

(i) That every weaver shall be ensured full and productive work for 240 days in a year;
(ii) That every handloom weaver individually will earn Rs. 50 per day as wages for a better living;
(iii) That the weaver shall not have to trek inconvenient distances for obtaining their days wages;

(iv) That the weavers shall not have to look towards the middlemen for work, material and wages;

(v) That to ensure this, the cooperative and other state organizations working for handloom development will be adequately strengthened and galvanized;

(vi) That as far as possible readymade warp beams, will be made available to weaver within a convenient distance along with a weaving programme;

(vii) That the modernized handlooms that will reduce the fatigue of the weavers will replace old looms to enhance production, productivity and quality;

(viii) That every weaver will be exposed to the intricacies of weaving more on paying styles in keeping with the market demands;

(ix) That the products of J&K State weavers will be promoted both within the country and abroad through an aggressive marketing strategy and establishment of a well-equipped dynamic marketing organisation supported with adequate financial backing, and above all;

(x) That handloom industry in the J&K State will be restored to its primitive glory by according it necessary priority in the planned development and policy for economic upliftment of rural and urban poor in the state.

To achieve the above suggested developmental goals, the state government may form its new policy towards the development of handloom industry in the state by including the following strategic measures which shall go a long way in achieving the cherished goals and also to improve the lot of the handloom weavers in the state:-

(i) Handloom be considered as an important and vital sector of the State's economy along with tourism and handicrafts by increasing its plan outlay.

(ii) The office of the Directorate of Handloom Development be suitably strengthened with adequate staff for auditing, supervision, technically guidance and for cooperative coverage.

(iii) The state government may plan to revive and develop 12,000 handlooms in the next five years, providing employment to 40,000 persons (30,000 directly and 10,000 indirectly).

(iv) The state government may consider opening of a yarn bank with its regional office in the state to provide yarn to weavers on a regular and continuous basis.

(v) In line with the national policy on handlooms the state government may recognize that the interests of the handloom weavers will be best served by bringing the maximum number of weavers in the cooperative fold within the next five years.

(vi) Steps may be taken by the state government for setting up a marketing federation to be known as J&K State Handloom Cooperative Marketing Federation to encourage sales and also develop an export potential to handloom cloth and marketing garments.

(vii) The proposed J&K State Handloom Cooperative Marketing Federation will cover not only those weavers who come under the cooperative fold and J&K State Handloom Development Corporation Ltd., but also those who are not affiliated to any of these agencies on account of their being in far flung areas not easily approachable. This federation will look into the requirements of raw material, work programme and wages to be paid to the weavers.

(viii) The exiting J&K State Handloom Development Corporation Ltd. will be retained to play its useful role and in addition cover new centres to be opened under the common facility.

(ix) The state government should expand the common facility centres to cover all the weavers under all segments.

(x) The state government may adopt certain measures to provide workshed type facilities to those weavers who come under the cooperative fold and J&K State Handloom Development Corporation Ltd. and also for those whose income is too small to purchase a loom on their own.

(xi) Steps may be taken to extend the role of cooperative banks to provide working capital assistance for purchase of raw material and for production activities to the primary cooperative societies and the repayment be linked up with the total produce of the primary societies.

(xii) The importance of imparting training to weavers and technical supervisory staff of primary cooperative societies be recognized as it will go a long way in achieving higher levels of efficiency.

(xiii) Adequate post-loom processing facilities be provided in the state in order to reduce dependence on agencies outside the state.

CONCLUSION

In conclusion it may be stated that the handloom industry in Jammu and Kashmir State has lot of potential for development. Old methods of work and the techniques employed have to substituted by the improved and new techniques of doing the work. This industry can play a very important role in the betterment of economic condition of the rural folk and may pave the way to overcome many of our economic ills. Already a good local market is there for most of the handloom products within the state. Certain products of the handloom industry of Jammu and Kashmir are increasingly gaining ground and popularity outside the state. In respect of some others especially the woollen and silk handloom products, it found a very good and ready market outside the state also. For strengthening the market of certain varieties of cotton handloom products in and outside the state, good amount of research on designs including traditional designs is well sought. The Jammu and Kashmir Government is conscious of this work.

In order to rebuilt the industry on sound footing, the J&K Government has made elaborate arrangements for imparting training to weavers and artisans on modern techniques in designing, dying, weaving, printing and finishing of handloom products. Handloom sector is the only industry in the J&K State having potential to generate greater employment opportunities as well as to check migration of people from rural to urban areas for seeking gainful employment. There is considerable scope for its further promotion and development in this state.

REFERENCES

Report on Socio-economic Survey of Handloom weavers and Development of Handloom Industry in J&K State, Vols. 1 and 11, Published by Agricultural Finance Corporation Ltd. Bombay 1987.

Research Study Report on Handloom Industry of J&K State prepared by J&K Industrial & Technical Consultant organization Ltd., July 1991.

Saraf, D.N., "Arts and Crafts of Jammu and Kashmir", Abhinav Publication, New Delhi, 1987.

Sudan, A.S, "Marketing Management of Handloom Products in Jammu and Kashmir", Anmol Publications Pvt. Ltd., New Delhi, 1997, pp. 90-127.

CHAPTER

29

Special Economic Zones in India—A Boon or Bane

S. LATHA AND DR. AMRIK SINGH SUDAN

INTRODUCTION

A **Special Economic Zone** (SEZ) is a geographical region that has economic laws that are more liberal than a country's typical economic laws.. A SEZ is a trade capacity development tool, with the goal to promote rapid economic growth by using tax and business incentives to attract foreign investment and technology.

A study conducted by Aradhna Aggarwal on the Impact of Special Economic Zones on Employment, Poverty and Human Development indicated that Employment generation, both direct and indirect, has thus, far been the most important channel, through which Special Economic Zones have impacted on human development and poverty reduction in India.

Today, there are approximately 3,000 SEZs operating in 120 countries, which account for over US$ 600 billion in exports and about 50 million jobs. By offering privileged terms, SEZs attract investment and foreign exchange, spur employment and boost the development of improved technologies and infrastructure.

SPECIAL ECONOMIC ZONE (SEZ) IN THE GLOBAL SCENARIO

One of the earliest and the most famous Special Economic Zones were founded by the government of the People's Republic of **China** under Deng Xiaoping in the early 1980s. The most successful Special Economic Zone in China, Shenzhen, has developed from a small village into a city with a population over 10 million within 20 years. The article Special Economic Zone of the People's Republic of China provides more information about the Special Economic Zones in China. Following the Chinese examples, Special Economic Zones have been established in several countries, including **Pakistan, India, Iran, Jordan, Poland, Kazakhstan, the Philippines, Russia, and Ukraine. North Korea has also attempted this to a degree, but failed. Currently, Puno, Peru has been slated to become a "Zona Ecomomica" by its President Alan Garcia. In the United States, SEZ are referred to as "Urban Enterprise Zones".**

After few years, China made experiment on Special Economic Zone (SEZ), and, embarked upon their SEZ experiment in 1979-80. It was a great impact in Chinese Economy, and from closed economy model (at that time), it registered 25.97 billion Yuan (about 3.13 billion US dollars) in gross domestic product (GDP) in the first six months of the year 2008-09, from the contributon given by Hainan Province, China's largest special conomic zone (SEZ).

Most developing countries in the world have recognized the importance of facilitating international trade for the sustained growth of the economy and increased contribution to the GDP of the nation.

SPECIAL ECONOMIC ZONE (SEZ) IN INDIAN SCENARIO

Indian Economy is one of the fastest growing economies of the world. Export contribution towards growth is the backbone of Indian economy. To strengthen the export, **India was one of the first Asian Country to recognize the effectiveness of Export Processing Zone (EPZ) model. In order to give colours to its dream, Asia's first EPZ was set-up in Kandla (India) in 1965.**

As part of its continuing commitment to liberalization, the Government of India has also, since the last decade, adopted a multi-pronged approach to promote foreign investment in India. The Government of India has pushed ahead with second-generation reforms and has made several policy changes to achieve this objective.

To provide a stable economic environment for the promotion of export-import of goods in a quick, efficient and hassle-free manner, Government of India enacted the SEZ Act, which received the assent of the President of India on June 23, 2005. The SEZ Act and the SEZ Rules, 2006 ("SEZ Rules") were notified on February 10, 2006.

REASONS FOR PROMOTING SPECIAL ECONOMIC ZONES IN INDIA

With a view to create an environment for **achieving rapid growth in exports,** a Special Economic Zone policy was announced in the Export and Import (EXIM) Policy, 2000. This policy intended to make **SEZs an engine for economic growth supported by quality infrastructure complemented by an attractive fiscal package, both at the Centre and the State level, with the minimum possible regulations.**

To provide an internationally competitive and hassle-free environment for exports, units were allowed be set-up in SEZ for manufacture of goods and rendering of services.

Under this policy, one of the main features is that the designated **duty free enclave** to be treated as foreign territory only for trade operations and duties and tariffs. **No licence** required for import. The manufacturing, trading or service activities are allowed.

All the import/export operations of the SEZ units is on **self-certification** basis. The units in the Zone are required to be a net foreign exchange earner but they would not be subjected to any pre-determined value addition or minimum export performance requirements.

Sales in the Domestic Tariff Area by SEZ units is subject to payment of full Custom Duty and as per import policy in force. Further **Offshore banking** units are being allowed to be set-up in the SEZs. There are 13 functional SEZs and about 61 SEZs, which have been approved and are under the process of establishment in India.

With a view to augmenting infrastructure facilities for export production it has been decided to permit the setting up of Special Economic Zones (SEZs) **in the public, private, joint sector or by the State Governments.**

The minimum size of the Special Economic Zone shall not be less than 1000 hectares. Minimum area requirement shall, however, not be applicable to product specific and port/airport-based SEZ. This measure is expected to promote self-contained areas supported by world-class infrastructure oriented towards export production.

THE SPECIAL ECONOMIC ZONE ACT, 2005

It came into force with effect from 10th February 2006, with SEZs Rules legally vetted and approved for notification. The SEZs Rules, inter-alia, provide for drastic simplification of procedures and for single window clearance on matters relating to central as well as state governments. Investment of the order of Rs. 100,000 crores over the next 3 years with an employment potential of over 5 lakh is expected from the new SEZs apart from indirect employment during the construction period of the SEZs. Heavy investments are expected in sectors like IT, Pharma, Bio-technology, Textiles, Petro-chemicals, Auto-components, etc.

The SEZ Act is expected to give a big thrust to exports and consequently to the **foreign direct investment ("FDI") inflows into India, and is considered to be one of the finest pieces of legislation that may well represent the future of the industrial development strategy in India**. The new law is aimed at encouraging public-private partnership to develop world-class infrastructure and attract private investment (domestic and foreign), boosting economic growth, exports and employment.

LIST OF SEZs IN INDIA

The special economic zones in India were established with a view to bring expertise for the country's exports sector. For this a policy was introduced on 01.04.2006. As to these policies, the Government has set-up SEZ's in the public, private, joint sector or by State Governments. In this regard some of the existing export processing zones were converted into Special economic zones.

Accordingly, the Government has converted Export Processing zones located at:

- Kandla and Surat (Gujarat)
- Cochin (Kerala)
- Santa Cruz (Mumbai–Maharashtra)
- Falta (West Bengal)
- Chennai (Tamil Nadu)
- Visakhapatnam (Andhra Pradesh)
- NOIDA (Uttar Pradesh)
- Nanguneri and Tirunelveli (Tamil Nadu)

In addition, 3 new Special Economic Zones were approved for establishment at Indore (Madhya Pradesh), Manikanchan—Salt Lake (Kolkata) and Jaipur and have already commenced operations.

At present, 1943 units are in operation in the SEZs. In the SEZs established prior to the Act coming into force, there are 1143 units providing direct employment to over 1.97 lakh persons; about 37% of whom are women. Private investment by entrepreneurs in these SEZs established prior to the SEZ Act is of the order of over Rs. 5626.24 crore.

Administrative Set-up for SEZs

SEZs is governed by a three-tier administrative set-up:

(a) The Board of Approval is the apex body in the Department, (b) The Unit Approval Committee at the Zonal level dealing with approval of units in the SEZs and other related issues, and (c) Each Zone is headed by a Development Commissioner, who also heads the Unit Approval Committee.

Approval Mechanism of SEZs

Any proposal for setting-up of SEZ in the Private/Joint/State Sector is routed through the concerned State government who in turn forwards the same to the Department of Commerce with its recommendations for consideration of the Board of Approval. On the other hand, any proposals for setting up of units in the SEZ are approved at the Zonal level by the Approval Committee consisting of Development Commissioner, Customs Authorities and representatives of State Government..

Approval given for Setting-up New SEZs in Private/Joint/State Sector

Approvals have so far been given for setting up of 117 new Special Economic Zones (including 3 Free Trade Warehousing Zones) spread over 15 States and 2 Union Territories in the Private/Joint Sector or by the State Governments and its agencies. Of the 117 SEZs approved for establishment, 7 SEZs have already become operational, 6 SEZs are now getting ready for operation and the other are at various stages of implementation.

RULES AND REGULATIONS REGARDING SEZs

The Ministry of Commerce and Industry lays down the regulations that govern the setting up and administering of the SEZs. The Central Government is functioning, while the State Governments play a significant lead role in the development of SEZs in their respective States by stipulating the conditions to be adhered to by SEZ and granting the necessary approvals. The policy framework for SEZs has been enacted in the SEZ Act and the supporting procedures are laid down in SEZ Rules.

The SEZ Rules provides the simplification of procedures for development, operation, and maintenance of the Special Economic Zones and for setting up and conducting business in SEZs. This includes simplified compliance procedures and documentation with an emphasis on self-certification; single window clearance for setting up of an SEZ, setting up a unit in SEZs and clearance on matters relating to Central as well as State Governments; no requirement for providing bank guarantees; contract manufacturing for foreign principals with option to obtain sub-contracting permission at the initial approval stage; and Import-Export of all items through personal baggage.

PRESENT SCENARIO

SEZs provide good infrastructure and make it **cost-effective for manufactures** to set-up manufacturing units. The environment is conducive the **timely production** of export goods and manufacturers receive exemptions from the payment of excise and income tax. Moreover, SEZs enable manufacturers to receive **faster clearances** from various government departments. Manufacturers in SEZs may also be able to use resting and **analytical facilities at concessional rates.**

At present, 1016 units are in operation in the SEZs, providing direct employment to over 3.49 lakh persons; about 40 per cent of whom are women. Private investment by entrepreneurs in the SEZs established prior to the SEZ Act is of the order of over Rs. 81,000 crore. In the 63 notified SEZs which have come up after 10th February 2006, investment of Rs. 13,435 crore has already been made in less than one year.

Physical exports in the year 2007-08 accounted for 84% of the total turnover of SEZs.

381% increase in exports over four years (2003-04—2007-08) year.

Value of Physical Exports from SEZs

(Rs. Crore)

	Growth Rate (over previous year)	
2003-2004	13,854	39%
2004-2005	18,314	32%
2005-2006	22,840	24.70%
2006-2007	34,615	52%
2007-2008	66,638	92%

TAX AND OTHER INCENTIVES BEING OFFERED TO THE SEZs

SEZ units are provided exemption from Income Tax under Section 10AA of the Income Tax Act, as given in the 2nd Schedule of the SEZ Act, 2005. Section 10AA of the Income Tax Act, as given in 2nd Schedule of the SEZ Act, 2005 has been amended by the Finance Bill, 2007.

The Finance Bill, 2007

Accordingly, Tax benefit has been provided only for new units in Special Economic Zones : Sections 10AA of the Income-tax Act, provides that in computing the total income of an entrepreneur, from his unit in the special economic zone, the following deduction shall be allowed:—

(i) hundred per cent. of profits and gains derived from the export made in eligible business for a period of five consecutive assessment years beginning from the year in which such business commences;

(ii) fifty per cent. of such profits and gains for further five assessment years and thereafter; and

(iii) an amount not exceeding fifty per cent of the profit debited to the profit and loss account of the previous year in respect of which the deduction is to be allowed and credited to a reserve account to be created and utilized for the purposes of the business in the specified manner, for the next five consecutive assessment years.

Section 10AA of the Indian Income Tax Act, 1961 allows exemption in respect of export profits of a unit located in a Special Economic Zone (SEZ), as per the provision, export profit required to be calculated for the purpose of exemption, with the reference of the total turnover of the assessee. This has resulted in discriminatory treatment of the assesses having units located both in SEZ and the Domestic Tariff Area (DTA), in compression with the assesses having units located in the SEZ only.

Benefits Available to the Developers

- SEZ Scheme allows private individuals or firms to participate by way of creating infrastructure in the zone.
- The developer of infrastructure is allowed duty free (customs and central excise) import/procurement of goods for development, operations and maintenance of SEZ.
- Income Tax exemption for 10 years in first 15 years is allowed.
- Full freedom in allocation of developed plots to approved SEZ units on purely commercial basis.
- Facility to develop township within SEZ with residential areas, markets, play grounds, clubs, recreation center, etc.

Incentives from Government

Investment subsidy of 20% of fixed capital invested subject to a maximum of Rs. 20 lakhs is given as subsidy. Of this, 50% is given in cash and the balance is given for the following purposes:

- An additional cash subsidy of 10% on capital investment subject to a limit of Rs. 10 lakhs will be given to SC/ST entrepreneurs in eligible SSI and tiny units.
- Exemption from payment of stamp duty and registration fee on registration of lease deeds for plots/built up space in the zone.
- 25% rebate on power tariff for eligible software and IT enabled service units.

Training of Entrepreneurs

- Training of workers.
- Import of new and second hand machinery.
- Testing and certificate facilities for raw materials and finished products..

Currently, India has 811 units in operations in 8 functional SEZ's, each an average size of 200 acres. 8 Export Processing Zones (EPZs) have been converted into SEZs. These are fully functional. All these SEZs are in various parts of the country in the private/joint sectors or by the State Government. But this process of planning and development is under question, as the states in which the SEZs have been approved are facing intense protests, from the farming community, accusing the government of forcibly snatching fertile land from them, at heavily discounted prices as against the prevailing prices in the commercial real estate industry. Also some reputed companies like Bajaj and others have commented against this policy and have suggested using barren and wasteland for setting up of SEZs.

SPECIAL ECONOMIC ZONE: A BOON FOR INDIAN ECONOMY

A Special Economic Zone (SEZ) is a geographical region that has economic laws more liberal than a country's typical economic laws. It is a trade capacity development tool, with a goal to promote rapid economic growth by using tax and business incentives to attract foreign investment and technology. By offering privileged terms, Special Economic Zones attract investment and foreign exchange, spur employment and boost the development of improved technologies and infrastructure.

In India, Special Economic Zones are being established in an attempt to deal with infrastructural deficiencies, procedural complexities, bureaucratic hassles and barriers raised by monetary, trade, fiscal, taxation, tariff and labour policies. Since country-wide development of the infrastructure is expensive and implementation of structural reforms would require time, (Special Economic Zones/Export Processing Zones) are being established as industrial enclaves for expediting the process of industrialization.

Special Economic Zones were announced by the Government of India in April 2000 as a part of the Export-Import policy of India. The government realized the need to enhance foreign investment, promote exports from the country and at the same time provide a level playing to the domestic enterprises, while ensuring manufacturers to be competitive globally.

The Special Economic Zones as announced by the Government of India in 2000 were deemed to be foreign territory for the purposes of trade operations, duties and tariffs. These zones were to provide an internationally competitive and **hassle-free environment for exports.** Units were allowed be set-up in Special Economic Zone for manufacture of goods and rendering of services.

All import/export operations of the Special Economic Zone units were on self-certification basis. Anything could be imported duty-free but sales in the Domestic Tariff Area by Special Economic Zone units were subject to payment of full Custom Duty and as per import policy in force.

Further **Offshore banking units were being allowed to be set-up in the Special Economic Zones.** The policy provided for setting-up of Special Economic Zones in the public, private, joint sector or by State Governments.

On 31st August 2004 the Department of Commerce announced the Foreign Trade Policy 2004-09 to create an appropriate institutional framework and policy environment for facilitation and growth of external trade. **The basic objective of this policy was to double India's share of global merchandise trade by 2009 and make exports an effective instrument of economic growth and employment generation.** The Special Economic Zone Act, 2005 and the Special Economic Zone Rules, 2006 were introduced under this policy, to regulate and promote the development of these industrial enclaves.

The Act designated the Special Economic Zones a duty free enclave to be treated as foreign territory only for trade operations and duties and tariffs. Under the Act, **no license is required for import and no routine examination is to be conducted by the custom authorities of the export/import cargo.**

To aid backward and forward integration of the economy, the **Act provides exemptions to Special Economic Zone units and**

Special Economic Zone developers from all indirect taxes, including basic customs duty, countervailing duty, education cess, and direct taxes while at the same time domestic sales are subject to full customs duty and import policy in force.

The Act provided the **freedom to subcontract.** It also permitted manufacturing, trading and service activities in the Special Economic Zones.

SPECIAL ECONOMIC ZONE: THE NEW CONFLICT GROUND (BANE) IN INDIA

More than a decade of opening of India, the Special Economic Zone (SEZ), probably has become the most controversial economic reforms announced in recent time. **While some consider it as India's supersonic engine of growth, others severely criticize it as the latest land grab instrument in the hands of the industrialists.**

Serious discourses on models of development, displacement and rehabilitation, employment generation, foreign investment, primacy of industry over agriculture are being raised in justification as well as against the whole concept of SEZ. Despite the heated argument as of January 2007, the Union Government has given final approvals to 237 SEZs and in-principle nod to about 165 zones. Besides, 300 more applications are in the process of approval.

While announcing its new Export-Import Policy in 2000 the Government of India also announced the introduction of SEZs; deemed to be foreign territory for the purposes of trade operations, duties and tariffs. As per the SEZ Act, 2005, SEZs are geographical regions that have economic laws different from a country's typical economic laws. Among all other incentives announced with the SEZ Act to woo the investment, the most lucrative are the 100 percent tax exemption for industries on export profits for first five years and the provision of 100 percent foreign direct investment in industries under automatic route.

However, in 2006 the Commerce Minister of India, Kamal Nath announced the detailed rules and regulations with regard to the SEZ. Since, millions of people from states like Haryana, Orissa and Maharashtra are engaged in a sustained agitation against the "unjust" acquisition of their land for many SEZ projects.

The threat of massive displacement from agricultural areas has been the main focus of these agitations. Strange but true the opposition to SEZs is taking the shape of a loose alliance of incongruent outfits which include activists like Medha Patkar, former Prime Minister V.P. Singh, many NGOs and even outfits owing allegiance to Left extremists.

The Indian corporate giant Reliance wants to develop a SEZ in Chirner, a suburb in Mumbai. Reliance wants the entire stretch from

Navi Mumbai to Dharamtar creek and plans to make it Mumbai's satellite city which would spread over 35,000 acres. The government on its part is acting as a perfect real estate agent and will eventually acquire 45 villages for the Reliance project. Although the government has issued notifications for land acquisition the local people have no clue how much they will get for their land. As per the local reports Reliance has been trying to buy up land at Rs. 3 lakhs to Rs. 4 lakhs an acre. In contrary, the market rate is said to be anywhere between Rs. 20 lakhs and Rs. 40 lakhs an acre. **The irony is that most of land will be acquired are owned by Agris and Kolis, traditional farmers and fisher folk who already have faced one displacement from Mumbai and Navi Mumbai. The reality is that even they would forgo their land they really don't have a place in the Reliance SEZ.**

Recently, Arcelor Mittal's Rs. 40,000 crore SEZ project planned at Patna Tehsil in Keonjhar in Orissa ran into rough weather when angry villagers from the 17 villages demonstrated a protest rally at the Keonjhar District Collector's office demanding that their land should not be acquired for the Mittal steel plant.

Thousands of protestors have given a clear **ultimatum to the Orissa government** to look for infertile lands as they are not going to hand over their land. The resistance to the SEZs has been particularly intense in left party governed **West Bengal**. In January at least five people were killed and dozens injured when communist supporters clashed with activists **opposed to the plan for a 14,500 acre economic zone set-up by TATA at Nandigram in West Bengal.**

The confusing signs emerging within the government departments have misled people from both the sides and which in return has led to violent protest and clash.

While the Commerce Ministry expects the zones to draw nearly 600 billion rupees ($13.5 billion) in investment by 2009 and create 890,000 jobs, **an internal assessment of the Finance Ministry has estimated that by 2010 the country will forgo about Rs. 1,60,000 crores in direct and indirect taxes because of the SEZs across the country.**

The Economic Survey of 2006-07 highlighted the fact that Special Economic Zones are testing grounds for the implementation of liberal market economy principles. At the same the survey emphasized on some apprehensions against the Special Economic Zones: Generation of little new activity as there may be **relocation of industries to take advantage of tax concessions and Revenue loss.**

Large-scale land acquisitions by the developers may lead to displacement of farmers with meager compensation.

Acquisition of prime agricultural land, having serious implications for food security.

Misuse of land by the developers for real estate. Uneven growth aggravating regional inequalities.

The Survey also mentioned that many of these apprehensions, however, could be addressed through appropriate policies and safeguards.

A major controversy surrounding the implementation of the Special Economic Zone scheme has been the ruthless manner adopted for acquiring land. News reports highlighted protests across the country against acquisition of lands for the purpose of establishing Special Economic Zones. The "SEZ No More" campaign gained momentum after the bloody chapter in Nandigram.

India and China: Whether on a Level Platform

Government of India wants to follow China way but while in China there are only six SEZs, in India twenty-four SEZs have been established just in 11 months and over 200 SEZs are in pipeline. **This shows that proper ground work has not been done before implementing the concept of SEZ.**

Success stories of large and small developing countries can be explained by the growth of world trade and opening up of these economies with market-based deregulation. But from any in-depth scrutiny one finds that the reform package under the broad heading of "liberalization" is very different from country to country. There is no standard recipe of a "reform package". A lot of factors influence the performance of Special Economic Zones in a country, e.g. economic history, location, industries, state policy, etc. Since India has adopted the idea of the special economic zones from China, it becomes pertinent to study the history of economic development in India and China.

China's success can be ascribed largely because of its effective population control. In the pre-reform days, both in China and India top priority was given to equity, removing poverty and increasing the social aspects of standards of living. This, however, was attempted in China under a total state-controlled economy and in India with the public sector playing a dominant role along with the market forces. Both the economies adopted a strategy of import substitution and heavy industry growth. China over time, realized that maintaining high standards of living becomes difficult unless efficiency in the use of resources is increased. Its attempt to maintain equity through forced saving and administered directives resulted in social unrest, which came to a breaking point after the controversial Cultural Revolution. **The key objective of present reform in China is to bring incentives back in the economy by increasing the role of the market with minimum changes in their political institutions. This is defined in China as an experiment in a socialistic market economy.**

In India, heavy import substitution lead to increased inefficiency in production and generation of surplus for maintaining the tempo of equity measures as a result social development became impossible. This led to heavy borrowing, culminating in a balance of payments crisis. To meet the crisis, this new economic policy in India was initiated.

China's success in attracting foreign direct investment and becoming one of the top exporting countries of the world hinged on the careful implementation of its Special Economic Zone policy. Size, location, flexible labour laws and stable policies were the factors primarily responsible for making Chinese Special Economic Zones attractive to foreign investors. In India, the fiscal concessions being offered to developers and units are the primary driving force.

Chinese government started building Special Economic Zones way back in 1979. The idea behind the Special Economic Zones was to experiment with liberal policies in certain ear-marked regions while insulating the rest of the economy from their influence. The government identified huge tracts of land, near the coastal region, and started building mega cities with all required infrastructure. Stringent labour laws applicable in China were relaxed in these regions and foreign investment was encouraged by offering concessions and promising of stability. One of the earliest and most famous Special Economic Zone was founded by the government of the People's Republic of China under Deng Xiaoping in the early 1980s.

Though India had a head-start in the direction by building its first export processing zone in 1969 with certain minimum infrastructure and fiscal sops, it could not muster enough political will to build full-fledged Special Economic Zones with foreign territory status in the matters of international trade till the turn of the century. As opposed to five mega Special Economic Zones built by the Chinese government (the largest being Shenzhen built over 49,500 hectares), India opened its doors to private players and allowed sector-specific Special Economic Zones to develop on just 10 hectares of land. As a result, more than 500 Special Economic Zones have been proposed, 220 of which have already been created. **The economies of scale, which seems to have worked so well in China by reducing production costs, may not have the same effect in the Indian Special Economic Zones.**

In China, the government chose the location for Special Economic Zones with a lot of thought with all five located near the coastal region. This made it easier for the Special Economic Zone units to export their products and import inputs. **In India, Special Economic Zones are being built all over the country, wherever land can be acquired by the developers.** This has also led to allegations of land-grabbing and conversion of productive agricultural land by developers. As a result, Centre has made it mandatory that all proposals should have a certificate from the state governments notifying that the land being used is non-agricultural for at least 90%.

Flexibility in labour laws, which played an important role in attracting foreign investors, is absent in the Indian Special Economic Zones. This is one of the prices India has to pay for the advantages of a federal democratic government. India has, however, tried to make up for all the disadvantages by offering attractive fiscal sops.

Tax holidays for Special Economic Zones in India are longer and steeper than those given by China. This had given rise to some dissent from the finance ministry which had complained that the fiscal loss would be immense. In fact the scheme has generated a difference of opinion between the Finance and Commerce Ministries. While the former is voicing its concerns about possible revenue loss from the tax privileges for the Special Economic Zones, the latter is stoutly defending the policy with statistics suggesting minimal losses and highlighting eventual gains in terms of employment and revenues

Reserve Bank of India has also expressed its concerns about the revenue losses and the uneven pattern of development. Reserve Bank of India insisted on factoring the revenue implications of the taxation benefits. The revenue loss for the Government in providing incentives may be justified only if the Special Economic Zone units ensure forward and backward linkages with the domestic economy. Also, as resources are being diverted from the less developed, growth will not be uniform.

One of the most basic difference between the Special Economic Zone model adopted in China and India is that the **Chinese Special Economic Zone initiative is government driven, whereas Indian Special Economic Zones are driven by private sector.** In China, the State acquires the land and develops the required infrastructure, while private enterprises are invited to set-up units. Under such a system, land continues to be under the ownership of the State. In India, however, private entities are being involved in developing the Special Economic Zone infrastructure. As a result, Land is being acquired by the State and handed over to private developers.

VALUABLE SUGGESTIONS

India is one of the fastest growing economy of the world. **Concept of SEZ is becoming a major pillar of the economic structure.** So far, SEZs have provided job opportunity, contributed in foreign exchange reserves, developed infrastructure, etc.

In future, it is advisable to provide m**ore fiscal and non-fiscal incentives to the developers as well as entrepreneurs**. Tax incentives and subsidies, should be provided on rational basis.

Continuous planning and execution of the plans, periodic review and appraisal of the plans are required to be more strengthened. Overall cooperation and dedicated services are required from merchant exporters and manufacturer exporters, business and industry as partners of Government in the achievement of its stated objectives and goals.

In order to obviate the need for litigation and nurture a constructive and conducive atmosphere, a **suitable Grievance Redressal Mechanism should be established,** which would, hopefully reduce litigation and further, it will create a healthy atmosphere among developers, entrepreneurs and government.

Since, developing Special Economic Zones involves massive displacement of farmers, it is essential that a systematic approach should be followed for ensuring balance of interests. Consequently, state governments have been advised that in land acquisition for Special Economic Zones, **first priority should be for acquisition of waste and barren land and if necessary single crop agricultural land.** If perforce a portion of double-cropped agricultural land has to be acquired to meet the minimum area requirements, especially for multi-product Special Economic Zone, the same should not exceed 10% of the total land required for the Special Economic Zone

The government has also announced the new National Policy on Rehabilitation and Resettlement, 2007. This policy would provide land-for-land compensation for acquisition of land for development purposes, including Special Economic Zones, and **employment to at least one person from each affected family.**

A National Rehabilitation Commission would be set-up by the Central Government, which would be duly empowered to exercise independent oversight over the rehabilitation and resettlement of the affected families. Further, wage employment would be provided to the willing affected persons in the construction work in the project. The policy also ensures housing benefits including houses to the landless affected families in both rural and urban areas.

Adequate provisions have been made for financial support to the affected families for construction of cattle sheds, shops, working sheds, transportation costs, temporary and transitional accommodation, comprehensive infrastructural facilities and amenities in the resettlement area including education, health care, drinking water, roads, electricity, sanitation, religious activities, cattle grazing, and other community resources.

The benefits expressed in monetary terms have been linked to the Consumer Price Index, and the same shall also be revised suitably at appropriate intervals.

Special provision has been made for providing life-time

monthly pension to the vulnerable persons, such as the disabled, destitute, orphans, widows, unmarried girls, abandoned women, or persons above 50 years of age (who are not provided or cannot immediately be provided with alternative livelihood).

A strong grievance redressal mechanism has been prescribed, which includes standing R&R Committees at the district level, R&R Committees at the project level, and an Ombudsman duly empowered in this regard. The R&R Committees shall have representatives from the affected families including women, voluntary organizations, Panchayats, local elected representatives, etc. Provision has also been made for post-implementation social audits of the rehabilitation and resettlement schemes and plans.

For effective monitoring of the progress of implementation of R&R plans, provisions have been made for a National Monitoring Committee, a National Monitoring Cell, mandatory information sharing by the States and Union Territories with the National Monitoring Cell, and Oversight Committees in the Ministries/ Departments concerned for each major project.

For ensuring transparency, provision has been made for mandatory dissemination of information on displacement, rehabilitation and resettlement, with names of the affected persons and details of the rehabilitation packages. Such information shall be placed in the public domain on the **Internet as well as shared with the concerned Gram Sabhas and Panchayats by the project authorities.** This policy aims at striking a balance between the need for land for development purposes and protecting the interests of land owners and other displaced people.

RECOMMENDATIONS

A study conducted by Aradhna Aggarwal on the Impact of Special Economic Zones on Employment, Poverty and Human Development indicated, that Employment generation, both direct and indirect, has thus, far been the most important channel, through which Special Economic Zones have impacted on human development and poverty reduction in India.

However, the role of Special Economic Zones in human capital formation and as an engine for promoting new knowledge, technologies and innovations through technology transfers and technology creation appears to be relatively limited. **With new generation Special Economic Zones emerging, the scope of human capital formation and technology upgrading effects will widen. It is therefore important for the government to play a pro-active role in strengthening these effects.**

For the contribution of Special Economic Zones to various aspects of human development to be realized, it is important to forge linkages between the domestic economy and Special Economic Zones. Systematic efforts need to be made to help zone units forge links with the outside units. Also, the effects of Special Economic Zones are contingent upon the success of these zones in attracting investment, in particular, Foreign Direct Investment. A comprehensive policy framework is required to attain this.

CONCLUSION

To conclude, in words of **Dr. M.S Swaminathan,** in building a paradigm of sustainable development, it is necessary to link concepts and procedures at the macro level with field-level practices to arrive at a way of translating a conceptual response into a reality".

Government should revise and update its policy not only to sustain its share in the development but also to increase it in the interest of the control and development of country as a whole. **The government has to ensure that strategies are developed in a timely manner to strengthen the opportunities that are likely to emerge, protect interests of the Special Economic Zones workers, and forge linkages between Special Economic Zones and the domestic economy. Such a regulated and monitored approach is the only means of attaining the actual potential of these Special Economic Zones.**

To overcome the **shortcomings** experienced on account of the multiplicity of controls and clearances; absence of world-class infrastructure, and an unstable fiscal regime and with a view to attract larger foreign investments in India, **it is necessary to continue study considering the international scenario of this concept. This is the time for unity of purpose and resolute action.** Thereby we may strengthen our strength and weaken out weakness by means of converting the global economic challenges into opportunities to make our nation into supreme financial position.

As countries with substantial domestic markets lead the way through sizeable stimulus packages, there should be indeed a light at the end of the tunnel. We must use this opportunity to better balance the sources of growth, and move steadfastly and vigorously toward greater regional cooperation and economic integration. That way, **India will not only fulfil its growing role as a global economic player, but will also move more quickly towards prosperity and freedom from poverty. With strong hearts, enlightened minds and willing hands, we will have to overcome all odds and remove all obstacles to create a brave new India of our dreams.**

CHAPTER

30

Technical Education in India: Myths and Reality and a Catalyst for Development

Dr. Amrik Singh Sudan, Rohini Gupta Suri and Dr. Rais Ahmad

INTRODUCTION

Education is consistently becoming the prime factor in measuring the growth of the country. Thus, it becomes very important for a nation to have not only an efficient but also a flexible educational framework that allows the students to go for their interested field in education and provides complete support to them from all perspectives. Because its only one follows his/her own interest and passion that will make him able to help the nation in future become better in all areas.

Educational system in India has been rooted deeply from ancient traditions. In fact, India is said to be the very origin of the Traditional educational system. Sanskrit, the traditional ancient Indian Language is said to be very rich in knowledge. In the current scenario, India is becoming the Hub for Global Education, in fact, a research says that India has now become the fourth most desirable Hub in the world.

India is now attracting more and more students from around the world for offering more desired courses at world class—B-Schools and technical colleges. In fact, India's Indian Institutes of Technology (IIMs) are said to be one of the top Management Institutes World-wide. Also, India's Indian Institutes of Technology (IITs) are among the most desired Technical Institutes around the world. So many students compete every year for admission into these world-class institutes to pursue their desired technical or Management Courses. As these institutes provide a consistent and fabulous growth rate in the carrier of a student and make him/her a better capable and productive person in future.

Education is indeed the very foundation of a society which brings economic wealth, social prosperity and political stability. A modern and well-developed society can become a possibility when quality of education improved. In other words, education is the very basis of development for a modern society and in this connection, the role of technical education is very important and significant.

TECHNICAL EDUCATION IN INDIA

The history of imparting formal technical education in India can be traced back to mid-19th century, although it got momentum in 20th century with the set-up Constitution of Technical Education Committee of the Central University Board of Education (CABE) in 1943; Preparation of Sergeant Report in 1944 and Formation of All India Council of Technical Education (AICTE) in 1945. With the country gaining independence in 1947, the development of technical education had become a major concern for the Government of India to face the new challenges and move the country forward.

The set-up of Indian Institutes of Technology, Indian Institutes of Management and Indian Institutes of Science was a major step in the development of technical education in the country. The quality of education of these institutes have managed to change the outlook of India so much that this ancient country which was earlier known for yoga and mediation is now known for computer engineers. However, it does not mean that the challenge of making technical education accessible to the rural populace and other under-developed sections of the society has been overcome.

In order to maintain the standard of technical education, a statutory authority—The All India Council for Technical Education (AICTE)—was set-up in 1945. AICTE is responsible for planning, formulation and maintenance of norms and standards, quality assurance through accreditation, funding in priority areas, monitoring and evaluation, maintaining parity of certification and awards and ensuring coordinated and integrated development and management of technical education in the country.

Technical Education Courses in India

The courses, which are known as 'technical' in India and therefore come under the purview of All India Council of Technical Education are—degree and diploma courses in Engineering, Master degree courses in Engineering, Master of Computer Application (MCA), Master of Business Administration (MBA), Pharmacy Courses in Architecture and Applied Arts and Hotel Management and Catering Technology Courses.

Institutes Offering Technical Education in India

As the technical education courses in India are quite diverse, the number of institutes providing technical courses in India is also huge. The number of AICTE approved institutes that offer engineering degree courses in India is 4,39,689. There are around 1244 institutes in India that offer diploma courses in engineering, 415 institutes offer diploma courses in pharmacy, 63 institutes offer diploma courses in Hotel Management and Catering Technology courses and 25 AICTE approved institutes that offer diploma courses in Architecture. The number of AICTE approved institutes that offer Master of Computer Application courses in India is 1012.

Likewise the AICTE also approves institutes from time to time institutes that offer MBA courses, M.E./M.Tech., Architecture and Applied Arts Courses., Hotel Management and Catering Technology courses.

SOME BASIC FACTS ABOUT TECHNICAL EDUCATION IN INDIA

The technical education in India has achieved a great success. It all happened in 19th century and received momentum by the end of twentieth century along with the origin of CABE. **CABE** well abbreviated as the **Constitutional Advisory Board of Education**. After India attained Independence, this gave the initiative to **technical education** and came out as a main concern for Government of India in turn to face future challenges and move the country forward.

The birth of "Indian Institute of Science (IISc)", "Indian Institute of Technology (IIT)" and the "Indian Institute of Management (IIMS)" was a fundamental move in the growth of practical and technological education in India. The main aim of these institutes is to produce hard core and competent engineers and scientists, who can manage and alter the viewpoint of Indians in the global prospective. Earlier, India was well-known for mediation, holy places and yoga, but today it is considered to be the highest produce of engineers in the world.

So, to go on with the level of practical and technogical education all over the nation, a constitutional authority that is- The AICTE came into existence in 1945. AICTE is abbreviated as All India Council for Technical Education. It is well-known for planning, sustaining and formulating analogous standards through the assistance of official recognition, by providing financial support in universities or colleges, evaluation and awards and monitoring thereby determining the organization in administering of technological and practical education in India.

The key intention of these establishments is to make sure that each and every selection criteria, admission procedure, entrance examination and details concerning allied research material are performed appropriately all over the country. In the word "Technical", there come loads of course which consist of diploma and degrees in engineering, i.e. 'Bachelor of Technology', 'Master Degrees such as the (MBA) 'Master of Business Administration' or the (MCA) 'Mater of Computer Applications' (MCA), 'Master of Business Administration' (MBA), courses in Applied Arts and Architecture and Catering Technology and Pharmacy Courses.

Technical and Engineering education is accessible at 5 levels. At the least level is the education in different courses provided by a huge number of Training Institutes (it is) across India. The eligibility and duration requirements vary depending on the nature of the business. The next higher level is the 3 year diploma courses offered in branches of technology and engineering existing in polytechnics. The primary educational eligibility for diploma engineering is 10th or 12th pass. Coming to the degrees offered by technical education of India are B.Tech. and BE courses. The educational eligibility to these degree courses is 12th pass. There are even **Post-graduate Courses** offered under technical education of India. The main aim of these technical courses is to produce hard core and competent engineers, who can manage and change the outlook of India in the global prospective.

NATIONAL POLICY ON EDUCATION 1986 AS UPDATED IN 1992

The National Policy on Education-1986 laid specific guidelines for the qualitative and quantitative development of technical and management education sectors, establishment of linkages between the concerned sectors and agencies, manpower assessment and technical education forecasting, increasing effectiveness of technical education management systems, proper delivery systems, measures to achieve greater cost-effectiveness and generation of resources through suitable means. In view of the rigid entry requirements to formal courses denying a large number of people to technical and managerial education programmes through distance-education system were recommended. Education in polytechnics was also made flexible. A strong guiding and counseling service was provided. It was under this policy that AITCE was given a statutory status and polytechnic education was modernized with the help of a massive assistance of Rs. 1650 crores from World Bank. Five hundred Government Polytechnics in 16 states and one UT were modernized, in fact transformed. The Project was implemented in two phases from 1992-98.

Women Polytechnics

The Modified National Policy on Education-1992 stressed the need for launching formal and non-formal technical education for women, the economically and socially weaker sections and the disabled. Additional women polytechnics and women hostels were to be provided. Guidance seminars for girls were to be organized by state governments at +2 stage, Boards of Apprenticeship Training and other selected institutions were to make the women aware of the opportunities in technical education and the potential for employment and self-employment. The result was that enrolment of girls in polytechnic and engineering courses has considerably increased. There are 83 Women's Polytechnics functioning in the country.

The up-dated policy of 1992 also envisaged programmes to meet the demands of teachers and professional for vocational education and its expansion, vocational technology and its use and curriculum development. To consider self-employment as a career option training in entrepreneurship for both men and women was to be provided through modular or optional courses, in degree or diploma programmes.

COMMUNITY POLYTECHNICS PULSE-BEAT OF RURAL INDIA

One scheme which has transformed the lives of rural people throughout India is the scheme of Community polytechnics. The scheme which was started in 1978-79 aims at sustainable community development through transfer of technology in rural areas, with people's participation at grassroot level. Out of about 500 Government Polytechnics, 110 are Community Polytechnics, administrated and financed by Central Government. The New Education Policy has laid special emphasis on the development of skills and trades which come natural to certain environments and communities thereby enhancing their employability and the values of their products.

Community Polytechnics in Minority Concentration Areas

The scheme of Community Polytechnics has covered all the 41 minority concentration districts under the 15-Point Directive on the Welfare of Minorities. At Government Polytechnic Ajmer trades and skills such as meter reading, tailoring, scooter repairing, house wiring stenography and radio repairing are taught. At Women's polytechnic New Delhi cutting, brass work, wire winding, tailoring, machine embroidery, textile designing, beauty culture, brass work, computer training, home craft are included in the courses. At university Polytechnic Aligarh Muslim University tailoring, knitting, Electrician's work, Moped repair, welding, plumbing, are taught. The Lucknow polytechnic specializes in carpentry, garment knitting, Chikan work, tailoring, etc. Moradabad polytechnic concentrates on brass utensils. The Polytechnic at Kilakarai in Tamil Nadu trains people in leather work while the one Aurangabad excels in other polytechnics in minority concentration areas such as Bhopal Goa, Ranchi, etc. have trained people in spray painting, soap and detergent making.

Apprenticeship Training

Four Regional Boards of Apprenticeship training have been set-up at Kanpur, Kolkata, Mumbai and Chennai by the Government of India and are functioning since 1969. Their jobs are to provide industrial training to engineering graduates and diploma holders passing out of engineering colleges and polytechnics, under Practical Training Stipendary Scheme. Under the provisions of the Act the industrial establishments are obliged to engage apprentices every year. The Central Government reimburses to the establishments who engaged these appretices half of the minimum amount of the stipend paid. The 1986 amendment has brought the 10+2 vocational stream under the purview of the Act. About sixty subject fields have been notified for training of technician (vocational) apprentices. About 30,000 trainees are engaged every year as apprentices by different industrial establishments. To improve the quality of apprenticeship training a number of programmes have been developed by the Boards. Recently, additional vocational subjects have been included under the purview of the Act.

Industry-Institute Interactions and Entrepreneurship Development

Linkages between technical institutions and industry have been strengthened in terms of curriculum development, resource sharing, joint research projects, apprenticeship training for students, exchange of faculty and experts on mutual basis, consultancy and sponsored research, counting education programmes, sandwich cooperative programmes and the concept of adjunct professorship. In order to meet the challenge of unemployment and to create an environment of self-employment, entrepreneurship development, entrepreneurship development programmes have been encouraged. For developing women entrepreneurship special programmes have been launched by the central and state governments at diploma and degree level.

Other Technical Institutions

Besides the above-mentioned vast network which exists at technician/technical education level, under-graduate level and post-graduate level including research, we have some special technical institutions which are taking the country towards a technological society. These include NITIE (National Institute of Training in Industrial Engineering) at Mumbai, set-up by the Central Government in 1963 as an autonomous body for training in industrial engineering and evolving and based programmes for the industry. We have also NIFFT (National Institute of Foundry and Forge Technology) at Ranchi, set-up in 1966 by the central government as an autonomous body to provide training in foundry and forge technology to the concerned industrial personnel. The Institute offers diploma, degree and M.Tech. courses in the field of foundry and forge skills/competencies. The School of Planning and Architecture is another important technical institution, established by the Government of India in 1955, to provide New Delhi, a Deemed University, Education & training facilities in the field, rural, urban and regional planning. The school conducts a bachelor's degree course in Architecture, Master's degree course in planning with specialization in urban and regional planning, transport planning, housing, landscape and architecture. In 1993-94 a new programme of industrial design has been started. Of equal importance are the four TTTIs (Technical Teachers' Training Institutes) at Bhopal, Kolkata, Chandigarh and Chennai, set-up in mid-sixties, to provide in-service training to polytechnic teachers and also to undertake various kinds of services to improve polytechnic education in India. In fact, these TTTI's have become Resource Institutions for technician education. They offer post-graduate courses in technical teaching and have done quality work in technical manpower planning and studies in it inter-action with industry. The 17 RECs (Regional Engineering Colleges) have also acquired a national status because of their expertise in engineering and technology. The Indian School of Mines, Dhanbad is a unique institution while the National Institute of Design is an institution of excellence. Sant Langowal Institute of Engineering and Technology was set in 1991-92 to meet the special technical manpower needs of the state of Punjab. The Institutes offers 12 Certificate, 10 Diploma and 8 Degree Courses.

Education is indeed the very foundation of society which brings economic wealth, social prosperity and political stability. A modern and well-developed society can become a possibility. When quality of education is improved. In other words, education is the very basis of development for a modern society and in this connection the role of technical education is very important.

Technical Education plays a pivotal role in human resource development of the country by creating skilled manpower, enhancing industrial productivity and improving the quality of life. It offers many courses and programmes in engineering, technology, management, architecture, town planning, pharmacy and applied arts and crafts, hotel management and catering technology.

In our country there are many institutes which offer technical education. These institutions are funded by the Central Government, State Government/State-funded Institutions and Self-financed Institutions. Among them, there are 79 centrally funded Institutes of Technical and Science Education.

Several new measures have been taken to implement the Government vision of providing increased access with equity and excellence. Eight new IITs, four new IIMs and ten new NITs have been set-up and are functional. IIMs at Udaypur and Kashipur would become functional from 2011-12.

Technical Education through different Programme envisages strengthening the Institutions to produce high quality engineers for better employability, establish Centers of Excellence for focused applicable research, training of faculty for effective teaching, enhancing Institutional and System Management effectiveness.

The Central founded technical Institutions have also implemented Central Education Institutions (Reservations in Admission) Act, 2006 from the year 2007-08 which provides 15%, 7-1/2% and 27% reservation in admission for SCs, STs and OBCs respectively. In the Scheme of Community Development through polytechnics, preference is given to training of rural youths/SCs/STs, women, school drop-outs and other disadvantaged groups and helps them to obtain need-based gainful employment. Scheme for upgrading existing polytechnics to integrate the physically disabled persons into the mainstream through technical and vocational education. The objective of the Polytechnics under coordinated action for skill development is to enhance employment-oriented skilled manpower through them. Under the scheme, financial assistance is provided to the State/UT Governments for setting up of 300 new Polytechnics. Out of 300 Polytechnics, financial assistance has been provided to the existing Government/Government-aided Polytechnics for strengthening of infrastructure facility and also for construction of women's hostel. Under the scheme of construction of women's hostel in Polytechnics, financial assistance has been provided to the existing Government/Government-aided polytechnics in the state of Jammu and Kashmir and North-East region. Financial assistance has also been provided to 18 districts of Jammu and Kashmir and 27 districts of North-East Region for establishment of new Polytechnics under the scheme of establishment of new Polytechnics in the country in unserved and under-served districts.

In the North-East Region of the country several technical institutions like Indian Institute of Technology, Guwahati (Assam), Rajiv Gandhi Indian Institute of Management (RGIIM), Shillong (Meghalaya); National Institute of Technology (NIT) Silchar (Assam), National Institute of Technology (NIT), Agartala (Tripura); North-Eastern Regional Institute of Science and Technology (NERIST), Itanagar (Arunachal Pradesh); and Central Institute of Technology (CIT), Kokrajhar (Assam) are taking care of the higher education.

Since technical education is a key enabler of growth in the country's economy, it has to adopt a facilitating approach to make technical education accessible, affordable and accountable. A need was felt to introduce flexibility and mobility in the system by investing public-private partnership for funding. The focus of AICTE is now on decentralized decision-making the various reforms introduced in the administrative procedures include strengthening of office procedures improvement in security measures and introduction of e-Governance for enhanced transparency, clarity, easy and assured communication.

Technical Education provides various facilities and promotes development in the country in a coordinated and integrated manner. All India Council for Technical Education (AICTE) was set-up in November 1945 as a national level Apex Advisory Body. In the National Policy of Education (1986), AICTE was vested with statutory authority for planning, formulation and maintenance of norms and standards, quality assurance through accreditation, funding in priority areas, monitoring and evaluation, maintaining parity of certification and awards an ensuring coordinated and integrated development and management of technical education in the country. The purview of AICTE (the council) covers programmes of technical education including training and research in Engineering and Technology, Architecture & Town Planning, Management, Pharmacy, Applied Arts and Crafts, Hotel Management and Catering Technology (HMCT), etc. at different level.

There are several schemes for faculty development, namely, Quality Improvement Programme (QIP), Career Award for Young Teachers (CAYT), Emeritus Fellowship, Visiting Professorship, Seminar Grant, Travel Grant, Staff Development Programmes, National Doctoral

Fellowship, AICTE-INAE Distinguished Visiting Professorship and Financial Assistance to Professional Societies/Bodies. Under the quality Improvement Programme (QIP), faculty members of degree level institutions are given opportunity to upgrade their qualifications to Master's and Ph.D. levels. Under the scheme Quality Improvement Programs (QIP) (polytechnic), polytechnic teachers can pursue Master's degree programme.

In an effort to bring in transparency, accountability, efficiency and swiftness in its decision-making process AICTE has launched the web portals of the AICTE and National Board of Accreditation (NBA) at URL://www.aicteindia.org and URL://www.nbaindia.org. Some other significant initiatives taken by the AICTE are Enhancement in seats in Engineering and Management Institutions, Reduction in land requirement norms in Metro and Mega Cities; Establishment of National Board of Accreditation (NBA), as an independent Body of AICTE, for making it eligible for full membership of Washington Accord; Providing 25% flexibility to Management Institutions in allocation of seats amongst different disciplines of Post-graduate Diploma in Management (PGDM), Cooption of Foreign Experts on Academic Boards; Conducting of first Gradual Pharmacy Aptitude Test (GPAT) for Pharmacy Graduates through MS University, Baroda.

CONCLUSION

Given the importance of technical education in the further development of the nation, the Government of India is keen on developing some more institutes in the line of Indian Institutes of Technology (IITs), Indian Institutes of Management (IIMs) and Indian Institutes of Science (IIScs). The Prime Minister of India Dr. Manmohan Singh has unleashed a plan to establish 8 Indian Institutes of Technology (IITs), 7 Indian Institutes of Management (IIMs) and 5 Indian Institutes of Science (IIScs) to improve the spread and quality of technical education in the country. These institutes along with various private institutes and foreign technical colleges have the potential of making technical education accessible to all sections of society in India without compromising on the quality of education.

Thus, we find that there is a technical upsurge in the country, that India is one the greatest technical education system and its products rub shoulders with the best and the brightest any and everywhere, yet it is not a perfect system. There is palpable imbalance between the rural areas and the urban settlements in so far as technical and vocational facilities are concerned.

But inspite of all that there is a change in the outlook of the society. While the general education is leading the country towards a learning society, technical education is moving towards a technological society. The superstitions Indian is moving towards rationally although his temper is still not scientific. Tomorrows' education is going to be different. To quote from the International Commission for Education (UNESCO): It must form a coordinated totality in which all sectors of society are structurally integrated. It will be universalized and continual, it will be the bulwark and the driving force in culture, as well as in promoting professional activity. This movement is irresistible and irreversible. It is the cultural revolution of our time.

CHAPTER

31

Cross Cultural Challenges while doing Business in India

LEKHA NAGAR, AND DR. AMRIK SINGH SUDAN

ABSTRACT

Technology has been the pace setter for structural changes in socio-economic With the globalisation of world business, India has become an appealing market for foreign investors. The problem of cross-cultural management arises as the cooperation between India and its culturally different Western partners continues to increase at an unprecedented rate. The 21st century is an era of the globalization of world economy. Cross-national business is facing great challenges in cultural differences.

In one survey entitled what is the major barrier in doing business in the world market, cultural differences ranked first in all eight items including "law, prize, information, competition, languages, delivery, foreign currency, time differences, and cultural differences." Hofstede (1993) believes that the spread of businesses into the global stage brings the issue national and regional differences to the fore. "There is something in all countries called 'management', but its meaning differs to a larger or smaller extent from one country to another" (Hofstede, 1993).

This paper presents an understanding on the general cultural differences between India, America and China by applying the cultural dimensions of Hofstede and Bond. It also discusses the impact of these cultural differences on their management practice from five aspects; cooperative strategies, conflict management, decision-making, work-group characteristics, and motivation systems.

INTRODUCTION

The 21st century is an era of the globalization of world economy. Cross-national business is facing great challenges in cultural differences.

In one survey entitled what is the major barrier in doing business in the world market, cultural differences ranked first in all eight items including "law, prize, information, competition, languages, delivery, foreign currency, time differences, and cultural differences." Hofstede (1993) believes that the spread of businesses into the global stage brings the issue of national and regional differences to the fore. "There is something in all countries called 'management', but its meaning differs to a larger or smaller extent from one country to another" (Hofstede, 1993).

It can also be observed that most of the failures faced by cross-national companies are caused by neglect of cultural differences. The globalization of the world economy, on one hand, has created tremendous opportunities for global collaboration among different countries; on the other hand, however, it has also created a unique set of problems and issues relating to the effective management of partnership with different cultures.

With the increasing importance of the India market in the world economy, many international companies rushed and planned to enter India to explore business opportunities. They enter the huge market by forming joint ventures or participating in mergers and acquisitions. This has spurred the need for cross-cultural research in India. It was reported that the great barriers caused by cultural differences like difficulty of communication, higher potential transaction costs, different objectives and means of cooperation and operating methods, have led to the failure of many Sino-foreign cooperation projects. The questions like "how to understand India" and "how to do business with Indian people" have occupied the minds of international business people who are planning to enter India.

GENERAL CULTURAL DIFFERENCES BETWEEN THE WEST AND INDIA

India, as the largest market and possibly the most appealing market in Asia, is entering into global collaboration with a wide range of foreign partners. It seems necessary to investigate the cultural differences between India and its international business partners in North America and Europe. To clarify the differences between India and the West, we will refer to Hofstede's four cultural dimensions and Bond's fifth dimension. Among researchers who have given a variety to definitions of culture. Hofstede is one of the first to adopt a pragmatic problem-solving approach in the field and relates culture to management. He defines culture as a kind of "collective programming of the mind, which distinguishes the members of one category of people from another" (Hofstede, 1980). He explained that culturally-based values systems comprised four dimensions: power distance, individualism/collectivism, masculinity/femininity, and uncertainty avoidance. Further research by Michael Bond (1989) identified a fifth "Eastern" dimension called long-term/short-term orientation.

By comparing some Western countries and India along these five dimensions according to their cultural dimension scores (Data source: Hofstede, 1991), some tentative conclusions may be drawn. First, Western countries seem to be generally lower (United States of America 40, Canada 39, United Kingdom 35, Germany 35, and France 68) than India (80) in power distance.

Second, in terms of individualism, Western countries are generally much higher (United States of America 91, Canada 80, United Kingdom 89, Germany 67, and France 71) than India (20). Third, Western countries seem to have short-term orientation while India is considered to be long-term-oriented.

Among these Western countries, America is frequently investigated in cross-cultural research, partly because of its economic power, partly because of its cultural representativeness. To a certain degree, the United States is considered as representing the so-called "Western Culture". Therefore, a comparison between USA and India seems to help clarify the cultural differences between the West and India and related cross-cultural challenges.

India and United States of America (USA) differ greatly with regard to their economic systems, political systems, social values, and laws, despite the substantial changes that have occurred in India during recent years. Table 1 shows the cultural dimension scores of United States of America (USA) and India (Data source: Hofstede, 1993). Some differences can be found. First, in terms of power distance, the scores of India are twice as those of United States of America (USA), which indicates that India is centralized (though it has shown some tendency toward decentralized power) while United States of America (USA) is relatively decentralised. Second, United States of America (USA) ranks first in individualism (strong Individualism) while India is low in individualism (strong Collectivism). Third, United States of America (USA) has higher value than India in masculinity, which indicates that United States of America (USA) is medium masculinity while India is medium femininity. Fourth, India has higher values for uncertainty avoidance than United States of America (USA), which shows that Indian are relatively risk-avoiding while Americans are relatively risk-taking. Last, United States of America (USA) has a short-term orientation while India has a long-term orientation.

It has been widely accepted that cultural differences greatly affect human thinking and behaviour and thus, business organizations in which people interact on the basis of shared values. Management is embedded in a wider societal setting, and is heavily influenced by local historical and cultural norms (DiMaggio and Powell, 1983). The significant differences between United States of America (USA) and India seem to some affect of their management.

IMPACT OF CULTURAL DIFFERENCES ON COOPERATIVE STRATEGIES

Because entrepreneurs mature within a societal context, their attitudes toward cooperation are likely to be influenced by the underlying values of their society (Weaver, 2000). As discussed, United States of America (USA) is strong in individualism and medium masculine. They rely on their own view to determine what they should do. They tend to work alone and are reluctant to cooperate because their individualism and masculine culture view cooperation in general as a sign of weakness and place a high value on independence and control. India is strong in collectivism and medium feminism. The Indians depend more on groups or institutions to determine what they should do and emphasis loyalty to the group. They are more likely to cooperate with others to avoid risks and reduce responsibilities. Their value systems appreciate duty to the group and harmony among its members while pursuing personal goals is viewed rather negatively in India.

In addition, in the process of cooperation, Americans place greater importance on contractual safeguards than the Indian. They believe that contracts can ensure that their partners' tendencies to focus on individual goals and aspirations do not interfere with their own individual goals and aspirations. But the Indians don't consider contracts as seriously as the Americans. They think there will always be changes and the contracts can be reasonably modified according to changes. Instead, they tend to pay more attention to relationships than contracts.

The two phenomena seem to be consistent with Weaver's findings. In his study of a group of seven-nation entrepreneurs about their attitudes toward cooperative strategies, he found that entrepreneurs from societies that are masculine and individualistic have a lower appreciations for cooperative strategies as compared to entrepreneurs from societies that are feminine and collectivist in nature. He also found that entrepreneurs from individualistic societies placed greater importance on contractual safeguards for maintaining effective cooperation than did those from collective cultures (Weaver, 2000).

IMPACT OF CULTURAL DIFFERENCES ON CONFLICT MANAGEMENT

The Indian and Americans tend to resolve conflicts in different ways. Since the Indian come from a strong collectivism and medium feminine society in which harmony and personal relationship are emphasized, they will try to use indirect ways to avoid direct and open conflict. When they face conflict, they prefer to use authority to suppress it, or settle things in private. They prefer to resolve conflict through negotiation and compromise. Individualistic and medium masculine American managers are used to confronting problems directly and bringing things out in the open. To resolve differences, American managers will prefer to use tactics that involve directly confronting others with rational arguments, factual evidence, and suggested solutions (Ting-Toomey, 1985). It is also consistent with the pragmatic short-term orientation and moderately low power distance in United States of America (USA). Indian managers use those tactics less than American managers, because using the tactics will provoke overt disagreement, which is considered highly undesirable.

In addition, American managers are reluctant to *invest* the time and effort required to enlist the help of other people (Yuki, Falbe and Youn, 1993), when they have conflicts or problems with another party. In contrast, the strong collective orientation and uncertainty avoidance values in India encourage Indian managers to use indirect forms of influence that involve the assistance of a third party (Bond, 1991).

To deal with a difficult or controversial request, indirect forms of influence are preferred by Indian managers to avoid losing face and damaging others. When their Western partners propose to use direct and open ways to deal with the conflict, they may feel embarrassed. On the other hand, Western partners may get totally confused by the roundabout why the Indian use to solve seemingly simple problems. The different ways that Indian and American managers resolve conflicts seem to find support from Weaver's finding that feminine societies prefer to resolve conflict through negotiation and compromise (Weaver, 2000).

IMPACT OF CULTURAL DIFFERENCES ON DECISION-MAKING

Risk-taking/Risk-avoiding

Indian and American managers differ greatly in the attitudes toward risks when they make decisions for their different values in uncertainty-avoidance. High uncertainty-avoidance Indian managers usually lack the adventurous spirit and the sence of risks. They dare not make immediate decisions if they feel the circumstance is uncertain, which may deprive them of the opportunity to compete in the market. In most cases, they would like to make comparatively safer and less risky decisions at the expense of the business opportunity. In contrast, low uncertainty-avoidance American managers are more likely to consider risks as natural and are volunteer to take the risks, especially in terms of developing new products, open a *new market* and applying new technology.

Levels of Participation in Decision-making

Managers from an individualistic or a collectivistic country prefer different levels of participation in decision-making. Another related cultural dimension is power distance. Individualistic American managers prefer making decisions individually or deferring to their supervisors instead of consulting with others. They come from a low power distance culture, and they value personal equality and believe in decentralization and empowered subordinates. But the situation in India is a little complicated. Some researches propose that decisions are typically participatory in collectivistic-high power distance countries (Smith *et al*, 1994). Other researchers think that employees in those countries should readily accept decisions handed down by their supervisors, and even resist participation in decision-making because of their unquestioning attitudes toward their supervisors (Graf *et al*, 1990).

In fact, the second phenomenon described by Graf *et al* (1990) predominates in Indian society while the first opinion is also true to a certain extent. That depends on many factors, the most important of which is "what kind of decisions are going to be made". Many Indian managers especially those in state-owned enterprises, adopt the non-participatory approach to decision-making. Final decisions are usually made by higher level superiors without consulting their subordinates. Since some characteristics like high formality, low disclosure and openness as well as centralization are common in most companies. Indian employees seldom have the chance to really participate in the decision-making process. Fortunately, things are getting better after 20 years of reform in India. Participatory decision-making is beginning to be adopted in a number of modern Indian companies.

Impact of Cultural Differences on Work-group Characteristics

India is a relational-oriented country in which people place great importance on personal relationship. Indian managers may initially focus more effort on building social and interpersonal relations before entering into business or contractual relationship. They would like to spend time developing and maintaining gunanxi during the process of interaction and consider it as a prerequisite to do business. In contrast, American managers may encourage their group members learn from each other, to focus on task rather than on social and interpersonal relations, and to build the confidence required for superior performance (Sosik and Jung, 2002). They place a much higher importance on the risk or business deal and hope to focus very quickly on specific business matters. They are achievement oriented, that is "work first". They don't think establishing personal relationship is necessarily involved in the work.

Living in a collectivism society, the Indian view differently as "in-group" and "out-group". They have much higher confidence in "in-group" members than "out-group" members. But the passing of time, and the development of gunanxi, the out-group members may turn into in-group members. If American managers want to do business with Indian managers, they should give their Indian partners enough time to know themselves and develop a personal relationship with them. Only when Indian managers believe that they can trust their partners and consider them as part of the in-group, will the business deal follow smoothly?

In contrast, individualists who exercise short-term thinking place more emphasis on pursuing hedonistic goals and may be unwilling to commit to long-term relationships developed through group interactions (Shamir, 1990). Many American managers are not patient enough to build personal relationships with Indian partners when they invest in India. They cannot understand how important the relations are to the Indian and their business.

Impact of Cultural Differences on Motivation Systems

The method of distributing pay to motivate employees may also depend on cultural values (Leung, 1997). As Aguinis (2002) stated, employees can be rewarded according to their performance (principle of equity), equally (principle of equality), or based on their needs (principle of need). In general, the equity principle is common in individualistic cultures while the equality principle is widely used in collectivistic cultures.

According to Triandis (1995), the basic distinction between collectivism and individualism is based on four universal dimensions. One is that the definition of the self is independent in individualism and interdependent in collectivism. Another is that personal and communal goals are not aligned in individualism but closely aligned in collectivism. These two may account for the great difference in motivation systems generally preferred by American companies and Indian companies, pay for performance in American companies, and pay equity in Indian companies.

In American value systems, great emphasis is placed on individual achievement, and they are expected to achieve success only by their individual efforts. They value competition, achievement and personal goals, and therefore, desire to have plans that recognize individual contributions. They generally see success as contingent upon their own efforts, so they prefer "pay for performance" systems which imply that an individual is solely responsible for what he has accomplished even though he may have had help from others. They consider these systems as effective means to motivate employees. The talents and job performance employees will be considered first by their superiors for salary increases and promotion.

Collectivistic cultures emphasise cooperation interdependence, and group goals, and thus, prefer plans that support group harmony (Aguinis, 2002). The Indian believe that the sense of belonging to the group and devotion to the group are important. They hold the view that one's success is mainly based on group work, so one cannot claim the reward just for oneself. In that case, the equality principle is reflected in the motivation system. When setting salaries, Indian managers will pay more attention to the working experience and academic qualifications of employees. In terms of promotion, one's political quality, history, and interpersonal relationship are given primary consideration.

CONCLUSION

By applying Hofstede's and Bond's cultural dimensions, a cultural comparison between America and India is made. The two countries differ greatly in nearly all the aspects. Based on the cultural difference, inferences are postulated on the impact of Sino-American cultural differences on some aspects of management is made in terms of characteristics, and motivation systems. There are several findings from this cultural comparison. First, Indian managers are more likely to favour cooperative strategies than American managers and American managers place greater importance on contractual safeguards than Indian managers. Second, when faced with conflicts, Indian managers tend to use indirect forms of influence that involve the assistance of a third party, while Americans prefer to use direct and open forms. Third, Indian managers tend to make less risky decisions than American managers. They tend to adopt the non-participatory approach to decision-making. Fourth, the Indian pay more attention to build social and interpersonal relations than Americans. Last, the equity principle is common in American companies while the equality principle is widely used in Indian companies.

TABLE 1

Cultural Dimension Scores of United States and India

Cultural Dimension Scores			
	Power Distance	*Individualism*	*Masculinity*
United States of America	40	91	62
India	80	20	50
Cultural Dimension Scores			
	Uncertainty Avoidance		*Long-term Orientation*
United States of America	46		29
India	60		118

TABLE 2

Cultural Differences Between America and India

Cultural Dimension	*USA*	*India*
Individualism/Collectivism Power Distance Uncertainty Avoidance Masculinity/Femininity Long/short-term Orientation	Strong Individualism Medium Risk-taking Medium Masculinity Short-term Orientation	Strong Collectivism Cantralised, Tendency toward democracy Risk-avoiding Medium Femininity Long-term Orientation

REFERENCES

Aguinis, Herman, 2002, "The Search for Universals in Cross-Cultural Organisational Behaviour", Chapter to appear in J. Greenberg (ed.), Organisational Behaviour: The State of the Science (2nd ed.) Mahwah, NJ: Lawrence Erlbaum Associates.

Bond, M. and G. Hofstede, 1989, "The Cash Value of Confucian Values", *Human System Management*, 8, pp. 195-200.

Chang, S.J., 1995, "International expansion strategy of Japanese firms: Capability building through sequential entry", *Academy of Management Journal*, 38(2), pp 383-407.

DiMaggio, P. and W.W. Powell, 1983, "The iron cage revisited: Institutional isomorphism and collective rationality in organizational field", *American Sociological Review*, 48: pp. 147-80.

Graf, L.A., M. Hemmasi, J.A. Lust and Y. Liang, 1990, "Perception of desirable organizational reforms in Indian state enterprises", International Studies of Management and Organisation, 20, pp. 47-56.

Hofstede, G., 1980, Culture's consequences: International differences in work-related values, Beverly Hills, CA: Sage.

———, 1991, Cultures and Organisations, New York: McGraw-Hill.

———, 1993, "Cultural constraints in management theories", *The Academy of Management Executives*, 7(1): pp. 81-94.

Leung, K., 1997, "Negotiation and reward allocations across cultures" in PC Early and M. Erez (eds.) New Perspectives on International Industrial/Organisational Psychology, pp. 640-75. San Francisco: the New Lexington Press.

Shamir, B., 1990, "Calculations, values and identities: The sources of collectivistic work motivation", *Human Relations*, 43, pp. 313-32.

Smith, P.B., M.F. Peterson, D. Akande, V. Callan, and N.G. Cho, 1994, "Organisational management in 14 countires: A comparison with Hofstede's dimensions" in A.M. Bouvy, F.J.R. van de Vijver, P. Boski and Schmitz (eds.), *Journey into Cross-cultural Psychology*, pp. 364-73, Amsterdam: Swets and Zeitlinger.

Sosik, John and Jung Dong, 2002, "Work-group characteristics and performance in collectivistic and individualistic cultures", *The Journal of Social Psychology*, Washington, pp. 5-23.

Ting-Toomey, S., 1985. Toward a theory of conflict and culture. In W.B. Gudykunst, L.P. Stewart, S. Ting-Toomey (eds.), *Communication. Culture and Organizational Processes*, pp. 71-86, Beverly Hills, CA: Sage.

Triandis, Harry C., 1995, Individualism and Collectivism, pp. 43-44, Boulder, Colorado: Westview Press.

Weaver, K. Mark, 2000, "Attitudes toward Cooperative Strategies: A Cross-cultural Analysis of Entrepreneurs", *Journal of International Business Studies*, 31, Fourth Quarter 2000: pp. 591-609.

Yukl, G., C.M. Falbe, J.Y. Youn, 1993, "Patterns of influence behaviour for managers", *Group and Organisation Management*, 18, 5-28.

CHAPTER

32

Women Entrepreneurship: Problems and Suggestions

Dr. Amirk Singh Sudan, Dr. Rais Ahmad and Dr. Radha Gupta

1. INTRODUCTION

In modern era, when women are showing their capabilities in various walks of life, the entrepreneurial class of women are playing an increasingly prominent role in the various sectors of Industrial growth and economic development. Entrepreneurship amongst women is a relatively recent phenomenon, which is gradually changing the growing sensitivity to the roles, responsibility and economic status of women in the society in general and family in particular. At the same time, it is also recognised that their challenges are immense and complex. For women entrepreneurs starting and operating a business involved considerable risk and difficulties, because in the Indian social environment women have always lived as subordinate to men. There has been noticeable changes in the socio-psycho-cultural and economic norms of our society due to liberalized policies of the Government of India, increase in education level of women and increased social awareness in respect of the role women plays. It has now been recognized that to promote self-employment and to reduce the incidences of poverty, some drastic efforts have to be made to accelerate self-employment in women in various sectors.

The contribution of women to the process of development is widely accepted. Liberalisation and modernization have provided major break-through for women to participate in the process of development. Keeping these into mind, policy-makers are of opinion that the women in the household and non-house hold sector are major emerging force in the areas of economic prosperity of her family and the state as well. Hence, it has become need of hour to participate the women in major decision-making process. The entrepreneurship activity is such in which women both in rural and urban educated and non-educated can play important decisive role.

Since the unemployment position in India specially among educated youths has been alarming and is getting worse with addition of number of graduates and post graduates from Colleges, Universities and Technical/Professional Institutions. It is very serious in rural areas where people are idle about six months in a year and women contribute only household activities. Hence, unemployment situation is severe both in rural and urban areas. To provide employment opportunities instead of government jobs, Government of India established in July 1983, National Institute of Entrepreneurship and Small Business Developments (NIESBUD), an apex body for coordinating the activities of various entrepreneurship development organization in India. In this time, about 1500 organisation are providing necessary vision and guidance for entrepreneurship development. In view of mounting unemployment National Institute for Entrepreneurship and Small Business Development (NIESBUD) is offering entrepreneurship as an alternative carrier profession and gradually introducing it not only in colleges but also in school level as a subject. Besides guidance and training, National Institute for Entrepreneurship and Small Business Development has instituted a Post-graduate Diploma in Entrepreneurship (PGDE) with a view to prepare professionally qualified trainers.

As far as women are concerned, National Institute for Entrepreneurship and Small Business Development had taken special attention to promote capabilities and capacity of women for initiating our business enterprises. It has become necessary in the modern context due to following factors:

(a) Social roles and responsibilities of women;
(b) Family responsibilities; and
(c) Psychological factors that are specific to women.

Now-a-days, women are playing a very important role in all walks of life, women entered into all developmental areas not confirmed to the house. The government has initiated many policies exclusively for women development. Many women proved that they are more capable of doing things with more efficiency than men, especially, in decision-making thanks to their sincerely, patience and hard work. In view of the above qualities, they are also becoming good entrepreneurs for the development of business and industry.

The main objective of this paper is to examine and highlighted in detail the following issues of Women Entrepreneurs in India:-

1. Introduction;
2. Scenario of women Entrepreneurs at International level;
3. Growth of Women Entrepreneurs in India;
4. Participation of women in Entrepreneurial activities in India;
5. Schemes for the Development and promotion of women Entrepreneurs;
6. Government policies and programmes;
7. Problems of women entrepreneurs;
8. Suggestions for the development and promotion of women entrepreneurs;
9. Conclusion; and
10. Reference.

2. SCENARIO OF WOMEN ENTREPRENEURS AT INTERNATIONAL LEVEL

The development of entrepreneurship among women is a major step to increase women participation in economic development. It will enhance economic growth and provide employment opportunities for women entrepreneurs providing economic opportunities for women can also improve the social, educational and health status of women and their families. In the advanced countries, there is a phenomenal increase in the number of self-employed women, after World War II. In the U.S.A. women own 30 percent of all business, even though their sales on an average are less than two-fifths of those of other small business. In Canada, one-third of small business is owned by women and in France is one-fifth. In the United Kingdom (U.K) since 1980, the number of self-employed women has increased three times as fast as the member of self-employed men (Vasant Desai, 1991). A similar trend is noticed even among the developing countries/nations in Asia. Women make for 40 per cent of the total workforce in these countries. It has been seen that women out number men by at least two times particularly when it comes to starting business in China. There are over five million women entrepreneurs constituting one fourth of all the entrepreneurs in China. In Japan too a similar trend has been noticed. The percentage of women entrepreneurs increased from 2.4 percent in 1980 to 5.2 percent in 1995 and 8.4 percent in 2010.

According to an International Labour Organization (ILO) report 2000, women are 50 percent of the worlds' population, do the two thirds of the worlds' income and own less than one percent of would property. In most countries, average earnings of women are lower than those of men. In several developing countries, marriage is the only career for most women. Women have confined their activities to selected professions such as education, nursing, medicine and office work, very few women enter professions like industry, engineering, trade, etc.

In the United States of America (USA) women entrepreneurs have emerged as the fastest growing group of entrepreneurs. According to the U.S. International Revenue Service (IRS), the number of female-owned enterprises increased from 7,02,000 in 1977 to 3.8 million in 2011. These women own textiles mills trucking firms highway and construction firms, beauty parlors, dry-cleaning shops, photographic studies, Restaurants, Malls, Shopping Centres, General Stores, Development Stores and dozen of other businesses.

In the advanced countries of the world, there is a phenomenon increase in the number of self-employed women after the World War II. In the United States of America (USA) women own 25 percent of all business. In Canada, one-third of small business owned by women and in France it is one-fifth. In United Kingdom (UK), since 1980, the increased three times as fast as the member of self-employed men.

3. GROWTH OF WOMEN ENTREPRENEURS IN INDIA

There has been a rapid growth of women entrepreneurs in India. In the world of fashion, beauty and career counseling, many fashion designers are women with their own labels. To name a few, Ritu Kumar known as the queen of Indian textiles, Ritu Beri who has had a stint in the French fashion industry and Rina Dhaka among others. The Indian fashion industry is now conscious of the opportunities available in this industry abroad and the need to be globally competitive. The world of beauty culture has opened up many opportunities for women entrepreneurs. Women are setting up successfully small establishments as beauticians and hair stylists. Apart from clothes, women are successful in producing design accessories as well. Indian women has established brands in the world. Vinita Jain of Boutique and Shahnaz Hussain of Shahnaz Herbals are global brands today. It is the time to collect data on small and big business run by women. In the field of education, counselors are mostly women, primarily because men did not see it as a lucrative sector. As disposable income increases with the opening up the economy, women are taking to opening up restaurants as well.

In India, there is a growing awareness of the need for women entrepreneurs in the nation's development. The growing awareness among women in India about profitability of entrepreneurship is mainly due to the fact that profile of Indian Women has undergone perceptible change in the post-independence period.

Women entrepreneurs have been making a significant impact in all segments of the economy. However, in India women entrepreneurs constitute a negligible proportion of the total entrepreneurs as compared to advance countries of the world. Attitudinal constraints, social traditions and kinship system inhibit the emergence of women entrepreneurs, out of total women populations of the country (i.e. 46.5%) only 4.5 percent self-employed are recorded. Majority of them are engaged in the unorganised sector like agriculture, agro-based industry, handicrafts, handlooms and cottage industries. Participation of women as industrial entrepreneurs however is comparatively a recent phenomenon continuing from 70s onwards.

In India women have contributed mainly in household industries due to the less technical know-how required and little competition from men in these industries. With the spread of education and the growing awareness among women, they have entered into engineering, electronics, energy and many other industries. Although, the number of such units are very small, women are putting up units to manufacturing solar cookers, Television (T.V.) capacitors, electronic ancillaries and small foundaries is, readymade garments, fabrics, handicrafts and handloom products, dairy, plastics ceramics, printing textile designing, Jewellery design, etc. Various government agencies and voluntary bodies like Mahila Mandals have accelerated the growth of women entrepreneurs in the country. Over the years women have become more task-oriented, career mind and economically independent. They aim to widen their sphere of work and enjoy achievement.

In the recent days, women are getting educated and undertaking any job at par with men driven by a zeal to start an industry and provide employment to some people while earnings their own livelihood.

4. PARTICIPATION OF WOMEN IN ENTREPRENEURIAL ACTIVITIES IN INDIA

The emergence of women entrepreneurs in a society depends to a great extent on the economic social, religion, cultural and psychological factors prevailing in the society, women participation in India is relatively a recent phenomenon. There is a noticeable increase in the number of women entrepreneurs and they are becoming ready to reach out for new jobs, new responsibilities and new experience with a high amount of challenging sprit. Thus, women entrepreneurs are the women or a group of women, who initiate organize and operate a business enterprise and are expected to innovate, imitate or adopt an economic activity.

For centuries, women have been victim of social prejudices and discrimination. Even today women are generally presumed to be weak, passive, dependent and people-oriented. These assumptions inhabit development of self-confidence, innovativeness achievement, motivation and risk-taking ability which are very much essential for an entrepreneurial career. In business, the entry of women is a relatively new phenomenon on account of this break-up of the joint family system and the need for additional income to maintain living standard on account of inflation. Women began to enter the competitive world of business. A woman may start her own business due to several reasons. She may not be able to find a job in the market place or she may not be able to work out of her house. Some women may start their own business as they are stagnating near the top of male owned firms.

In recent years, women have made their mark in different walks of life and are competing successfully with men despite the social, psychological and economic barriers. This has been possible due to education, political awakening, urbanization, legal safeguards, social reforms, etc. Some of the women have distinguished themselves in many unconventional field as Prime Ministers, Ambassadors. Governors, Chief Minister, Minister of Space Scientists, Supreme Court Judges, High Court Judges, Advocates, Vice-Chancellors, Pilot, Administrators, Professors, Doctors and entrepreneurs.

Women entrepreneurs are observed dealing with production of both consumer goods and intermediate goods/Articles, like a male entrepreneurs, a women entrepreneurs also require to perform the functions of exploring the prospects of starting new enterprise undertaking of risks and the handling of economic uncertainties, introduction of innovations. Coordinating control change management.

Table 1 shows that the total number of women enterprises in the SSI Sector was estimated at 10,63,721 (10.11%). The estimated number of enterprises actually managed by women was 9,95,141 (9.46%). About 13% of the women enterprises were in the registered SSI Category, the rest being unregistered. Of the enterprises managed by women, 11.5% were in the registered SSI category. The share of the units managed by women in terms of employment was 7.14%. The employment generated per Rs. 1 lakh investment in the units managed by women 2.49. The total number of female employees in the SSI sector is estimated at 33,17,496. About 57.62% of the women were employed in SSI units located in the states of Tamil Naidu, Kerala, Karnataka, West Bengal and Andhra Pradesh.

The table under reference further reveal that the proportions of female employees in the total employment in the SSI sector was

TABLE I

Participation of Women in Small Scale Industrial (SSI) Sector State-wise in India

S. No.	Name of State/UT	Number of Female Employees			Number of Enterprises managed by Women	Number of Women Enterprises
		Registered SSI Sector	Unregistered SSI Sector	Total		
1.	Jammu & Kashmir	3313	8153	11466	5640	5742
2.	Himachal Pradesh	4016	4134	8150	3515	3722
3.	Punjab	11757	36013	47770	30190	29088
4.	Chandigarh	659	1975	2634	2059	2243
5.	Uttarakhand	2940	13240	16180	8706	8804
6.	Haryana	15651	14889	30540	10087	9620
7.	Delhi	6306	35798	42104	13368	14383
8.	Rajasthan	15003	46065	61068	29785	38371
9.	Uttar Pradesh	23506	180918	20442	54491	72667
10.	Bihar	8353	68908	77261	38170	49443
11.	Sikkim	212	09	221	30	98
12.	Arunachal Pradesh	342	104	446	131	150
13.	Nagaland	637	2574	3211	207	179
14.	Manipur	3853	19485	23338	9168	10745
15.	Mizoram	2188	4636	6824	3076	10745
16.	Tripura	4294	3325	7619	631	863
17.	Meghalaya	3139	5664	8803	3658	3580
18.	Assam	9077	16988	26065	11189	11757
19.	West Bengal	26549	304969	331518	71847	69625
20.	Jharkhand	5105	8907	14012	7271	7885
21.	Orissa	11723	213123	224846	33274	38233
22.	Chhattisgarh	10177	52476	62653	11766	10034
23.	MadhyaPradesh	29612	111703	141315	622351	68823
24.	Gujarat	41189	79930	121179	55361	53703
25.	Daman & Diu and Dadra & Nagar Haveli	6106	126	6232	167	213
26.	Maharashtra	78731	162700	241431	80662	100670
27.	Andhra Pradesh	60693	270026	330719	77347	77166
28.	Karnataka	117993	223142	341076	101264	103169
29.	Goa	5309	1833	7142	677	810
30.	Lakshadweep	26	405	431	61	67
31.	Kerala	189640	224491	414131	137561	139225
32.	Tamilnadu	270936	223050	493986	130289	129808
33.	Pondicherry	5613	2670	8383	1089	1065
34.	Andaman & Nicobar Islands	124	294	418	53	110
	Total	974713	2343783	3317496	995141	1063721

Source: 3rd Small Scale Industrial Sector (SSI) Census, 2001-02, Ministry of MSME, Government of India, New Delhi.

13.315%. In the states/union Territories of Mizoram, Orissa, Karnataka, Goa, Lakshadweep, Kerala, Tamil Naidu and Poncherry, the share of women employment was significantly higher (more than 20%) compared with the total employment in the respective states. In India, women entrepreneurs have been in business for quite some time and achieved remarkable success.

5. SCHEMES FOR THE DEVELOPMENT AND PROMOTION OF WOMEN ENTREPRENEURS

According to the Third All India Census of Small Scale Industries conducted in 2001-02 and subsequent estimates made by Government of India, only 10.11% of the Micro and Small Enterprises in India are owned by women while 9.4% of the Micro and small enterprises are managed by women. Currently in 2006-07, these estimated number is 12.99 lakh women owned enterprises and 12.15 lakh women managed enterprises.

In orde to encourage more and more women enterprises in the Micro and Small Enterprises (MSE) Sector, Several schemes have been formulated by this ministry and some more are in the process of being finalized, targeted only at the development of women enterprises in India. These schemes are as follows:

(1) Trade-related Entrepreneurship Assistance and Development Scheme for Women (TREAD);
(2) Micro and Small Enterprises cluster Development Programmes (MSE-CDP);
(3) Credit Guarantee Fund Scheme for Micro and Small Enterprises;
(4) Support for Entrepreneurial and Managerial Development; and
(5) Exhibitions for Women Under Promotional package for Micro & Small Enterprises Approved by CCEA under Marketing Support.

6. GOVERNMENT POLICIES AND PROGRAMMES

In order to facilitate and motivate the growth of entrepreneurship, the central and state governments have created an elaborate institutional framework in the country. There are several schemes launched by the government to motivate women entrepreneurs and changed to role of women from a housewife to a businessmen women. For example:

(i) SBI Store Shakti package;

(ii) IDBI Mahila Udyana Scheme;
(iii) SIDBI's assistance for Women Entrepreneurs;
(iv) Bank of India's Pryadarshine Yojana;
(v) Prime Minister Rojgar Yojana (PMRY);
(vi) Self-Employment Programme for Urban Poor (SEPUP);
(vii) Swarnajayanti Gram Swarojgar Yojana (SGSY); and
(viii) Integrated Rural Development Programme (IRDP)

Micro fiancé through Rashtriya Mahila Kosh (RMK) are some of the schemes launched for the development of women entrepreneurs.

7. PROBLEMS OF WOMEN ENTREPRENEURS

There are umpteen problems faced by women right from the idea of initiating an enterprise till the enterprise lives. Various problems faced by Indian women are as follows:

(a) Patriarchal Society

Entrepreneurship has traditionally been seen as a male preserve and idea of women taking up any entrepreneurial activities is considered as distant dream. Any deviation from the norm is frowned upon and, if possible, immediately curbed. Women also have to face role conflicts as soon as they initiate any entrepreneurial activity. Women have had many ambitions during their student days, but most of them usually lose track of their ambitions in the period of post-education, i.e. during marriage and settling down as a housewife. Family is always as major influencing factor for married women entrepreneur and husband being the main constituent of the family, has a major influencing role. Thus, it is an uphill task for women to face such conflicts and cope with the multiple roles.

(b) Absence of Entrepreneurial Aptitude

Many women take the training by attending the Entrepreneurship Development Programmes without an entrepreneurial bent of mind. As per a study, involvement of women in the small scale sector as owners stands at mere 7 percent. Women who are imparted training by various institutes must be verified on account of aptitude through the tests, interviews, etc.

(c) Marketing Problems

Access to market is as important as access to finance. In order to make the company successful and profitable, the entrepreneur must be able to sell his products or services. Women entrepreneurs, by and large continue to face the problems of marketing their products. Every entrepreneurs, man or woman, does a market survey before launching his or her product or service, however small, the initial size of the business would be. But this is just the beginning of marketing. Once the product or service is launched and is available in the market, access to markets does pose a very big challenge in some cases. Such access sometimes is more difficult than access to finance. It is one of the core problems as this area is mainly dominated by males and even women with adequate experience fail to make a dent.

For, marketing the products women entrepreneurs have to be at the mercy of middlemen who pocket the chunk of profit. Although the middlemen exploit the women entrepreneurs, the elimination of middlemen is difficult, because it involves a lot of running about. Women entrepreneurs also find it difficult to capture the market and make their products popular.

(d) Financial Problems

Obtaining the support of bankers, managing the working capital, lack of credit resources are the problems which still remain in the domain of males. According to a report by the United Nations Industrial Development Organisation (UNIDO) "despite evidence that women's loan repayments rates are higher than men's, women still face more difficulties in obtaining credit", often due to discriminatory attitudes of banks and informal lending groups. Studies have shown that even educated modern women entrepreneurs in the advanced and developing economies experienced difficulties in obtaining loans. Although the procedure of obtaining the loans has been simplified to a great extent but for women it still remains a challenge. Women are yet to make a significant mark in quantitative terms.

(e) Family Conflicts

Women also face the conflict of performing of home role as they are not available to spend enough time with their families. They spend long hours in business and as a result, they find it difficult to meet the demands of their family members and society as well. Their inability to attend to domestic work, time for education of children, personal hobbies, entertainment add to their conflicts.

(f) Credit Facilities

Though women constitute about 50 percent of population, the percentage of small scale enterprise where women own 51 percent of share capital is less than 5 percent. Women are often denied credit by bankers on the ground of lack of collateral security. Therefore, women's access to risk capital is limited.

The complicated procedure of bank loans, the inordinate delay in obtaining the loans and running about involved do deter many women from venturing out. At the same time, a good deal of self-employment programme has been promoted by the government and commercial banks.

(g) Shortage of Raw-Materials

Women entrepreneurs encounter the problems of shortage of raw-materials. The failure of many women co-operations in 1971 such as these engaged in basket materials.

(h) Severe Competition

Many of the women enterprises have imperfect organizational set-up. But they have to face severe competition from organized industries.

(i) High Cost of Production

High cost of production undermines the efficiency and stands in the way of development and expansion of women's enterprises. Government assistance in the form of grant and subsidies to some extent enables them to tide over the difficult situations. However, in the long-run, it would be necessary to increase efficiency and expand productive capacity and thereby reduce cost to make their ultimate survival possible, other than these, women entrepreneurs also face the problems of labour, human resources, infrastructure, legal formalities, overload of work, lack of family support, mistrust, etc.

(j) Social Barriers

Women entrepreneurs in India are always seen with suspicious eyes, particularly in rural areas, they face more social barriers. Not only this, so many castes and religions dominate with one another and hinders women entrepreneurs.

(k) Problems of Middlemen

Women entrepreneurs depend on the middlemen for the sale of their product. Middlemen add their own profit margin and hence cause for higher selling price. In such a competitive, customer's attraction towards the product produced by women entrepreneurs decrease. This results in less sales and lesser profit.

(l) Lack of Information

Women entrepreneurs lack knowledge and information of availability of raw materials, financial facilities, government help and various kinds of subsidies available. They are not, sometimes, aware of developments of information technology. Lack of advanced knowledge hinders the expansion of their markets.

(m) Lack of Self-Confidence

Women lack confidence and they even hesitate taking risks. They are conservative in taking risks and their risk-bearing capacity is also less. Since she has to play dual role, she has to strive hard to strike a balance between her dual roles. Many times, she has to sacrifice or suppress her entrepreneurial urge in order to strike balance.

(n) Lack of Access to Technology

Women-controlled business are often small and it is not always, innovative schemes, concessions, alternative markets, etc. They do attend and meetings of various associations, speak to other entrepreneurs but that is not sufficient. While technology is being used to assist in delivering information to women-owned businesses, but all businesses don't have access to such technology. Just a small percentage of women entrepreneurs avail this technology an they too remain confined to word processing software in the computer. They hardly make use to advanced software available like statistical software SAP, Accounting Package TALLY, Animation software 3D MAX, internet, etc. Women entrepreneurs will not feel empowered unless they make use of technology and update themselves accordingly.

(o) Lack of Training

Many women become entrepreneurs out of change and not choice. They also have to play dual role of business women and a care take of her family as well. As a result, they often lack access to training that will go a long way in helping them in developing business skills. Regular training is a necessity in this competitive environment.

8. SUGGESTIONS FOR THE GROWTH OF WOMEN ENTREPRENEURS

Empirical evidence shows that women contribute significantly to the running of family businesses mostly in the form of unpaid efforts and skills. Their efforts and contribution is always undervalued. Even many of the enterprises defined as being run by women are in fact run in their names by men who control the operations and decision-making. Various programmes introduce by the government can succeed only, if independence and mobility is not denied to women. Programmes started are bound to fall or succeed partially if taken up in isolation. Entrepreneurship basically implies being in control of one's life and activities and women entrepreneurs need to be given confidence, independence and mobility to come out of their paradoxes. The following measures are suggested to empower the women to seize various opportunities and face challenges in business.

(a) Change in Attitudes

Promoting entrepreneurship for women requires reversal of traditional attitudes than the mere creation of jobs for women would. Societal change is not an overnight phenomenon, it is a gradual and slow process. Unless attitudes are changed, subsidies and credit allocation to women will not bring much change. The various programmes initiated by the government have to go beyond incentives and exemption to change social outlook, attitudes, group formation, training and other support services.

(b) Training

Training in entrepreneurial attitudes have to be imparted which should start right at the high school level through well designed courses which will instill confidence. Various case studies and behavioural games can be formed part of curriculum. In addition, practical knowledge can be imparted regarding financial management, cost accounting, preparation of project report and various legal aspects related to business. The curriculum should be designed to give hands on experience of assessing the marketability of a commodity or a service. The syllabus should introduce this subject on a compulsory basis and should also be reviewed continuously on the basis of the feedback received.

(c) Increase in Mobility

The constraints on mobility should be released as mobility contributes to raise the confidence levels to an incredible extent. Girls should be compulsorily trained to learn cycling, play games, participate in extra-curricular activities and youth festivals and also trained to learn the techniques of self defence (karate, etc.)

(d) Initiation into Professional Work

Efforts made by the government to train the girls and to increase their mobility will turn out to be futile unless women are initiated into independent professional work. Families usually provide financial and emotional support to sons but don't extend similar support to the daughters. Parents and daughters together need to be convinced that the skills acquired in school and polytechnics could provide them with profitable occupations. In women's institutes therefore, an additional year of training should be introduced where women are put to work in training-*cum*-production workshops, whose produce is sold and income earned. The training programmes should be viewed constantly and the spirit of entrepreneurship spread among women.

(e) Self-recognition and Decision-making Authority

Any programmes started for women could be misused by using the truly deserving beneficiaries, i.e. women, s fronts for their personal interests. This practice can be curbed only if women beneficiaries are induced to claim greater decision making authority in family businesses, whether run in their names or not. This can come only from greater experience and motivation included by greater knowledge and experience of dealing with external world and from moving with other successful women entrepreneurs.

The women must be pulled out of normal restrictive environment and should taught to recognise her own psychological needs and give her needs an expression. Women must be motivated to involve themselves actively in business being run in their names but actively driven by menfolk of their families. They should not be contended by just lending their names in the business and putting their signatures as and when required.

(f) Provision of Amenities

Infrastructure to set-up industries is provided in the form of industrial plots and sheds. Apart from allotting units to women entrepreneurs on priority basis, it is desirable to provide amenities particularly required by them in industrial areas.

Following amenities can be arranged:

- Easy mobility and closeness to homes;
- Child care facilities;
- Regular bus service inside industrial areas;
- Creches for children; and
- Allotment of sheds on priority basis.

Whenever sheds or land is allotted on priority basis or at concessional rates to women, the scope of misuse is higher than in the case of any other disadvantaged group. Therefore, the screening committee should take precautionary measures to evaluate whether the beneficiary is indeed going to be in full charge of the enterprise especially in case of allotments made for highly sought after areas. Rather training in running an enterprise form quality approved institutions should be insisted upon as a precondition for such priority allottees.

The same condition can be applied while granting special investment subsidies or other facilities being extended for women.

(g) Inculcating Marketing Skills

Women generally prefer to undertake those projects which ensure almost total marketing support. Women lack in confidence to seek out and develop markets and are usually short of time due to their family obligations. In many cases, they depend on the males of the family for marketing. Marketing requires mobility and confidence in dealing with the external world and women have been discouraged in both the requirements.

As a remedy to this ailment, the long-term strategy should be to inculcate marketing skills in women entrepreneurs. Women should be equipped to conduct (i) market survey to select the product, (ii) training and advice should be given to expose her enterprise to the market, (iii) in addition, professional marketing expertise should be provided to identify marketing channels for the products made by women entrepreneurs, and (iv) industrial estates could also provide marketing outlets for the display and sale of products made by women.

A Women Entrepreneur's Guidance cell has been set-up by Karnataka State Financial Corporation (KSFC) to handle various problems of women entrepreneurs. The experiment has proved to be successful to familiarize the women with external world and has assisted women in their projects. Similar cells can be setup in the District Industrial Centres (DICs) and through Single Window Agencies.

(h) Causes of Stagnation in Growth

Women entrepreneurs usually suffer from stagnation in growth of their enterprise. The stagnation is caused due to various reasons like household responsibilities, lack of mobility, fear of expansion of business and staff, lack of self-confidence, fear of success, resentment and hostility from family, etc. In addition, lack of managerial and technical skills also act as a barrier to the growth of women's businesses. Women entrepreneurs should also understand the need for constant training, development and counseling in order to cope up with advancement taking place in their respective enterprises. Women entrepreneurs can be counseled through NGOs, psychologists behavioural experts, managerial experts and technical persons so as to enable them to adopt appropriate strategies to full growth.

9. CONCLUSION

In conclusion, it may be stated that there has been a positive growth in women entrepreneurship, more and more encouragement and support be provided to them as it would certainly enhance their Socio-economic status, which is pre-requisite for economic development.

The future of women entrepreneurship is very bright and hopeful. There is a radical and admirable change in the attitude of society. The help from government and the financial institutions is bound to increase their attainments in life. Their participation in building an industrialised India will be a reality. All their dreams will come true through their education, labour and endless efforts for their empowerment and betterment of mankind. There is no doubt that tomorrow will bring a beautiful dawn for women entrepreneurs. The potentiality within women must be channelized appropriately. The inborn capacity of Indian women to manage a home can be extended to the level of managing a firm and ultimately building a nation or country. In the new century, the women have stood with a self-dependent, self-confident, assertive personality to fight against all constraints. They are spreading their wings to the sky unlimited and this height would lead to a better human resource development and strengthen nation's prosperity. In the present age of globalization, women entrepreneurs in India will play a greater role and emerge themselves in yet another new dimension to the centre Universe by enlightening the society with a glorious future.

In India, women entrepreneurs are playing a very important and significant role in the development of the country. But they have started a responsibility of making the country to be a develop one. There is a greater awakening among women. If the opportunities are provided to women, they can show the better results. There is a need of motivation and awareness to the young women for coming in the entrepreneurial stream and to extend support with technical and financial assistance. The integral approach from the government and non-governmental organizations is necessary for the success of the movement of women entrepreneurship.

It is said when a man starts a business ventures, he only becomes the entrepreneur, while when a women takes to entrepreneurship the whole family becomes entrepreneurial in its behaviour. Hence, if the aim is to entrepreneurise the society, one should promote entrepreneurship among women. Either because of this reason or just to raise the economic level of the women folk or to raise their social status in the society, various government agencies, Non-governmental Organizations (NGOs) are working to promote and develop entrepreneurship amongst women. Small Industries Service Industries (SISIs), State Level Institutes of Entrepreneurship Development (IEDs), Technical Counseling Organisations (TCOs) and Entrepreneurship Development Institute of India (EDI) and various Non-governmental Organsiations (NGOs) regularly conduct 4 to 6 weeks training programmes to develop women entrepreneurs. Since its in caption Small Industries Development Bank of India has sponsored more than 400 Entrepreneurship Development Training Programmes exclusively for women.

Indian women was like a well-defined predictable master plan. They were to act as per the liking and disliking of their male counterparts. Slowly changes started creeping in women migrated to cities and towns, got education and economic independence. Here role as an earning woman was no more looked down upon women entrepreneurship is the process whereby women organize the business and provide employment to others. A women entrepreneur enterprise is termed as an small scale industrial (SSI) unit/industry-related service or business enterprise, managed by one or more women entrepreneurs in proprietary concerns in which she/they individually or jointly have a share capital of not less than 51 percent as shareholders/partners/ directors of a private limited company/members of a cooperative society. Unlike other countries, Indian women run just 8 percent of Small Scale Industries (SSI) while in USA, their number is as high as 50 percent. Indian women have to face a number of problems beginning right from the commencement of the enterprise. Male dominated society, heavy competition, non-cooperative attitude of financial institutions, marketing problems, family conflicts, etc. are just some problems faced by women entrepreneurs. However, we can say that the traditional phase has set in since 1990's, India has become a major centre for business outsourcing. Besides, world of fashion has opened and education has develop rapidly as an industry. Government has come forward with umpteen facilities concession and incentives exclusively for women entrepreneurs. Various schemes like PMRY, TRYSEM, DWCRA, IRDP, TREAD, SGSY, SIDBI developmental schemes to has been launched for women whereby they are provided various relaxations as compared to their male counterparts. Inspite of all these incentives, women have to go a long way fulfilling their multiple role of a mother, wife and a business woman as well.

Promoting entrepreneurship for women will require an even greater reversal of traditional attitudes than the mere creation of jobs for women world. This doesn't mean that we should wait for societal change to take lace first. But it does imply that the programme should go beyond subsidyes and credit allocation to attitudinal change, group formation, training and other support services but also practical application of the academic knowledge regarding management like marketing and finance of a business enterprise.

Activities in which women are trained should focus on their marketability and profitability and not be routinely restricted to making pickles and garments. A high power and professionally involved committee must continuously review the courses and the curriculum in the institutions on the basis of evaluation studies and current market developments. In addition to skill development these institutes should also provides practical management inputs. Appropriate training is still the key to a successful programme to develop entrepreneurship among women cult of entrepreneurship among young women. A women or any continuous monitoring an improvement of training programmes should eventually spared the small business has not access to institutionalized trade finance and must operate only through very expensive informal channels. Any move to make banks and institutions enter this area is likely to benefits women more than men. Apart from allotting industrial plots and sheds by state-run agencies to women entrepreneurs on priority basis, it is desirable to provide amenities particularly required by them in industrial areas. Imparting training by experts in different fields namely, production, marketing, banking, accounts, industrial and consumer relations legal issues, etc. is an added value to women entrepreneurs.

Training and development of skills are essential for the development of women entrepreneurs. Training schemes should be so designed that women can take full advantage. Women have the potential and the will to establish and manage enterprises of their own. What they need is encouragement and support. Government should offer an salary units to women entrepreneurs on priority basis. Tax concessions market subsidies, low interest rate, are the push factors with which women can join the mainstream of national economy and thereby excel well and gets the capacity of contributing significantly to the country's economic progress even under changing turbulent global business environmental conditions. When women moves forward, the family moves, the village moves, and the nation moves. These were the words of Pandit Jawahar Lal Nehru which is fact. Employment gives economic status to women. Economic status gives a way to social status and thereby empowerment to women.

References

Arora, Renu and Sood, S.K, "Fundamental of Entrepreneurship and Small Business", Kalyani Publisher, Ludhiana, Sixth Edition 2011.

Desai, Vasant, "Entrepreneurship and Technology", Himalya Publishing House, Mumbai, 1999.

Gupta, C.B. and Khanka, S.S., "Entrepreneurship and Small Business Management", Sultan Chand & Sons, New Delhi, 2000.

Gupta, M.C., "Entrepreneurship in Small Scale Industries", Anmol Publication (P) Ltd., New Delhi, 1987.

Indian Journal of Marketing, Volume XXXIV; Number 5, May 2004.

Kamulun, N. and Kumar, A., "Entrepreneurship Development in Orissa: Some Issues", *Yojna*, 16(2), 1992, (18-19).

Kapoor, K., "Entrepreneurial Behaviour: A study of selected Women Entrepreneurs, 1998", M.Sc. Dissertation/Thesis, Punjabi Agricultural University, Ludhiana, (Pb.).

Khanka, S.S., "Entrepreneurship Development in India", S. Chand & Company, New Delhi, 2010.

Shukla, M.B. and Gupta, Ranjita, "Entrepreneurship Development in India", Book Land India, Varanasi (U.P), 2003.

Shukla, M.B., "Entrepreneurship and Small Business Management", Kitab Mahal, Allhabad, 2003.

CHAPTER

33

Rural Tourism in India: Issues and Challenges in Marketing Strategy of Community Tourism

VINAY RAJ R.

ABSTRACT

Rural Markets are defined as those segments of overall market of any economy, which are distinct from the other types of markets like stock market, commodity markets or labour economics. Rural markets constitute an important segment of overall economy. In recent years, rural markets have acquired significance in countries like China and India, as the overall growth of the economy has resulted into substantial increase in the purchasing power of the rural communities. On account of the green revolution in India, the rural areas are consuming a large quantity of industrial and urban manufactured products. In this context, a special marketing strategy, namely, rural marketing has taken shape. Sometimes, rural marketing is confused with agricultural marketing—the later denotes marketing of produce of the rural areas to the urban consumers or industrial consumers, whereas rural marketing involves delivering manufactured or processed inputs or services to rural producers or consumers. Also, when we consider the scenario of India and China, there is a picture that comes out, huge market for the developed products as well as the labour support. This has led to the change in the mindset of the marketers to move to these parts of the world. Aggressive marketing campaigns launched by the Kerala Tourism Development Corporation—the government agency that oversees tourism prospects of the state—laid the foundation for the growth of the tourism industry. In the decades that followed, Kerala Tourism was able to transform itself into one of the niche holiday destinations in India. The tag line Kerala—God's Own Country was adopted in its tourism promotions and became synonymous with the state. Today, Kerala Tourism is a global super brand and regarded as one of the destinations with the highest brand recall.

Keywords: Freebies, USP,

TOURISM IN ERNAKULAM

Ernakulam District is a district of the state of Kerala in southern India. Ernakulam, the commercial capital of Kerala, served by an international airport, fine communication facilities, excellent accommodation, modern amenities and a lot of attractive tourist places gives a breathtaking experience for its visitors and settlers. The district took its form on 1st April 1958, by carving out the areas of erstwhile Travancore, Kochi & Malabar kingdoms. The word Ernakulam was derived from a Tamil word 'Erayanarkulam' which means 'Lord Shiva's abode'. The commercial capital of Kerala and the most cosmopolitan of Kerala's cities, Cochin is a fascinating blend of cultures. From time immemorial, it has been the favorite destination of globe-trotters and explorers. Blessed by one of the finest natural harbors in the world, Cochin has been eulogised as the Queen of the Arabian Sea.In and around Cochin are several cities of monumental value, but step into Fort Cochin, and you are in a totally different world. The Chinese fishing nets, the Jewish Synagogue, the Dutch palace, the Portuguese and British Architecture, all finely blend with the rich heritage of the land, intricately woven folklore and fact. With an international airport, an international sea port and a busy network of road and rail transport, the city is well connected to the rest of the world.

The Importance of Community-based Tourism (CBT)

The importance of Community-based Tourism (CBT) increases over time, while countries and public being aware of positive and negative impacts of tourism industry. The objectives of CBT are suitable to the aims of UN Commission on Sussainable Development and the "Tourism Enriches" campaign and declaration of WTO which can be summarized as "to gain local economic development, reach some forms of participation, provide socially and environmentally responsible experience for visitors, and bring positive effect on the conservation of natural and cultural sources. Community Tourism is a new phenomenon. It is a spirit that permeates all sectors of the tourism industry. In other words it is a philosophy around which tourism products are developed. Community Tourism is about new levels of relationships between the host country and the visitor. What is appealing to the visitor is found among the varied natural attractions, local resources and talents, and indigenous attributes of a community or area. Through visitor-community interaction, respective cultures are explored, ideas and information are exchanged, and new friends are made. Community Tourism fosters opportunities at the community level for local people wishing to participate more fully in the tourism industry. This may range from establishing bed & breakfast accommodation in a rural home to creating income-generating tourism opportunities for an entire village. Community Tourism is managed tourism in its profoundest sense, taking into account evolving travel trends in such areas as the environment, study and research, seniors travel and other special interests. This study is mainly based on Community Tourism and Natural Resource Conservation the wise use of the earth's resources by humanity.

RURAL MARKET: AN OVERVIEW

Is the Shift due to the Growth in Indian Agriculture

For the last ten consecutive years, we have had good monsoons. So, agriculture is prospering. Of course, there have been setbacks in the last couple of years. Another interesting aspect is, today rural India is not 100 per cent dependent on an agrarian economy. Unlike in the past where the ratio between those who involved in agriculture and in other business was 75-25, today the estimated ratio is 50:50, if not 60:40. So today, 50-60 per cent of the rural population is involved in other businesses. A lot of people belonging to the second generation are getting white-collar jobs in nearby towns. So, there is a growing middle class with a monthly income in rural India and it is a drastic change from the past where their income was totally dependent on the monsoon, cropping season, etc. This has resulted in a definite growth in the prosperity level in rural India. Of course, there are still a lot of poor people, especially the agricultural labourers. But there is a growing middle class with regular income and the rural rich are becoming richer. Inspite of the crucial and pervasive role of agriculture in Indian economy, its two main partners—farmers and the labourers are not paid that much attention as they deserve. While 75 percent of farmers are marginal and small farmers with less than two hectares of holding more than 50 percent of agricultural labour is below poverty line. Much has to be sown in the country's agriculture before a bountiful harvest can be reaped. It is not as if targets of 4 percent annual growth for agriculture have been realized today in keeping with contemporary needs. Such threshold has been the mandated target from the ninth plan onwards, but hardly ever achieved.

Is the Urban-Rural Divide in India Thinning Now

The urban-rural divide is still there, but the divide between urban and rural India is thinning among the top segment of rural India. The rural rich are almost like urban India. Rural India is like a pyramid. The rich farmers and businessmen occupy the top of the pyramid. They may constitute around 5 per cent of the population. The next level belongs to those with a regular income and the vast majority of the people who are daily wage labourers occupy the base of the pyramid. So we cannot say that the urban-rural divide has melted. It is still there. But there is hope with the growing emphasis on education.

How would you Categorise Different Parts of Rural

In India, we have the developed rural India and undeveloped rural India. Punjab, Haryana, Tamil Nadu, Andhra Pradesh, Kerala and parts of Maharashtra come under the developed rural India but the rest of the states are undeveloped where power, infrastructure, etc. are big problems. The prosperity of Kerala has come from the NRI income and not from agriculture. Today, there is hardly any village in Kerala. Tamil Nadu is prosperous as power and good roads are available. All the villages with proper infrastructure have developed. In such villages, people also have better access to towns and cities.

What are the Major Reasons for the Change in the Lifestyle of the Developed of Rural India

Television has done wonders to rural India. Today, especially in the south, the penetration of satellite television is very high, which is around 50 per cent unlike 25-30 per cent in the rest of the country. These people may not be literate in the true sense but they know what is happening around the world because of television. They know how the rest of the country live. Another influence is due to the growth of rural youth. The new forces of social change, politicization and media exposure have changed youth's values and social psychology throughout the country from villages to the town and to the metropolis. The globalization of the economy and the decentralization of the production of goods and services by the multinational corporations have totally altered for these youth the project of their future and the perception of their Life-World.

Rural-Urban Disparities

Of the total estimated gross income generated in the household sector, the share of rural income accounted for 55.6 percent with 74.6

percent of country's population which was 66.8 percent in 1975-76 with 79.1 percent of population. The rural share of income seems to have gone down during the last two decades resulting in wider disparities in income distributions between rural and urban India. On average one in every 25 households in rural areas has an income of more than Rs. 72,000 while every fourth household in urban India rose and in the highest income brackets more than two-thirds of the households reside in cities and towns.

Strategies

Dynamics of rural markets differ from other market types, and similarly rural marketing strategies are also significantly different from the marketing strategies aimed at an urban or industrial consumer. This, along with several other related issues, have been subject matter of intense discussions and debate in countries like India and China and focus of even international symposia organized in these countries

Rural Markets and Rural Marketing Involve a Number of Strategies, which include:

- Client and location specific promotion,
- Joint or cooperative promotion,
- Bundling of inputs,
- Management of demand,
- Developmental marketing, and
- Unique Selling Proposition (USP).

Client and Location Specific Promotion

Involves a strategy designed to be suitable to the location and the client.

Joint or Co-operative Promotion

Strategy involves participation between the marketing agencies and the client.

'Bundling of Inputs'

Denote a marketing strategy, in which several related items are sold to the target client, including arrangements of credit, after-sale service, and so on.

Management of demand involve continuous market research of buyer's needs and problems at various levels so that continuous improvements and innovations can be undertaken for a sustainable market performance.

Developmental Marketing

Refer to taking up marketing programmes keeping the development objective in mind and using various managerial and other inputs of marketing to achieve these objectives.

Media

Both traditional as well as the modern media, is used as a marketing strategy.

Unique Selling Propositions (USP)

Involve presenting a theme with the product to attract the client to buy that particular product. For examples, some of famous Indian Farm equipment manufactures have coined catchy themes, which they display along with the products, to attract the target client, that is the farmers. English version of some of such themes would read like:

- The heartbeats of rural India,
- With new technique for a life time of company, and
- For the sake of progress and prosperity.

The Basics of Rural Marketing

Marketing in rural areas has a lot in common with warfare. Both are challenging in nature and one cannot win over unless he has a sound knowledge of the terrain weather conditions and above all the main target, the mind of the consumer or the enemy as the case may be.

- *Simplicity in product strategy and designing.*
- *Simple and well defined goals.* What is your ultimate aim and correct idea of the aim?
- *Theme Selection.* A simple and deep penetration theme aimed at the mind of the rural consumer.
- *Synergistic survey and research.* Marketing research as we may term it.
- *Direct relationship with mass.* Rural marketing has to be a system that the mass beckons you whole-heartedly.
- *An effective public relation.* Can be linked to the above point as well.
- *Feedback.* Do it before you launch yourself, they will give you the actual on ground requirement and not in thin air.

A Key to Rural Marketing

Rural marketing by and large is simple provided you have the aptitude for it without utlising high-sounding tooth breaking termonoligies. In Hindi there is a popular word called *TARIQA* meaning the way to execute.

T—Truth-based; A—Assurance; R—Reliability; I—Ingenuity; Q—Quality; and A—Acceptability.

Challenges in Rural Marketing

Though rural markets are a huge attraction to marketers, it is not easy to enter the market and take a sizeable share of the market, in the short time due to the following reasons

Population Spread—The spread of population is to a great extent and even at times a house of ten forms a hamlet leading to problems for the marketer to reach out.

Infrastructure—Most of the rural areas are devoid of proper roads and lacks of connectivity.

Income—this is once again comparable and area specific.

Terrain Factors—Terrain plays an important role in rural marketing and poses substantial challenges to a marketer in execution of task. This attribute of terrain availability in India has to be clearly understood before venturing out to these areas.

Low Literacy

There are not enough opportunities for education in rural areas. The literacy level is as low (36%) when compared to all-India average of 52%.

Seasonal Demand

Demand for goods in rural markets depends upon agricultural situation, as agriculture is the main source of income. Agriculture to a large extent depends upon monsoon and, therefore, the demand or buying capacity is not stable or regular

Transportation

Many rural areas are not connected by rail transport. Kacha roads become unserviceable during the monsoon and interior villages get isolated

Distribution

An effective distribution system requires village-level shopkeeper, Mandal/Taluka-level wholesaler or preferred dealer, distributor or stockiest at district level and company-owned depot or consignment distribution at state-level. The presence of too many tiers in the distribution system increases the cost of distribution

Communication Problems

Facilities such as telephone, fax and telegram are rather poor in rural areas.

Traditional Life

Life in rural areas is still governed by customs and traditions and people do not easily adapt new practices. For example, even rich and educated class of farmers does not wear jeans or branded shoes.

Buying Decisions

Rural consumers are cautious in buying and decisions are slow and delayed. They like to give a trial and only after being personally satisfied, do they buy the product.

Media for Promotions

Television has made a great impact and large audience has been exposed to this medium. Radio reaches large population in rural areas at a relatively low cost. However, reach of formal media is low in rural households; therefore, the market has to undertake specific sales promotion activities in rural areas like participating in *melas* or fairs.

Cultural Factors

Culture is a system of shared values, beliefs and perceptions that influence the behaviour of consumers. There are different groups based on religion, caste, occupation, income, age, education and politics and each group exerts influence on the behaviour of people in villages, there is a belief among rural people that experience is more important than formal education and they respect salespersons who can offer practical solutions to their problems. Therefore, it is desirable that sales persons, especially those who have been brought up in cities are given a thorough training consisting of both theory and practical aspects of village life. The training will help these sales persons to align themselves with the market realities and settle down smoothly in their jobs. Rural market has a tremendous potential that is yet to be tapped. A small increase in rural income, results in an exponential increase in buying power.

Prevailing Constraints for Growth

- Inherent Weakness,
- Lack of latest Updated Technology,
- Poor Infrastructure,
- Labour Policy,
- Access to Foreign Markets,
- Different approaches in tapping foreign goods,
- Purchasing power of domestic market, and
- Political intervention in corporate decision.

CONCLUSION

Tourism plays an increasingly important role in the development of communities. The benefits of tourism include both tangible (e.g. job creation, state and local tax revenue, etc.) and less tangible (e.g. social structure, quality-of-life, etc.) community effects. In addition, tourism can, and often does, result in less desirable effects on the economic, social, and environmental fabrics of communities. These benefits and costs provide ample opportunity for creative public policy debate. But still we can see that most of the rural areas are untapped. Thanks to television, today a customer in a rural area is quite literate about myriad products that are on offer in the market place. An important tool to reach out to the rural audience is through effective communication. A rural consumer is brand loyal and understands symbols better.

REFERENCES

A New Approach to Rural Marketing by "Lt. Col. Kaushik Sircar, First Edition, 2004", Himalaya Publishing House.

Article Source: http://EzineArticles.com

Article Source: written by an Rtd. Civil Servant, Lalit Mathur—Director General of the National Rural Development, Ministry of Rural Development, Govt. of India, Published in *The Hindu Magazine*, Dated 5-4-2008.

Information from KTDC (Ernakulam).

Journal Source: Understanding the Rural Markets and Exploring the Scope for Consumer Market Segmentation in Rural India, by Omkumar Krishnan and Barada Panigrahi (Faculty Members, ICFAI Business School), *ICFAI Journal*, February 2006.

Rural Marketing: Text and Cases by "C.S.G. Krishnamacharyulu, and Lalitha Ramakrishnan", Published by Pearson Education (Singapore) Pte. Ltd., Indian Branch, 482, F.I.E. Patparganj, Delhi, Second Indian Print, 2003.

http://ernakulam.nic.in/

http://www.indianetzone.com/9/tourism_ernakulum_district.htm

CHAPTER

34

Entrepreneurship Development in J&K State

DR. AMRIK SINGH SUDAN AND DR. RAIS AHMAD

INTRODUCTION

Entrepreneurship is the most important factor in the economic development of the country. It plays a very important and catalytic role in activating the factors of production leading to the overall economic development. The Entrepreneur organizes and utilize the various factors and sets productive machinery in action. The availability of entrepreneurs, therefore, is the precondition for economic growth. The supply of entrepreneurs depends on the rate of profit and social climate. It is profit that induces the prospective entrepreneurs to get into the business and start new activities or expand the existing activities. Profit, therefore, is a factor which induces the entrepreneur to organize and utilise the factors of production for development. Entrepreneur is an innovator with a certain achievement motive and that achievement motive may also mean something more than money. Social climate also plays its role in attracting prospective entrepreneur to innovate, assume risk and face the music of uncertainty. There is high degree of risk and uncertainty in economic world and in this context entrepreneurial behaviour is important as development is a discontinuous process which itself proceeds with spurts.

The entrepreneurship plays an important role in the economic development of the country. It creates wealth for the nation. It provides means of livelihood by creating job opportunities. The standard of living of the people is also linked to the growth of entrepreneurship. Entrepreneurship plays a dominant role in the growth and development of an economy.

Entrepreneurship can solve the problems like unemployment, concentration of economic power in few hands, in balanced regional development, increasing wastage of youth power in destructive activities. Entrepreneurship is the ability to start a new enterprises to earn profits by producing or marketing of goods and services to meet the needs and requirements of customers. It is the ability and quality of entrepreneur to identify an investment opportunity and to organise an enterprise in order to contribute for the real economic growth. Entrepreneurship lies more in the ability to minimize the use of factors of production such as land, capital labour and to exploit them to maximize advantage. Entrepreneurship largely depends upon the personal qualities like accepting the challenge, bearing the risk, creativity organizing and coordinating process of entrepreneur.

The word entrepreneurship has been derived from a French language which means to undertakes. Today it is called by various names, e.g. adventurism, risk taking, thrill seeking, innovating, etc. by different people.

Entrepreneurship is, thus, a cycle of actions of an entrepreneur undertakes to start a new enterprise. It is the process of giving birth to new enterprise.

Entrepreneur play a key role in the process of economic development of a nation. They are, in fact, the engine of growth and considered to be catalytic agents for promotion and expansion of productive activities in every sphere of economic life of a developing country. The entrepreneurs, with the help of their strong initiative, drive skills and spirit of innovation can bring significant changes in the process of economic development. They not only look for opportunities but identify opportunities and capture opportunities mainly for economic gains. They are highly motivated individuals who are always action-oriented and prepared to take all possible risks to attain their economic goals. The process of economic development is closely connected with the process of social change in a country and the role of entrepreneurs in this context cannot be overemphasized. The word entrepreneur is derived from the French word entreprendre which means to undertake. This word was first used by a Frenchman in the early sixteenth century and was applied to persons engaged in military expeditions. It was further extended to cover construction and civil engineering activities in the 17th century. It was only in 18th century that the word was used to refer to economic activities. Richard Cantillon, a French economist used the term of "entrepreneur" to business for a person who bears uncertainty and risks. According to Cantillon, "an entrepreneur is the agent who buys factors of production at certain prices in order to combine them into a product with a view to selling it at uncertain prices in future".

Development in the last two decades specially in the economic sphere has knitted the world economic order in a new fashion holding a paradigm shift in trade equations across the globe. The world is witnessing acceleration of integration of the global economy and the emergence of an increasingly complex production system.

Entrepreneurship has its unique role to play in the economic development of a nation. It is specially so, in case of a developing economy where unemployment, poverty and inequality, plague of process of economic development. The only alternative is self-employment and its contribution for a better future. In India, until recently the small scale industries were considered to be a panacea to socio-economic ills, and accordingly items were exclusively reserved for this category of industries.

The government of India also felt the need in envisaging a greater role for entrepreneurship development activities in nation-building in the Eighth Five Year Plan, and the various state governments and financial institutions, etc. are also in the process of setting up a few more entrepreneurship development organizations in the states where they do not exist.

Since the unemployment position is India. Specially among educated youths has been alarming and is getting worse with addition of number of graduates and post-graduates from Colleges, Universities and Technical Institutions. It is very serious in rural areas where people are idle about six months in a year and women contribute only household activities. Hence, unemployment situation is severe both in rural and urban areas. To provide employment opportunities instead of governments jobs, Government of India established in July 1983 National Institute for Entrepreneurship and Small Business Development (NIESBUD) an apex body for coordinating the activities of various entrepreneurship development organizations in India. In this regard, about 1500 organisations are providing necessary vision and guidance for entrepreneurship development. In view of mounting unemployment, National Institute for Entrepreneurship and Small Business Development is offering entrepreneurship as an alternative carrier/profession and gradually introducing it not only in colleges but also in school-level as a subject. Besides, guidance and training, National

Institute for Entrepreneurship and Small Business Development had instituted a Post-graduate Diploma in Entrepreneurship (PRDE) with a view to prepare professionally qualified trainers.

During the last three decades many countries of the world have resorted to entrepreneurs development as a means to rapid economic development resulting in creating of gainful employment for masses. The myth hat entrepreneurs are born and not created seems to be slowly fading away.

Entrepreneurship is an important determinant of industrialization process and entrepreneurs have significant contribution in the development of modern but innovative society. The changing entrepreneurial behaviour plays a vital role redefining the social structure in an economy. The Indian entrepreneurial environment is shaply going through the state of transition from the traditional pattern to the highly diversified global level over the last decade, there has been a great shaft in national pride for the better. In the year 1990, national pride emphasized our past achievements. Today, its about our confidence in the future. In the past, we asked in the glory of our forefathers, but today we perceive ourselves as creators of new history directing our own destines. The "Made in India" Label is no longer something that we need to defend.

Despite the difficult environment, the Indian entrepreneurs are bravely setting up new standard in the country and contributing to the drastic change in the modern Indian Society. Entrepreneurship is an emerging movement that will hope fully continue to grow and prosper. According to U.S. National Intelligence Council's report that India will become global player in the 21st century and that this change will be manifest by 2020. Future will see entrepreneurship as the key driver of economic development because it helps to create jobs, even if self-employment only. This may help to increase the graph of employment and create new ventures for overall development of the country. All forms of entrepreneurial activity increase the exposure of the entrepreneurial to market opportunities and give them valuable entrepreneurial experience what we need is the scaling up of experiment, a mass movement take India to a developed country status by 2020.

There is a great need of entrepreneurship to mitigate the ever increasing menace of unemployment confronting the whole country including the State of J&K. India faces the challenge of generating 200 million new employment opportunities over the next two decades. This is a Herculean task on the part of any state or Central Government to take serious initiative in this regard. Private Sector in general and individual entrepreneurs in particular have to shoulder lot of responsibility in this fight against the monster of unemployment.

Entrepreneurial ecosystem in J&K is at a very nascent stage. The long drawn militancy and the disturbed political environment have eroded the skill base of the youth in the State. J&K economy is a high cost mountain economy and has a number of characteristics that pose special development challenges. The State is especially vulnerable to external events, natural disasters, that cause high volatility in its State Domestic Product; and it suffers from limited capacity in the public and private sectors. More specifically, the following characteristics define the special development challenges and vulnerabilities that J&K faces:

(a) Remoteness and Isolation

Landlocked, and located far from major markets. For J&K high transport costs make it harder to turn to major markets to compensate for the drawbacks of the small size of its domestic markets.

(b) Limited Capacity

While a weakness in both public and private sector capacity is a key problem for J&K, size has added a further dimension to the challenge. This is further compounded in J&K where the internal distances are large and the population is scattered. J&K does not possess sufficient institutional capacity to participate fully in national finance and trade activities, the outcomes of which can profoundly affect its economy. In the private sector, as already noted, lack of diversification and domestic competition has held back successful development. Many of these factors combine to make J&K economy especially vulnerable, and in particular they affect:

(i) Income Volatility

Overall, the range of per capita incomes and rates of growth are not significantly different across Indian state economies. However, the residents of J&K experience higher volatility in their incomes.

(ii) Access to External Capital

The access to national capital markets is very limited for J&K. There is just one listed corporate in the State. Most private sector enterprises are still in the primitive stage of organization. The fact is that private markets tend to see J&K as more risky than other States, as the spreads are wider and market access more difficult.

(c) Limited Diversification

Because of its small domestic markets, J&K is necessarily relatively un-diversified in its production structure and exports. This adds to vulnerability to changes in the external environment.

(d) Susceptibility to Natural Disasters and Environmental Changes

Of late, J&K has become susceptible to natural disasters such as earthquakes and drought which typically affect large parts of the population and economy. These events cause income volatility which creates additional hardship as the poor are less able to withstand negative shocks to their incomes.

While examining the characteristic features of the economy of J&K State, one comes to the conclusion that the State is economically backward as compared to the other States of the Country. In view of this state of affairs, it is highly essential to diversify its economic structure by developing industries so that the socio-economic conditions of the state are improved. It will not only play a significant role in providing employment to a large number of educated youths, but also give solution to the unemployment crisis through which the state is passing these days.

Unemployment is a social issue of serious concern in the present times, both at National as well as state-level. J&K State has been facing the problem of unemployment since long. According to 2001 census, the number of total workers stands at 37.54 lakh of which the main workers constitute 26.09 lakh or 69.49% and the number of marginal workers constitutes 11.45 lakh or 30.51%. Out of 11.45 lakh of marginal workers, half of this working force suffers from disguised unemployment.

At state-level, the number of registered illiterate unemployed in the year 2006 was 9363 where as the number in the year 2007 has been decreased to 4030 thereby registering negative growth (retardation) of 56.98 percent. The number of literate unemployed (literate below Matric) has recorded a marginal decrease of about 1.8 percent during the year 2007. Unemployment among the registered educated people had recorded a rising trend. The educated unemployed population has been recorded as 74,630 in the year 2006 and 82,618 in the year 2007 recording an increase of 10.70 percent. The overall number of registered unemployed in the state during 2006 was 1,09,383 persons which have increased to 1,11,564 persons in the year 2007, showing an increase of about 1.99 percent. This clearly indicates the growing problem of educated unemployed population in the State.

The total population of the state has now increased to 1.25 crore with 6 lakh unemployed youth registered in District Employment and Counseling Centers. The objective of the State Government of providing at least one productive job per family is quite hard to achieve primarily through providing at least one productive job per family is quite hard to achieve primarily through providing Government jobs. Thus, employment opportunities are greatly required to be created through entrepreneurship development initiatives. The state Government has clearly envisaged entrepreneurship development in the state as a tool to provide opportunities to educated unemployed. Lack of entrepreneurial skills of the local youth has been identified as an impediment for sustained industrial growth. Thus, to cope up with the unemployment crisis in the State, the Government of J&K has entrusted Jammu and Kashmir Entrepreneurship Development Institute (JKEDI) with the responsibility of inculcating entrepreneurship spirit in educated youth and pave way for creation of large scale job avenues. Underlining significance of Sher-e-Kashmir Employment and Welfare Programme for youth (SKEWPY) in addressing the unemployment crisis in the State, Chief Minister, Omar Abdullah has said that, "We see our youth beset with the fashion of pondering for Government employment, even if it might be of lower standard", he said adding that this fashion has to give way to Micro Small and Medium Entrepreneurship (MSME) culture to evolve a viable and suitable solution to unemployment problem.

PROBLEMS

Lack of entrepreneurship is generally attribute to economic pressures and socio-economic inflexibilities. Factors like lack of aptitude, family background, kinship, customs, absence of creative spirits, lack of freedom to choose profession, lack of education, etc. act as obstacles in the growth of entrepreneurship. There are some extrinsic factors which also obstruct the growth of entrepreneurship in the State and these are:

- Poor public infrastructures;
- Unstable economic environment;
- Poor Marketing of goods and services;
- Sourcing raw material from outside state;
- Non-Availability of finance;
- Lack of Training and Development Program of budding entrepreneurs;
- Lack of Research and Development Activities; and
- Poor institutional support system for entrepreneurship development.

Moreover, J&K is a Land Locked State and is severely handicapped due to locational disadvantages. This result in increase in the cost of production as compared to our neighbouring states. Thus, geographical location of the state is detrimental in venture set-up. The shocks of the industrial sickness experienced by the entrepreneurs in the past have also contributed in suppressing the morale and confidence of the unemployed youth to venture for in the risky and innovative projects. Procedural hassle in the setting up of enterprises is one more pebble in the growth of entrepreneurship. Even today the single-man unit is subjected to an ample number of procedures and formalities which unnecessarily de-motivate the potential entrepreneurs to go venture set-up.

Our national policy for boosting the economic growth takes into account the promotion of indigenous industries as first and the foremost task. Rapid growth of the entrepreneurship and thus, indigenous enterprises is the need of the hour. Industrialization in J&K State has not many feathers in the cap agreed. However, gradually it is pitching up despite having topographical disadvantages. With the help of Central Government incentives along with state incentives given to industries, the scenario is fast changing.

The objective of the State Government of providing at least one productive job per family is hard to achieve primarily through providing government jobs but employment opportunities need to be created primarily through two routes;

- Accelerating the rate of economic growth of the State.
- More effective implementation of employment-oriented schemes and programmes (broad-based inclusive growth schemes).

The following 8 sectors of economy have been identified for generation of employment:

- Agriculture (including Horticulture, Floriculture, Food Processing and Animal Husbandry),
- Handlooms and Handicrafts,
- Industries (including Small Scale Industries and Rural Industries),
- Tourism & travels,
- Education and health,
- Large infrastructure projects (Roads & Railways),
- Information Technology & Telecommunication, and
- Construction Sector.

The economy of Jammu & Kashmir has suffered from disturbed conditions prevailing in the State for almost two decades. It would, therefore, be necessary to put the economy back on the rails to enable the average person to get employment opportunities. This would require giving fillip to the economic activities that have traditionally been the mainstay of the State's economy and continue to hold significant potential for growth and employment. Such activities include Agriculture (including Horticulture), Food Processing, Handicrafts and Handlooms, Tourism, etc. It would be equally necessary to ensure diversification of the State economy, especially expanding the industrial base by promoting private capital inflows into the state through various incentives and concessions in the initial stage.

Agriculture constitutes an important sector of the State Economy as more than half of the population of J&K derive its livelihood directly or indirectly from this sector. As per census 2001, 18.389 lakh persons comprising 15.92 lakh as cultivators and 2.46 lakh as agricultural labours depend directly on agriculture for their livelihood forming 49% of the total working force (37.54 lakh) of the State. There is a tremendous scope for setting up agro-based industries in the state for those products where raw material require is locally available. This sector can also provide additional employment for the local youth in packaging, marketing and transportation of the products.

Horticulture is another potential sector for income as well as employment generation. With an annual turnover of fruits and dry fruits of the order of Rs. 2000 crore, horticulture plays a significant role in the economy of the state. Around 6 lakh farm families comprising around 25 lakh people are connected directly or indirectly with this sector. Area under horticulture at the end of 2007-08 was 2.95 lakh hectares in the state. Almost 45 percent of economic returns under Agriculture sector are contributed by Horticulture sector, thereby adding 7 percent to the Gross State Domestic Product. Developing this sector will not only increase direct employment but also increase employment in related activities such as servicing of inputs, packaging, transportation of fruits, etc.

Handloom and handicrafts is one of the traditional sectors of the state employing more than three lakh people. Handicrafts Sector engages approximately 3.50 lakh artisans in manufacturing of Handicrafts in the state. Production of Handicrafts goods for the year 2007-08 stood at Rs. 1614.59 crore out of which carpet alone has a contribution of 47% with Rs. 761.27 crore. During 2007-08, handicraft goods valuing Rs. 1200.47 crore were exported of which carpet accounted for 54%. As on March 2008, there were 598 handloom units in the state providing employment 23.84 percent. The department runs 100 training centers. About 500 trainees are provided technical skills annually. Production of Handloom fabrics during the year 2007-08 in the Cooperative Sector was 5.84 m/mtrs., approximately valuing Rs. 31.59 crore. Emphasis on improving designs, colour combinations and marketing can generate additional income and employment in the State.

Tourism has been a major source of income and employment for the local people in the state. Tourist inflow to J&K has continuously been increasing steadily from 69.12 lakh in 2004 to 79.28 lakh in 2007. About 77.07 lakh tourists visited J&K during 2008. Out of the total tourist visitors to J&K, about 85% to 90% tourists visit the shrine of Shri Mata

Vaishno Devi Ji annually and hence pilgrimage tourism is an important sector. The share of tourist visitors to Shri Amarnath Ji Shrine cave is also quite significant and stood at 4.98 lakh for the year 2008 and 5.37 lakh for the year 2010 and it is expected to cross 8 lakhs in this year, i.e. 2011. It is roughly estimated that 30% of the people get direct or indirect employment from Tourism sector. Employment in Hotels and Restaurants is about 13,785 of which 9869 are employed in urban areas and 3916 are employed in rural areas. However, this sector has been the main casualty due to militancy. By reviving the tourism industry, the state can re-establish those rendered unemployed. Thus, to revive the tourism industry, 19 Tourism development authorities have been constituted which are functioning in the State. At present, 3 tourist circuits are being developed and 50 villages have been identified to promote tourist activities. These are proposed to be prepared on international standards with all ultramodern facilities and infrastructure.

Developing sericulture can also generate employment opportunities in the state. This sector has the potential of absorbing the entire family and becomes an important source of employment for female workers who otherwise would remain unemployed or under-employed. However, there is an urgent need to develop this sector through improved silkworm rearing, mulberry cultivation, etc. Apart from sericulture, there is a tremendous demand for dairy, poultry and fishery products. Developing this sector can also boost employment generation in the State.

Information Technology (IT) and Telecommunication are the new emerging knowledge-based industries which have high employment potential. Increased application of Information Technology (IT) in various sectors can enhance employment and income generation. Thus, the future success of J&K State will depend on creating excellence employment in the above mentioned sectors, both in the emerging and the existing categories.

CONCLUSION

In conclusion, it may be stated that the growing unemployment rate in the J&K State is due to our failure in not developing the spirit of entrepreneurship among the educated youth. The spirit needs to developed among the youth during their School, College and University days.

After a student comes out of College/University with a degree/certificate in hand, he/she does not know what to do with it. He/she has in his/her mind a Government job, which most of the youth do not get in the present times. Had he/she been imputed technical skills during his/her academic career, it would have become easy for him/her to start a job of his/her own liking. That system does not prevail in this state, students are taught such subjects which do not help him/her in getting a job in private sector, where a skilled worker is preferred over a non-skilled worker.

The Government must take such initiatives which may enable students to get self-dependent.

In the backdrop of above discussed sectors which can generate number of jobs for unemployed youth in the State it can be conclude that entrepreneurial development initiatives should be encouraged.

References

Desai, Vasant, "Dynamic of Entrepreneurship Development and Management, Himalaya Publishing House, Bombay, 1992.

Khanka, S.S., "Human Resources and Entrepreneurship in Economic Development," Himalaya Publishing House, Bombay, 1988.

Rao, N.G., "Entrepreneurship and Growth of Enterprise in Industrial Estates", Deep & Deep Publications, New Delhi, 1986.

Samiuddin, "Entrepreneurship Development in India", Mittal Publication, New Delhi, 1989.

Singh, Jagroop and Parray, M.A., "Entrepreneurship", Kalyani Publisher, Ludhiana, 2010.

"*The Indian Journal of Commerce*", The Quarterly of the Indian Commerce Association, Vol. L, Part IV, No. 193, December 1997.

CHAPTER

35

Impact of Tourism on Indian Economy

DR. AMRIK SINGH SUDAN AND MS. LEKHA NAGAR

INTRODUCTION

Tourism has been traditionally viewed as a great force in promoting understanding among nations and within the national boundaries facilitating emotional integration. Tourism is the movement of people from their normal place of residence and work, for a period of not less than twenty-four hours and not exceeding a year. Tourism is the most important industry in our economy without chimneys. Over the past three decades, tourism has emerged as the world's largest industry. Tourism is one of the most important items of export. It is well said by Husker and Kraft, that, "Tourism is the sum of the phenomenon and relationships arising from the travel and residents in so far as it does lead to permanent residence and is connected to any earning activity". Today, we all know about the tourism but before the independence, in India tourism industry was not promoted by the Britishers, but soon after independence, the importance of tourism in India was felt vigorously by the Indian Government. They set-up a committee under the chairman of Sir John Sergeant. This committee submitted its report in 1948. It was thought at that time, it was beneficial for India to encourage and develop the tourist traffic by all means. A small tourist section was set-up in the

Ministry of Transport in 1948. A separate department was set-up early in 1958 which is called Department of Tourism. Tourism at that time was a new activity for India. Its importance was not realize by the people and Government of India. This was so because the scope of tourism was low.

There was not proper infrastructure, such as the number of hotels was very less. There were no proper means of accommodations of the tourists, same was the case of transport also. But now-a-days, people have realized the value of Tourism and the economic benefits of tourism. In India, now tourism Industry is developing a lot and this has increased the nation's income also.

Tourism has become a popular activity not only in India but across the globe. It is one of the major revenue generating industries. If the international statistics are evaluated, it is seen that there were about 980 million international tourist arrival in the world, and the projections, says that this number would go upto 1.8 billion for 2030. As of the contribution of tourism to Gross Domestic Product (GDP) is concerned globally, it is about 10% and this industry is contributing about 8% to the employment (in absolute number, it generated about 235 million jobs in 2011)..But, if these figures are compared with Indian Scenario, it is seen that India's share in world tourist arrivals is just 0.6%, i.e. about 6 million tourists. And its contribution to Gross Domestic Product (GDP) in India is about 6.2% out of the global tourist arrivals. India is ranked at 41st place. Tourism is one of the third largest net earner of foreign exchange for the country and also one of the sectors, which employs the largest number of manpower. Looking at many benefits of tourism in India, the present union Minister of Tourism, Govt. of India has projected to increase in share of India in the world tourists arrivals from 0.6% to 1% by 2016. But this target cannot be achieved if SWOT analysis of India as a tourist destination is not done.

SWOT ANALYSIS OF INDIA AS A TOURIST DESTINATIONS

(A) Strengths

India as a tourist destination has got many strengths, particularly its diversity in culture, all kinds of products, be it religion, heritage, adventure, Sports, Cultural, Cuisine, ethnic, etc. All forms of tourism are possible in India. It has such climatic diversity also that at the same time in Kashmir, there would be snow but in Kanyakumari, there will be immense heat. It has such product diversity that in Rajasthan, one will find beautiful colours and in Assam, one would get the view of awesome tea gardens. Fairs and festivals, clusine, architecture, temples, monuments, museums, leisure, tourism Skikara, back waters, hill stations, adventure activities, beaches, wild life, music, varied attires, languages and what not is there in India and that is what which give the strength to our nation as a tourism destination.

(B) Weaknesses

If India as a country has so many strength, then what is making it weak to attract only a fraction of tourists among the total world tourist arrivals. There are two major factors, i.e.

(i) Safety and Security

India as a destination is not considered to be safe out of the 7S's framework, i.e. Swagat, Sochna, Safai, Suvidha, Sanrachna, Suraksha, and Sahyog. National tourism policy in 2002 has given special importance to Suraksha and for that Government of India has adopted the code of conduct for safe and honorable tourism on July 2010.

(ii) Poor Infrasturcture

India ranks 68th among 139 countries in the global travel and tourism competitiveness in den which is surveyed by the World Economic Forum. This forum conducts survey every after 2 years. India slipped sin positions to 68th from 62nd in 2009. And it is conducted that India is in this number mainly because of the poor infrastructure that it has quality of roads, quality of Airports, hotel rooms, etc. needs to be upgrade to the world class set-up if we have to attract more tourists.

(C) Opportunities

Since India has the diverse product range, there is an inmnervre opportunity for it to have new paradigms in tourism like MICE as India is already projected to become 5th faster growing business travel destinations from 2010. Nineteen (19) Medical tourism as India's share, in the global medical tourism industry will reach around 3% by 2013, Yoga and meditation and so on and so fourth.

(D) Threats

If India has got many strengths and opportunities, it has a lot threats too like threat of terrorism/militancy as already there is an issue in India of safety and security. Terrorism is always a major threat to tourism in the country. Better infrastructure in neighbouring countries like China which has already created a mark in the world tourism map, etc.

(E) Though on one side, Government of India is taking a lot of initiatives to boost tourism in India, yet on the other side not much is been considered while developing the budget. The various initiatives of

Government of India included tourist visa on arrival for 11 countries, effective marketing outside the country, more allocations of budget to Ministry of Tourism in Union budge, etc. Still tunes on accommodations and air travel would surely lower down the tourism in the country as the services tax on hotel accommodations in excess of declared tariff of Rs. 1,000 per day. Then the revision of service tax rates on air travel shows that there is a hike in domestic international fares.

In the early 1980s, Gantzer and Gantzer, wrote about an increasing awareness among government officials of the need to redefine tourism in order to make a clear distinction between... 'the functional business of travel and the essential escapism of tourism' (1983:125). Regretfully, this increasing awareness didn't lead to a change in definitions. Both on an international and national level, relatively straightforward all-travel-inclusive definitions have become the general norm.

The U.N. World Tourism Organization (WTO) and the World Travel and Tourism Council (WTTC) define tourists as people who 'travel to and stay in places outside their usual environment for more than twenty-four hours and not more than one consecutive year for leisure, business and other purposes not related to the exercise of an activity remunerated from within the place visited'. The Indian Department of Tourism's definition is almost the same. It explicitly includes people traveling for the following reasons: leisure-recreation, holiday, health, study, religion, sport, business, family mission and meeting (GOI, 2008: 257-60).

THE GROWTH POTENTIAL OF INDIAN (INTERNATIONAL) TOURISM

Let's briefly forget about the definitions and statistics, and simply start from the idea that inflated numbers are a world-wide phenomenon and, hence, that India received 5 million foreign 'tourists' in 2007. Is this a significant number? It equals the *monthly* number of tourists visiting Paris, the capital of France, and the yearly number of tourists who stayed in a hotel in Amsterdam, the capital of Holland (750,000 inhabitants). These findings serve to put things in perspective. Indeed, even with the help of inflated statistics, in 2007, the second most populated country in the world attracted only 0.56% of all international tourists.

The Indian Government wants us to believe that this situation can be structurally and significantly altered. While it argues that the variety of tourism activities need to be differentiated and improved, and that the tourism infrastructure requires augmentation, it seems convinced of the idea that the things India has to offer, merit a much larger influx of international tourists. There is, however, no evidence to substantiate this view. In most cases it is based on some sort of nationalist conviction, which wants 'India to attain its proper place in the global tourism scene' (e.g. Raguramam, 1998). It is true: India can be regarded as an old civilization, with a rich history, a number of important heritage sites, mountains, beached, deserts and the like. Offers superior sights; and (b) tourists would be drawn to such sights if they were properly informed and seduced?

In the early days of independent India, quite rationally, the Government didn't pay much attention to tourism. Worldwide, the number of international tourists was still limited and among those tourists there were only few who considered going to faraway places like India. Moreover, the Indian Government had more urgent matters at hand. The first public milestone in the history of tourism was the creation of the Indian Tourism Development Corporation (ITDC), in 1996. This federal organization was meant to develop tourist infrastructure and services. On a state-level, similar Corporations were established, albeit reluctantly and after considerable delay. Their budget were small and the scales of their operations were limited. Moreover, the tourist services they offered were generally considered substandard and indifferent.

This held true for different states to different degrees. In Kerala, Jammu and Kashmir, Goa and Himachal Pradesh tourism was much more consistently incorporated into state (five year) plans than in other states, even relatively early on—in the 1970s. At that time, these states already attracted a relatively large share of India's international tourists and were to become India's main tourist states and 'models for tourism development' (Singh, 2001:143). Still, even in these states, tourism was largely left to the private sector and to a comparatively unrestricted market. While there have been slight modifications in policy, this basic condition hasn't changed much until the present day.

TOURISM DEVELOPMENT IN THE 1990s

In the 1980s, the growing public interest in tourism resulted in the recognition of tourism as an export industry (including the implied tax exemptions) and the creation of a special public tourism finance corporation (1987). These initiatives were thought to openly and explicitly invite private investors and entrepreneurs to participate in tourism development. In 1997, the department of tourism published a (new) National Tourism Action Plan. Apart from identifying a few areas for 'integrated tourism development', along the lines of the aforementioned (thematic) tourism circuits, the aim of the plan was to achieve an overall growth and improvement of the tourism sector in India, by stepping up marketing, infrastructure building and human resource development. According to some, the plan didn't present anything new. It just was phrased in a more fashionable development sector jargon (Singh, 2001:144). Others maintained that the plan was over-ambitious and unrealistic. Funding by no means matched the challenging quantitative targets (Raguraman, 1998:535). In fact, from independence onwards the budget outlays for tourism have always been very small (less than 0.2%).

This goes for India's share in worldwide international tourism too. During the first fifty years of independent India, the portion of international tourists visiting India has not been more than 0.4%. It is claimed to have risen to some 0.45% in 2004 (GOI, 2004). Given the rise of international tourists in the world, this implies that there has been a steady increase in tourist arrivals which accelerated in 2004. Indeed, according to the Indian Tour Operators' Promotion Council (ITOPC), over the period 2001-06, the number of foreign tourists in India has nearly doubled (to some 4.4 million). While these figures are unrealistically high, they do who that in absolute terms the number of international tourists visiting India has increased considerably. The increase in tourism arrivals was aided by the growing popularity of far-off and exotic destinations among Western tourists and recent advances in the general outlook and international image of India. The opening its markets in the early 1990s and the recent years of high economic growth have profoundly changed its surface. In the past India was naturally associated with slums and poverty. Now-a-days it is often portrayed as a rapidly emerging economic superpower. The rising economic significance and potential of tourism has gone hand in hand with a growing public interest in the sector. With the opening up of the Indian market, a lively competition between the states emerged in attracting investors in industry and other sectors. Along similar lines, states started competing for their share of international and domestic tourists. Particularly in the traditional tourists states there was an urge to develop tourism to its full potential. A notable example is Kerala, a relatively small state with a rich variety of natural tourist settings. The Kerala government took up a largely enabling role, supporting and promoting a great number of different tourist activities.

In the governments of other states increasingly follow Kerala's example. They do so by: (a) recognizing the income earning and job generation potential of tourism; (b) (more) actively promoting tourism through publicity campaigns and giving support and incentives to the private sector; and (c) largely putting private entrepreneurs in charge of the provision of tourist services.

The example of Kerala shows that Indian tourist destinations can outgrow the phase of basic backpacker enclaves referred to above and offer a more differentiated mix of products to a socio-economically much more varied group of tourists. Amongst these tourists, there is a large section of people from India, i.e. domestic tourists. The importance of domestic tourism was recognized by public policy-makers in the 1990s. The included it as an important issue in the Tourism Action Plan of 1997 and decided that it was a state government (policy) issue. The central government was to take care of international tourists. Traditionally, domestic tourism mainly concerned pilgrimage and work-related travel. From the 1990s onwards there has been a steep rise in modern forms of domestic tourism. This new phenomenon is related to the booming Indian economy and the new susceptibility of the Indian middle and higher classes to rather alien, Western ideas of Holiday making. At present, an ever growing group of Indian tourists travels around the country for rather prosaic, leisure and sightseeing–related reasons. This new trend in underscored by the emergence of Indian travel magazines and growing explicit attention for domestic tourist destinations in leading newspapers.

THE NEW TOURISM POLICY (2002)

In 2002, when the action plan was finally translated into a tourism policy. Tourism policy itself, however, was designed by the central government. To a large extent, it concerns old wine in new bottles. It holds the kind of goals and expectations exemplary for the first policy. To start with, the policy document attempts to establish tourism's great contribution to national development and its role as an *engine of growth.* It suggests that tourism not only generates government revenue, foreign currency, but also provides an optimal use of India's scarce resources, sustainable development, high quality employment (especially to youngsters, women and disabled people), and finally, peace, understanding, national unity and stability (GOI, 2002: 8-9). The policy starts from the idea that tourism can be used as a development tool, e.g. that it can generate high quality, mass employment and prosperity among vulnerable groups in backward areas.

In more practical terms, the policy aims at increasing the number of domestic and international tourists. In order to do this, the government proposes to diversify the Indian tourism product and substantially improve the quality of (tourism) infrastructure, marketing, visa arrangements and air-travel. The aforementioned *tourism-as-a-development-tool* largely concerns domestic tourism, which in this capacity is conceptually linked to 'sustainable' rural development. As far as international tourism is concerned the Indian Government mainly wants to target the '*high-yielding variety*' of tourists.

BACKGROUND: DEVELOPMENT SECTOR IDEOLOGY

'The participation of different private and public agencies makes the tourism industry a complex phenomenon requiring a strong cooperation and coordination for its to be developed and expanded along lines that will contribute to the overall national development objectives. Left to itself, the industry will develop naturally, but not necessarily optimally or sustainable, and without clear links to the broad development objectives of the country. Uncontrolled tourism growth cloud damage India's socio-cultural structure, degrade its tangible and intangible cultural and natural heritage, and lead to adverse economic impacts such as high importation costs, and weakening inter-industry linkages, (GOI, 2002:12).

SUSTAINABLE TOURISM IN PRACTICE: THE ENDOGENOUS TOURISM PROJECT

One of the general objectives of the Indian Tourism Policy (2002) is to...'substantially increase the proportion of the urban resident leisure and pilgrimage tourism to rural areas...'. It proposes the development of tourist services in villages and rural regions located off the beaten tourist track. To this effect, in collaboration with the UNDP, it initiated a public rural tourism program geared towards all round, instantaneous village development: the Endogenous Tourism Project/ Rural Tourism Scheme (2003-07). The project was meant to be implemented in 31 villages across 20 states. Most of these villages harbored traditional artisans (weavers, potters, sculptors, block printers and the like) who were thought to attract tourists. Local or nearby Non-Government Organisations (NGOs) were identified as the most important implementing agents in the otherwise 'community-owned' project. According to the program's website its focus is to develop the culture, craft and sustainable dimensions of rural life, a means to viable livelihood opportunities for low-income rural communities...

CONCLUSION

Speaking of the contentions of the present tourism policy we suspect that the stress on the long haul high yielding variety of international tourism is somewhat misplaced. Putting this issue in a broad development context, one might suggest that this type of international tourism development is a function of economic development, rather than the other way around. One can't change India for the sake of 'world class tourism development'. Rather, India will develop economically an along with this development, opportunities for various kinds of tourism will emerge. No doubt, in the near future, the greatest opportunities will arise in the domestic sector. From a research and policy perspective, the present roles of the Tourism Development Corporations in the states, identifying and developing new destinations, merits much greater attention. We firmly believe that many of the new 20 destinations created by these public agencies attracting mainly domestic tourists, will ultimately emerge as new hubs for both domestic and international tourists. Some serious thinking on purposeful guidance of such development seems in place!

The various strengths and opportunities shows that there is a tremendous scope of developing tourism in India, but still, there are many challenges to be catered to in terms of the weakness and threats of India as a tourist destination. There is a dire need to evaluate the strength of India and overcome the weaknesses in order to bring it to the top list of the world tourist destination and the right strategies and initiatives in the right direction would bring our country to the targeted place.

References

Daily Excelsior, July 11, 2011.

Daily Excelsior, March 31, 2012.

Gantzer, H. and C. Gantzer (1983), Managing Tourist and Politicians in India. In: *Tourism Management*, June 1983, pp. 118-25.

Government of India (GOI), (2004), *India Tourism Statistics 2004*, Ministry of Tourism, Market Research Division, New Delhi.

Government of India (GOI), (2008), *India Tourism Statistic 2007*, Ministry of Tourism, Market Research Division, New Delhi.

Government of India (GOI), Ministry of Railways (2007), Indian Railways, *Yearbook*, 2006-07, New Delhi.

Government of India (GOI), (2002), *National Tourism Policy, 2002*, New Delhi, p. 21.

Ghosh, P. (1999), Name it, India has got it, except tourist; The Rediff Business Special/ tourism in doldrums. http://www.rediff.com/business/1999/mar/29tour1.htm.

Leiper, N. (1999), A conceptual analysis of tourism-supported employment which reduces the incidence of exaggerated, misleading statistics about jobs. In: *Tourism Management*, Vol. 20, pp. 605-13.

Leiper, N. (1995), Tourism Management, Melbourne, RMIT Press.

Mansukhani, B. (2003), Indian Tourism—Ready to Exhale?, In: *Express Travel and Tourism*, October 2003.

Sudan, A.S. (2004), "Encylopaedia of Tourism and Hospitality Management in 21st Century in Six Volumes", Anmol Publications (P) Ltd., New Delhi.

Index